Data-Driven Technology for Engineering Systems Health Management

Gang Niu

Data-Driven Technology for Engineering Systems Health Management

Design Approach, Feature Construction, Fault Diagnosis, Prognosis, Fusion and Decision

Science Press
Beijing

Springer

Gang Niu
Institute of Rail Transit
Tongji University
Shanghai
China

ISBN 978-981-10-2031-5 ISBN 978-981-10-2032-2 (eBook)
DOI 10.1007/978-981-10-2032-2

Jointly published with Science Press, Beijing, China

Library of Congress Control Number: 2016946012

Printed on acid-free paper

This Springer imprint is published by Springer Nature
The registered company is Springer Science+Business Media Singapore Pte Ltd.

Preface

Systems health management (SHM) has emerged over recent years as significant technologies that are making an impact on both military and commercial maintenance practices. This discipline links studies of failure mechanisms to system lifecycle management and is often referred to as prognostics and health management (PHM), or in transportation applications—vehicle health management (VHM). Technical approaches to building models in SHM/PHM can be categorized broadly into data-driven approach, model-based approach, and hybrid approach. The data-driven approach for SHM/PHM is also explained in condition-based maintenance (CBM). CBM can be applied as a technical architecture and engineering strategy of data-driven PHM. In this book, data-driven PHM/CBM is introduced in details, which mainly emphasis functions of condition monitoring, fault diagnosis, and prognosis.

Condition monitoring, fault diagnosis, and prognosis of engineering systems have received considerable attention in recent years and are increasingly becoming important in industry because of the need to increase reliability and decrease possible loss of production due to faulty systems. Early fault detection, diagnosis, and prognosis can increase system availability and performance, reduce consequential damage, prolong machine life, and reduce spare parts inventories and breakdown maintenance. With the development of artificial intelligence techniques, many intelligent systems have been employed to assist the maintenance management task to correctly interpret the fault data.

This book aims to provide latest research findings and advanced techniques for the fault diagnosis and prognosis area of engineering systems. It introduces the developments and applications of intelligent diagnosis and prognosis techniques in recent years.

This book details the technique for intelligent fault diagnosis and prognosis that implements data-driven approach. Data-driven methodology consists of data acquisition, feature extraction, feature selection, classification, prognosis and data fusion algorithms, etc. for decision making. Each step of data-driven strategy is reviewed including examples. It provides a foundation in the data acquisition, analysis, feature extraction and selection, classification of equipment faults and prognosis through text discussion, worked examples, applications, and use of modern computer tools. Most chapters include with examples showing how to use these tools to solve situations not easily amenable to analytic solutions. This book provides practice in identifying, formulating, and solving fault diagnosis and prognosis problems. An extensive set of worked examples offers the opportunity to apply concepts discussed in the book to analyze and solve a variety of problems.

The organization and basic subject matter for this book parallel a 17 intensive course entitled "Engineering Systems Health Management" offered by Dr. Gang Niu at Tongji University.

Shanghai, China Gang Niu

Acknowledgements

The integration task needed to pull together cohesive and up-to-date materials was enormous and would not have been possible without the unselfish and enthusiastic support I received from my previous colleagues and coworkers. The research work of faculty and students at the Intelligent Mechanical Laboratory (IML) of Pukyong National University and the Center for Advanced Life Cycle Engineering (CALCE) of University of Maryland form the backbone of many topics covered in this book. Special thanks also to the Intelligent Control Systems Laboratory of the Georgia Institute of Technology and the Center for Intelligent Maintenance Systems (IMS) of University of Cincinnati for valuable references. The author is grateful to Achmad Widodo, Van Tung Tran, Tian Han, Martha Arbayani Bin Zaidan, and Jing Tian, among others for their valuable contributions.

I particularly thank my graduate students for helping collect and integrate some of the original draft.

I am gratefully acknowledging the support received over the past years from many industrial sponsors. It is impossible to acknowledge by name many researchers whose work referred in this book.

The work presented in this book was supported by the National Natural Science Foundation of China (Grant No. 51575396 and 51205291).

Finally, I would like to express my gratitude to my parents, my wife, and my little son James, for their patience, understanding, and encouragement during the hard creation of this book.

Gang Niu

Contents

Chapter 1
Background of Systems Health Management

1.1 Introduction

As the global competition intensifies in the manufacturing sector, the players are forced to compute on several measures, such as costs, quality, on-time delivery, and time to market. This new environment has forced managers and engineers to optimize all systems involved in their organizations. The manufacturing facilities hence become very important engineering assets. The capital invested in an organization's assets requires maximum benefit to be obtained from them throughout their life cycle—an asset out of use is a direct cost to the bottom line. Thus, assets must be managed in order to minimize downtime and sweat the assets for maximum usage. To effectively manage the assets requires a full understanding of the asset from an engineering point of view:

- *Configuration control*: It is essential to know the build structure of an asset and the configuration standard of the constituent parts, to ensure that the overall configuration is maintained when replacement of a part becomes necessary.
- *Maintenance management*: Where an asset has predictable wear characteristics, or is subject to degradation of performance with time, the life of the asset can be prolonged by anticipating probable failure and conducting maintenance tasks to avoid the failure. It is also important to record the occurrences of random failure to assist in identification of recurring problems.
- *Component living*: If a component has a known life, this needs to be identified within any software tool that may be employed by the asset manager. Life usage can then be recorded against the component as it is operated, which in turn enables the replacement or reconditioning of the component to be planned into engineering activities for the component or its parent platform.

Maintenance management plays a key role in achieving organizational goals and objectives. The term maintenance is normally used to cover a broad range of

G. Niu, *Data-Driven Technology for Engineering Systems Health Management*,
DOI 10.1007/978-981-10-2032-2_1

planned or unplanned activities for preserving an asset in its original condition. Maintenance generally consists of the following (Vanier 2001):

- *Inspections* are carried out periodically to monitor and record how systems are performing;
- *Preventive maintenance* ensures that systems or components continue to perform their intended functions throughout their service life;
- *Repairs* are required when defects occur and unplanned intervention is required;
- *Rehabilitation* replaces one major component of a system when the system is reaching the end of its service life of the system.

An efficient and effective maintenance becomes crucial for reducing costs, ensuring on-time product deliveries, minimizing equipment downtime, improving quality, increasing productivity, and providing reliable equipment that is safe and well configured to achieve timely delivery of orders to customers.

Operational reliability of industrial machinery and production systems has a significant influence on the profitability and competitiveness of industrial companies. This emphasizes the increasing importance of online condition monitoring, diagnosis and prognosis of machinery, and production processes and system in industry (Helle 2006).

Over the past 20 years, maintenance also has greatly changed because of a huge increase in the number and variety of plants and equipments, which must be maintained throughout much more complex designs, new maintenance techniques, and changing views on maintenance organization and responsibilities.

Recently, DiStefano and Covino (2007) studied statistics from the US Department of Commerce, including their measurement of what they call *net stock of private fixed assets* in various industries. This measurement is a close proxy of replacement asset value (RAV). In 2003, there were $4.9 trillion of physical assets on the ground in the US industry. They have a proprietary benchmarking system that measures actual maintenance spends (as a percentage of RAV) in about 10

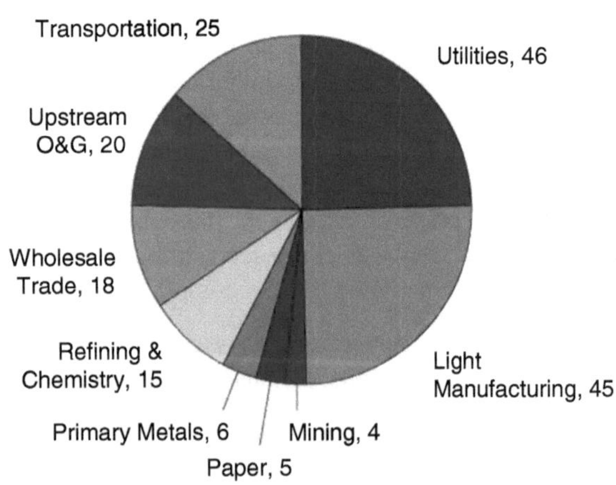

Fig. 1.1 Annual excess maintenance spend (in billion US $) by US industry

different industries. As it turns out, the average spends is between 5 and 8 %, with the best performers spending less than 2–3 %. They applied their proprietary benchmarking system to the US Department of Commerce Statistics and dollarized the value of elevating fourth, third, and second quartile plants to first quartile in maintenance spend. The result is shown in Fig. 1.1 and suggests that industry wastes more than $180 billion in excess maintenance spend annually.

1.2 Maintenance Strategy

In maintenance, there are three basic approaches: *corrective* maintenance, *scheduled* maintenance, and *preventive* maintenance. According to the way these basic approaches are applied, five basic maintenance policies need to be distinguished: failure-based maintenance, design-out maintenance, use-based maintenance, condition-based maintenance, and detection-based maintenance. The advantages and disadvantages of each maintenance policy are well documented by Waeyenbergh and Pintelon. From a time perspective, the evolution of maintenance is represented in Table 1.1.

The development of the maintenance concept started after the study of existing maintenance concepts: reliability-centered maintenance, total productive maintenance, business-centered maintenance, integrated logistic support, and logistic support analysis. These concepts are often very time-consuming to implement or only valid for a special class of equipment or a specific industry. Recently, a new generation of maintenance, e-maintenance, emerges with globalization and fast growth of the communication technologies, and computer and information technologies.

Table 1.1 Maintenance in a time perspective (Waeyenbergh and Pintelon 2002)

<1950	1950–1975	>1975	2000–
Manpower (simple)	Mechanization (complex)	Automation (more complex)	Globalization (crossing boundaries)
"Fix it when it breaks"	"I operate—you fix" (availability, longevity, cost) PM, WO management	RAM (safety, quality, environment), CBM, CM, DOM, multi-skilling, MMIS asset management	Optimal concept + Outsourcing and ICT
Maintenance is "a production task"	Maintenance is "a task of the maintenance department"	Maintenance is "(may be) not an isolated function integration efforts"	Maintenance is "external and internal partnerships" Maintenance meets production
"Necessary evil"	"Technical matter"	"Profit contributor"	"Partnership"

RAM reliability, available, maintainability; *PM* preventive maintenance; *ICT* information and communication technology; *CBM* condition-based maintenance; *CM* condition monitoring; *WO* work order

The concept of e-maintenance refers to a major pillar in modern industries that supports the success of the integration of production (e-manufacturing) and business (e-business) (Han and Yang 2006; Muller et al. 2007).

The evolution of maintenance strategies can be summarized from breakdown maintenance (BM) and preventive maintenance (PM) to predictive maintenance (PdM). BM is also known as reactive maintenance or corrective maintenance. BM is essentially the *run-it-till-it-breaks* mode and is performed when action is taken to restore the functional capabilities of failed or malfunctioned equipment or systems. There are no routine maintenance tasks to perform, and equipment is repaired or replaced only when obvious problems occur. This is a reactive approach to maintenance because the action is triggered by the unscheduled event of an equipment failure. Since companies do not incur any maintenance expense until something breaks, reactive maintenance may appear to be the least expensive approach. The result is more frequent replacement and higher capital costs. In this maintenance policy, the maintenance costs are usually higher because downtime events are often unplanned, more frequent, and longer in duration and also due to the following reasons:

- The high cost of restoring equipment to an operable condition under crisis situation,
- The secondary damage and safety/health hazards inflicted by the failure, and
- The penalty associated with lost production.

Reactive maintenance works well if equipment shutdowns do not affect product quality or revenue generation and if higher repair/replacement costs and a longer mean time to repair are within an acceptable range.

PM is the approach developed to avoid this kind of waste. PM is also known as time-based maintenance or periodic preventive maintenance. This assumes that the estimated failure behavior of the equipment, i.e., the mean time between failures (MTBF), has statistically or experimentally been known during equipment and machinery degrading within normal usage. PM takes mainly the form of equipment overhaul or item replacement. This practice is known as time-directed maintenance which is performed on a monthly, quarterly, biannual, or annual basis depending on the type of equipment, performance against specifications, and operating conditions. The reasons for performing PM tasks are to prevent the failure, to detect the onset of failure, and to discover the hidden failure. Most common type of PM task is done at hard time intervals regardless of other information that may be available when the preset time occurs. Also, it requires an intrusion into the equipment, thereby rendering it out of service until the task is completed. While this approach can help reduce equipment failure and extend component life, the process can be labor-intensive, maintenance is performed based on the preset schedule regardless of the condition of the equipment.

PdM is to monitor the equipment for changes that could be destructive in the future, but allows you to correct them before the destruction starts or to identify production equipment needing maintenance attention before product quality is reduced or an unplanned shutdown occurs (Thorp 2007). PdM is also known as on-condition maintenance, condition-based maintenance (CBM), or condition-directed maintenance. This approach involves monitoring the condition and operation of equipment to assess whether the equipment will fail during some future period, and then taking action to avoid the consequences of that failure. Unlike preventive maintenance, the need here is based on the actual condition of the asset rather than on some preset schedule. Although PdM require an investment in order to effectively implement, operate, and maintain them, the actual cost is substantially lower than the lost production resulting from failure. PdM is designed to detect the onset of a failure/fault and applied to critical equipment where a failure would interrupt a continuous process or impact quality. It is an appropriate option for PM when the following conditions apply:

- Either failure prevention is not feasible, or how it can be achieved is not yet known, as in cases where the event leading to failure occurs in a predominantly random manner;
- A measurable parameter which correlates with the onset of fault has been identified; for example, the vibration level is an indicator of the machine condition;
- It is possible to identify a value of that parameter when action may be taken before full fault occurs, such as the setting of warning limits for the vibration.

PdM does not normally involve an intrusion into the equipment, and the actual preventative action is taken only when it is believed that an incipient fault has been detected. This strategy works well if maintenance personnel are properly trained and have the time to perform the necessary maintenance work to address the potential problem.

Table 1.2 summarizes these different strategies of maintenance which being commonly practiced in the industry. Table 1.3 shows a survey conducted by Rockwell Automation and leading maintenance trade publications. Even though companies ideally want to spend most of their time on predictive maintenance, a large percentage of their time is being spent on preventive and reactive activities (Fig. 1.2).

According to the Federal Energy Management Program (FEMP) in USA, a functional PdM program can provide savings of 8–12 % over a program utilizing

Table 1.2 Types of maintenance strategies

Maintenance strategy	Maintenance approach	Signification
Breakdown maintenance (BM)	Fix it when broke	Large maintenance budget
Preventive maintenance (PM)	Scheduled maintenance	Periodic component replacement
Predictive maintenance (PdM)	Condition-based monitoring	Maintenance decision based on equipment condition

Table 1.3 Percentage of time spent on equipment maintenance activities

Maintenance strategy	Plant services		Maintenance technology	
	Ideal	Actual (%)	Ideal	Actual (%)
Breakdown maintenance (BM)	12	29	12	40
Preventive maintenance (PM)	35	33	44	32
Predictive maintenance (PdM)	18	10	33	15
Shutdown/turnaround	10	11	11	13

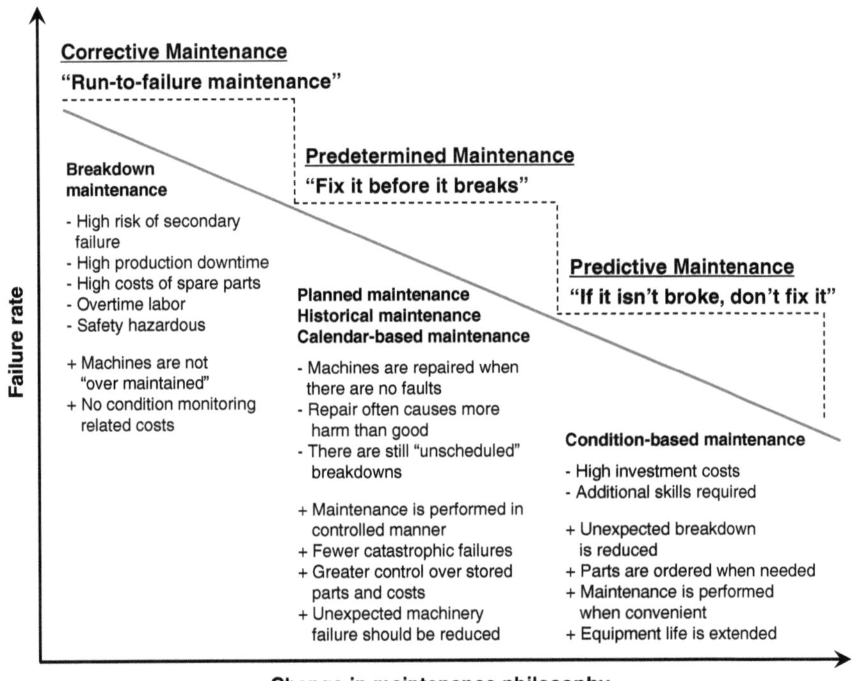

Fig. 1.2 Strengths and weaknesses of different maintenance types (Bengtsson et al. 2004)

only PM strategies. The FEMP also came up with the following industrial savings based on the initiation of a functional PdM program (Thorp 2007).

- Return on investment: 10 %,
- Reduction in maintenance costs: 25–30 %,
- Elimination of breakdowns: 70–75 %,
- Reduction in downtime: 35–45 %, and
- Increase in production: 20–25 %.

Figure 1.3 shows the maintenance cost by maintenance strategy which summarized by an Electric Power Research Institute (EPRI) study.

Fig. 1.3 Maintenance cost versus maintenance practices employed (EPRI)

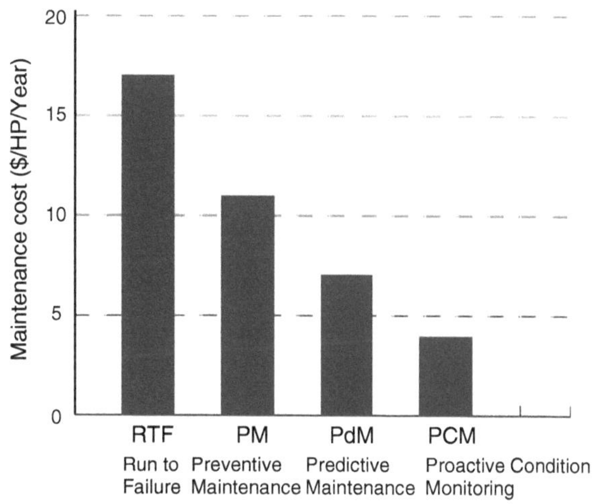

In general, maintenance strategy should be a mix of predictive, preventive, and reactive methods, depending on the desired goal and part of the process being maintained. In applications where the criticality of the equipment and the impact of unplanned downtime and quality are high, a maintenance strategy that includes preventive or predictive components offers numerous advantages. Independent studies indicate that preventive maintenance has a 5:1 cost advantage over reactive maintenance.

In practice, the choice of the optimum maintenance strategy is not as simple as noted above. Not all failures can be detected by monitoring. The economics of the situation may limit the number of components that can be monitored. There will also be a number of components and/or machines for which condition monitoring is not particularly appropriate. In many cases, these three strategies (BM, PM, and PdM) mentioned above are used simultaneously within an organization. Therefore, decision makers face such questions as follows (Mechefske and Wang 2003):

- Which strategy should be introduced for a specific type of machine?
- How to justify their decision?

Answering these questions is a difficult task. There are always a variety of objectives that a company wants to achieve through its maintenance strategy. Most traditional economic analysis methods are based upon a comparison of the initial investigation to the estimated cost savings and often focus on easily quantifiable factors. The majority of the recognized benefits of condition monitoring are considered to be "intangible" or "non-monetary." They are hard to quantify and therefore often ignored in traditional discounted cash flow analysis methods.

The decision elements with respect to maintenance strategy are shown in Table 1.4 (Pintelon et al. 2006). The first four decision elements in Table 1.4 can be viewed as the maintenance resources.

Table 1.4 Summary of maintenance strategy decision elements

Structural decision elements	
Maintenance capacity	Capacity in terms of work force, supervisory, and management staff. Shift patterns of work force, temporary hiring of work force
Maintenance facilities	Tools, equipment, spares, workforce specialization (mechanics, electricians, etc.), location of workforce
Maintenance technology	Predictive maintenance, or condition monitoring technology, expert systems, maintenance technology (intelligent maintenance)
Vertical integration	In-house maintenance versus outsourcing and relationship with suppliers
Infrastructure decision elements	
Maintenance organization	Organization structure (centralized, decentralized, or mixed), responsibilities
Maintenance policy and concepts	Policies such as corrective, preventive, and predictive maintenance. Concepts such as total productive maintenance (TPM) and reliability-centered maintenance (RCM)
Maintenance planning and control systems	Maintenance activity planning and scheduling. Control of spares, costs, etc. Computerized maintenance management systems (CMMS)
Human resources	Recruitment policies, training and development of workforce, and staff. Culture and management style
Maintenance modifications	Maintenance modifications, equipment design improvements, new equipment installations, and new machine design support
Maintenance performance measurement and reward systems	Performance recognition, reporting and reward systems, overall equipment effectiveness (OEE), and balanced score card (BSC)

They are termed as structural, because decisions made in those areas are generally assumed to be fixed. For instance, a company outsourcing its entire maintenance activities cannot revert immediately to in-house maintenance. The majority of the maintenance budget is consumed by these structural elements. The last six infrastructure elements can be viewed as maintenance management elements. These structural and infrastructure elements are interrelated. For instance, effective utilization of maintenance resources depends upon the decisions taken in the infrastructure elements. Over a period of time, decisions must be made in all of these maintenance strategy elements. The way these elements are managed or utilized can have a major impact on the maintenance function's ability to implement and support the company's manufacturing and business strategies.

Companies mainly differ in their maintenance strategies by the combination of decisions taken in these elements. Several operating aspects and business requirements influence these decisions. An effective maintenance strategy is one that fits the needs of the business. Its performance is judged based on certain measurable criteria. Cholasuke et al. (2004) studied the status of maintenance management in UK manufacturing organizations based on a pilot survey. They categorized the maintenance effectiveness measures into nine areas based on a literature study. Most of these measures fall under the structural and infrastructure elements shown in Table 1.4. The measures constitute the following:

- Policy deployment and organization,
- Human resources management,
- Financial aspects,
- Continuous improvement,
- Contracting out maintenance,
- Maintenance approach,
- Task planning and scheduling,
- Information management,
- Computerized maintenance management systems (CMMSs), and
- Spare parts management.

Maintenance management is an orderly and systematic approach to planning, organizing, monitoring, and evaluating maintenance activities and their costs. The maintenance management system is usually designed to optimize the management of deferred maintenance and capital improvement activities throughout the service. This can be achieved by using standardized procedures to document and prioritize the field facility and resources needed and to report accomplishments. Several activities should be supported by the maintenance management system, which can be summarized (Piotrowski 2001) as follows:

- Work order generation, work prioritization, and tracking by asset;
- Tracking of scheduled and unscheduled maintenance activities;
- Historical tracking of all work orders generated;
- Storing of all technical documents and reports;
- Generation of real-time reports of ongoing activities;
- Capital and labor cost tracking by component as well as shortest, median, and longest times to close a work order by component; and
- Complete asset, inventory, equipment, and material records.

Maintenance information systems, such as CMMS and enterprise resource planning systems (ERP), has been developed for data storage and handling. The main objectives of a CMMS can be summarized as follows (Wireman 1994):

- Improve maintenance efficiency,
- Reduce maintenance cost,
- Reduce asset downtime by scheduling preventive maintenance,
- Increase the design life of an asset,
- Provide historical records to assist in maintenance planning and budgeting, and
- Provide maintenance reports in a format that is required by the user.

CMMS is computer-based software program used to control work activities and resources used as well as to monitor and report work execution. CMMS is tools for data capture and data analysis. Managing and processing maintenance operations data constitutes probably the main contribution of current CMMS. CMMS transform maintenance records with data into proper information for decision making in maintenance. In order to do so, they need to have a suitable technical database structure and an efficient work order management system for easy data input, update and control.

With the rapid development of computer and advanced sensor technologies, data acquisition facilities and technologies have been more powerful and less expensive, making data acquisition for condition-based maintenance (CBM) implementation more affordable and feasible.

CBM and prognostics and health management (PHM) are essential components of a robust systems health management. The aim of CBM is diagnostic in nature. Given the current condition, it tries to identify appropriate maintenance actions to detect a fault condition before it turns into a failure. PHM is predictive in nature, aiming to determine how long from now will a fault happen in a system given the current operating conditions (Sreerupa et al. 2012). Both PHM and CBM analyses determine a set of maintenance actions based on real-time or near real-time assessment systems health.

1.3 From Maintenance to PHM

The concept and framework of PHM have been developed based on the well-known maintenance methodologies and diagnostics techniques, such as preventative maintenance (PM), reliability-centered maintenance (RCM), and condition-based maintenance (CBM). The future development of PHM will be mutually inspired and promoted from various fields besides engineering. The related research topics in meteorology/climatology, decision science/policy, financial/economic, and other fields can also be followed to expand the vision of PHM.

CBM consists of data acquisition and data processing (condition monitoring), resulting in actionable condition information on which maintenance decision making can be based, thus avoiding unnecessary maintenance tasks. Currently, more and more research effort has shifted toward prognostics and health management which focuses more on incipient failure detection, current health assessment, and remaining useful life prediction. However, in various maintenance scenarios with different system complexities and uncertainties, the maintenance strategies should be different. Figure 1.4 is the maintenance transformation map in which diverse maintenance strategies are shown with the system complexity and uncertainty.

CBM can be applied in systems that (1) can be regarded as being deterministic to some extent, (2) is stationary or static, and (3) for which signal variables that can be good health indicators can be extracted, despite low dimensionality. If the system is

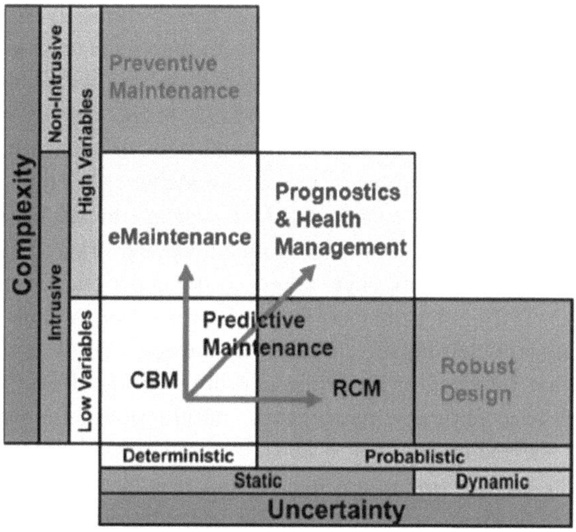

Fig. 1.4 Maintenance transformation map (Lee et al. 2013)

a probabilistic system for which the output cannot be easily determined with a known relationship model and is conditionally dependent upon the input, output data will not always be repeatable in instances observed at different timings. Hence, the future behavior cannot be predicted accurately based on the domain knowledge of the system and the historical observations. In such an uncertainty laden system as described, RCM is more suitable. RCM focuses on the ability of a system to have expected reliability in a certain period of time and utilizes statistical tools such as failure modes and effects criticality analysis (FMECA) to retrieve the information that can help identify failure modes and possible durations before each of the modes can happen. Since RCM relies on statistical estimation of the total operation life expectation, it can reduce unscheduled or unnecessary maintenance if the system is static and the failure modes are well studied. However, RCM is still prone to large deviation of the system dynamic and it lacks significant insight into the actual system performance. If the system uncertainty is more complicated, like in a highly dynamic system in which the behavior varies over time, robust design should be considered. In this scenario, resources should be allocated to the design processes, rather than relying on inspection to ensure quality. As well, the product performance, due to robust design being deployed, should have minimal sensitivity to material, manufacturing, or operation variations.

Along the other axis in Fig. 1.4, if a system has a high level of complexity and a large number of variables that represent different aspects of the system, such as vibration, position, velocity, stress, current, environment, and controller parameters, e-maintenance is preferred as CBM techniques usually deal with low dimension of data. By using Internet and tether-free communication technologies, e-maintenance enables a system to achieve near-zero-breakdown performance on a common platform to integrate information. Moreover, by linking with industrial business systems, e-maintenance is able to align the maintenance process with the business and operations processes to achieve optimally integrated production and asset management. But if the system complexity is even higher so that only non-intrusive approaches can be applied to the system, which means no or only very limited instrumentation can be applied due to the product package, seal, and the concern that externally added components will dramatically affect the precision and efficiency of the products, preventive maintenance may be chosen rather than CBM or e-maintenance which requires systems to be heavily instrumented. Preventive maintenance is a time-driven maintenance strategy that schedules maintenance for a machine or component based on the experience of the mean time between failures (MTBF). This method follows strong assumptions that the machine is working under deterministic and static conditions, and therefore cannot be applied to system that operates in dynamic working regimes. Hence, preventive maintenance may lead to untimely maintenance and non-optimal cost.

PHM can be treated as an evolved form of CBM. CBM techniques can be used to provide input for the prognostics models in PHM and support the timely, accurate decision making that prevents downtime and maximizes profit. For its capability to assess the health status and predict the occurrence of failure and downtime, PHM is considered to be the foundation, when complemented with other techniques, for advanced areas including self-maintenance, resilient system, and engineering immune systems. The disciplines of PHM need to be further developed and extended to help building these areas.

1.4 Definitions and Terms of Systems Health Management

Both diagnostics and prognostics originally come from the medical field. As machinery maintenance technology and engineering systems health management emerged, diagnostics and prognostics gradually permeated all areas of systems engineering. Nowadays, there are many kinds of professional instruments, such as sensors, meters, controllers, and computational devices, for conducting fault diagnostics. These instruments can be used to acquire and analyze signals from a machine or process. More and more sophisticated diagnostics methodologies are available to determine the root causes of component failure. However, *diagnostics*, which is conducted when a fault has already occurred, is a reactive process for maintenance decisions and cannot prevent downtime as well as corresponding expense from happening. In order to reduce the maintenance cost and maintain the machine uptime at the highest possible level, maintenance should be carried out in a proactive way that means a transformation of maintenance strategy from the traditional fail-and-fix practices (diagnostics) to a predict-and-prevent methodology (prognostics).

Prognostics has been applied to the field of maintenance for more than 10 years; however, most of these applications only address forecasting, or remaining useful life (RUL) prediction, which is just one facet of PHM.

As an engineering discipline, PHM aims to provide users with an integrated view of the health state of a machine or an overall system. Diagnostics is also included in prognostics and health management. Diagnostics can be summarized as the process of identifying and determining the relationship between cause and effect in that its function is to isolate faults and identify failure root causes. Prognostics can be interpreted as the process of health assessment and prediction, which includes detecting incipient failure and predicting RUL.

Health management is the process of taking timely, appropriate maintenance actions and making accurate logistics decisions based on the outputs from diagnostics and prognostics, available resources, and operational demand. It focuses on assessing impact of failures and minimizing impact and loss with maintenance management.

An effective PHM system is expected to provide early detection and isolation of the precursor and/or incipient fault of components or subelements; to have the means to monitor and predict the progression of the fault; and to aid in making or autonomously triggering maintenance schedule and asset management decisions or actions. The detected, incipient fault condition should be monitored, trended from a small fault as it progresses to a larger fault, until it warrants some maintenance action and/or replacement. By employing such a system, the health of a machine, component, or system can be known at any point of time, and the eventual occurrence of a failure can be predicted and prevented, enabling the achievement of near-zero downtime performance. Unnecessary and costly preventive maintenance can be eliminated, maintenance scheduling can be optimized, and lead time for spare parts and resources can be reduced—all of which can result in significant cost savings.

1.5 Preface to Book Chapters

This book details the technologies in data-driven PHM/CBM that have been introduced over the recent past by researchers and practitioners and are making significant inroads in such application domains such as mechanical, thermal, electromechanical, and, more recently, electrical and electronics systems.

This book is structured as follows: Chapter 2 introduces those fundamental system concepts that set the stage for the effective design of systems health management. We review systems-based methodologies that have a direct and significant impact on the design of PHM/CBM systems.

Chapter 3 gives an overview of data-driven PHM/CBM approach. OSA-CBM architecture is explained in detail which includes 7 layers: data acquisition, data manipulation, condition monitor, health assessment, prognostics, automatic decision reasoning, and human–computer interface.

Following the contents above, Chaps. 4–8 address the key issues regarding data acquisition processing and analysis, feature extraction and clustering, feature selection optimization, intelligent fault diagnosis and prognosis, and typical technical and methodology. Related case study will be exhibited.

Then, Chap. 9 emphasis utilizing data fusion strategy to improve the performance of data-driven PHM/CBM, fusion architectures, and classical fusion methodology in feature layer and decision layer, which will be described with engineering case study.

The tenth chapter answers the question of the following: How to make decisions based on the output results of anomaly detection, diagnosis, and prognosis? A data-driven PHM/CBM-based intelligent maintenance platform will be proposed; in addition, autonomous control and contingency management also will be introduced. Future development of systems health management toward engineering immune systems will be clarified at last.

References

Bengtsson M, Olsson E, Funk P, Jackson M (2004) Technical design of condition based maintenance system—a case study using sound analysis and case-based reasoning. In: Maintenance and reliability conference

Cholasuke C, Bhardwa R, Antony F (2004) The status of maintenance management in UKmanufacturing organizations: results from a pilot survey. J Qual Maint Eng 10(1):5–15

DiStefano R, Covino L (2007) Enterprise reliability: changing the game. Asset Manag Mainten J 20(2):22–31

Han T, Yang BS (2006) Development of an e-maintenance system integrating advanced techniques. Comput Ind 57(6):569–580

Helle A (2006) Development of prognostic concepts and tools. In: VTT symposium 243

Lee J et al (2013) Prognostics and health management design for rotary machinery systems-reviews, methodology and applications. Mech. Syst Signal Process

Mechefske CK, Wang Z (2003) Using fuzzy linguistics to select optimum maintenance and condition monitoring strategies. Mech Syst Signal Process 17(2):305–316

Muller A, Marquez AC, Iung B (2007) On the concept of e-maintenance: review and current research. Reliab Eng Syst Safety

Pintelon L, Pinjala SK, Vereecke A (2006) Evaluating the effectiveness of maintenance strategies. J Qual Mainten Eng 12(1):7–20

Piotrowski J (2001) Proactive maintenance for pumps. Archives. http://www.pump-zone.com

Sreerupa D, Richard H, Amar P et al (2012) An open architecture for enabling CBM/PHM capabilities in ground vehicles. In: IEEE conference on prognostics and health management

Thorp B (2007) What is predictive maintenance and how it has benefited Seminole Electric Coop. Inc. Asset Manag Mainten J 20(2):14–21

Vanier DJ (2001) Asset management: "A" to "Z." In: APWA international public works congress, pp 1–7

Waeyenbergh G, Pintelon L (2002) A frame for maintenance concept development. Int J Prod Econ 77:299–313

Waeyenbergh G, Pintelon L (2004) Maintenance concept development: a case study. Int J Prod Econ 77:395–405

Wireman T (1994) Computerized maintenance management systems. Industrial Press, New York

Chapter 2
Design Approach for Systems Health Management

2.1 Introduction

Over the last several decades, there has been a wide range of approaches and implementation strategies for performing manual, semiautomated, or fully automated fault diagnosis and prognosis (i.e., health management) on critical systems in commercial and defense markets. Associated with these diagnostic and prognostic systems, designs are an equally diverse number of philosophies and associated architectures used to implement them for particular application.

Following the evolution of diagnostic systems in the modern industry, prognostic initiatives stared to be introduced in order to try to take advantage of the maintenance planning and logistics benefits. However, the early prognostic initiatives often were driven by infield failures that resulted in critical safety or high-cost failures, and thus, retrofitted technology was hard to implement and costly to develop. Hence, diagnostic and prognostic systems developers found the need to analyze and describe the benefits associated with reducing infield failures and their positive impact on safety, reliability, and overall lifecycle cost reduction. This leads to many cost–benefit analyses and ensuing discussions and presentations to engineering management about why the diagnostic and prognostic technologies need to be included in the design process of the system and not simply an afterthought once field failures occur. This had us to the point where many complex vehicle/system designs such as advanced fighter and high-speed train are now developing "*designed in*" health management technologies that can be implemented within an integrated maintenance and logistics system that supports the equipment throughout its lifetime. This "*designed in*" approach to health management is performed with the hardware design itself and also acts as the process for systems validation and managing inevitable changes from infield experiences and evaluating systems design trade-offs, as shown in Fig. 2.1.

G. Niu, *Data-Driven Technology for Engineering Systems Health Management*,
DOI 10.1007/978-981-10-2032-2_2

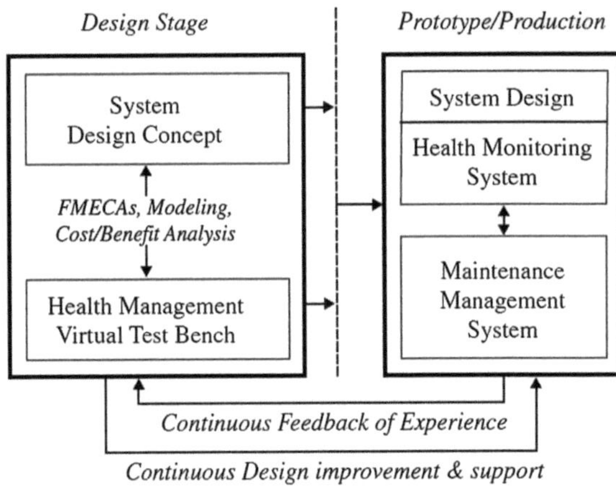

Fig. 2.1 The "designed in" approach to health management (Vachtsevanos et al. 2006)

Realizing such an approach will involve synergistic deployments of component health monitoring technologies, as well as integrated reasoning capabilities for the interpretation of fault-detect outputs. Further, it will involve the introduction of learning technologies to support the continuous improvement of the knowledge enabling these reasoning capabilities. Finally, it will involve organizing these elements into a maintenance and logistics architecture that governs integration and interoperation with the system, between its onboard elements and their ground-based support functions and between the health management system and external maintenance and operation functions.

Condition-based maintenance (CBM) is the use of machinery run-time data to determine the machinery condition, and hence, its current fault/failure condition, which can be used to schedule, required repair and maintenance prior to breakdown. *Prognostics and health management* (PHM) refers specifically to the phase involved with predicting future behavior, including remaining useful life (RUL), in terms of current operating state and the scheduling of required maintenance actions to maintain systems health. Detecting a component fault or incipient failure for a critical dynamic system (aircraft, gas turbine, pump, etc.) and predicting its remaining useful life necessitate a series of studies that are intended to familiarize the PHM/CBM designer with the physics of failure mechanisms associated with the particular system/component. Moreover, the designer must have a thorough understanding of methods for optimal selection of monitoring strategies, tools, and algorithms needed to detect, isolate, and predict the time evolution of the fault, as well as systems approaches for designing experiments and testing protocols, performance metrics, and means to verify and validate the effectiveness and performance of the selected models.

This chapter will introduce the concept of a system, as well as an engineering viewpoint for thinking about systems. The general consideration for systems health management (SHM) life cycle and analysis models is described. Then, two systems-based methodologies for the design of health management systems will be introduced.

2.2 Systems Engineering

Systems engineering is presented as a framework organized around stages of the product development life cycle. We specify a systems health management (SHM) process view of the systems engineering life cycle which provides a basis for understanding the broader issues of SHM and how they fit into the system life cycle. This supports the very important notion that SHM is an essential property of engineered systems that exerts considerable influence on system performance and affordability—it, therefore, must be addressed with an appropriate level of concern early on and throughout the product life cycle.

The term "*systems engineering*" was introduced by Bell Labs in the late 1940s, and by the latter part of the twentieth century, the field of systems engineering was recognized by most engineering organizations as a functional and integral part of the design process.

Systems engineering is concerned with the use of work products, processes, and tools to support and manage the design of complex engineering projects. A systems engineering implementation is typically organized around the system life cycle, addressing the identification and definition of customer requirements, creation, and analysis of alternative design artifacts by performing relevant trade studies, the implementation and integration of selected design approaches, verification and validation of the implementation, and then production support and evaluation/maturation of the embodiment of the design.

There is general agreement in the engineering community as to the nature and components of the systems engineering life cycle (ISO 2008). Figure 2.2 illustrates the most atomic stages of the systems engineering life cycle. These stages should not be construed as discrete events in a timeline, but rather as evolutionary phases with a necessary order of evolution, including the parallel utilization and support stages.

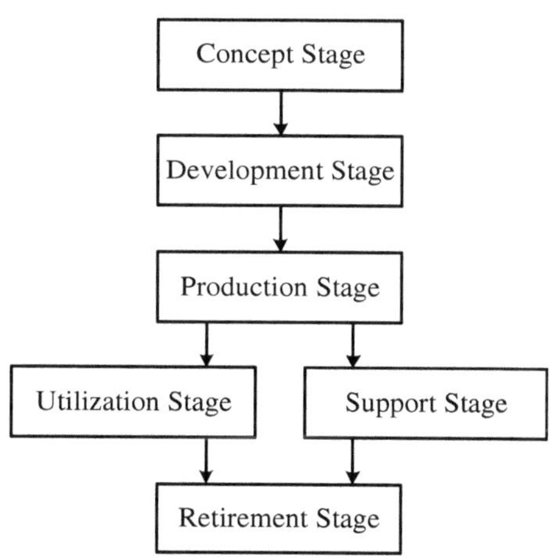

Fig. 2.2 The components of the systems engineering life cycle (Stephen et al. 2011)

2.3 Systems Engineering, Dependability, and Health Management

The creation of systems involves the application of methods from a variety of disciplines, coordinating and controlling the system creation process, and performing these functions under the influence of a number of external factors. Creating dependable systems requires that systems engineers develop an awareness of the holistic, interdependent nature of these processes and their effects on the dependability of the systems being created.

Dependable systems are those that perform their intended function when called upon to do so within their expected lifetime while not performing any unintended functions (Campbell et al. 1992). Dependability does not mean perfect, and while experienced engineers will tell you that you cannot build a perfect system, it is generally a critically important requirement that the system must be able to survive and recover from a failure condition (*mission-critical function* of SHM). This simple requirement has far-reaching implications, however, because systems of this nature typically do not exist in isolation. Engineered systems are generally hierarchical in nature, interact freely with each other, and in general exhibit behavior of an extremely complex nature. A less critical requirement may be that system failures be predicted or detected and isolated in a manner that supports efficient maintenance processes (*support-critical function* of SHM).

The notion of health management (HM) in complex systems, therefore, transcends engineering, management, and social processes and can only be obtained as an emergent property of a system that accounts for all of these issues. This "health" property is best viewed as the result of a dynamic process that changes based on the context of the lifecycle phase in which one is operating, the scale and complexity of the system being created, and the social interactions that take place between the individuals and organizations involved in the overall task of creating the system. The multi-organizational nature of the product development process adds a considerable degree of difficulty in understanding, analysis, and mitigation of system failures.

This systems engineering perspective provides us with a convenient framework for representing the SHM process. This perspective supports representation of the roles and interactions of system management methods, engineering activities, and cross-functional teams in the planning, implementation, and evaluation of the SHM process. SHM can be treated as a specialized view of the systems engineering process (ISO 2008). In this specialized process view, one can represent the HM design for a system as a system in its own right. In this view, HM system influences and interactions cut across multiple subsystems, serving to integrate the SHM perspective at each of the levels of hierarchy, as shown in Fig. 2.3.

Figure 2.3 expands upon the composite system view by including the notion of a health management system that encompasses all of the aircraft system components. This implies that each of the vehicle/system components participates in the health management system and that the health management system is itself hierarchical in

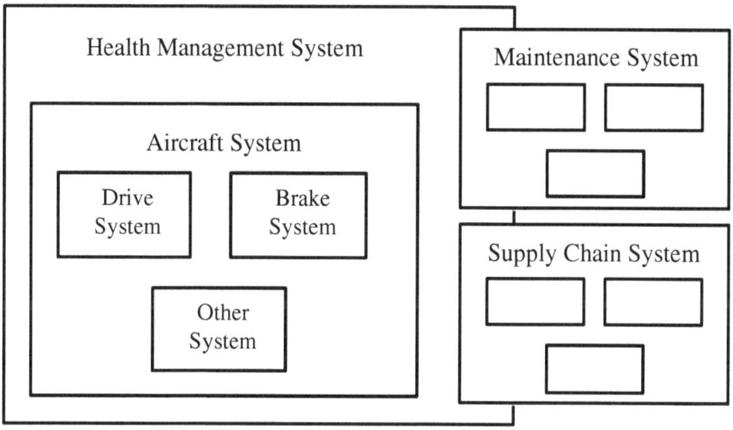

Fig. 2.3 Integrated SHM perspective (Stephen et al. 2011)

nature. The figure also shows that the health management system encompasses elements of the maintenance and supply chain systems. Health management system functions provide critical decision support functions to maintenance and supply chain systems, and in some applications, maintenance and supply chain system states may inform the health management systems. As described above, there are numerous other (unrepresented) internal and external elements, not shown here, which may influence all of the elements in this SHM hierarchy in unanticipated ways, providing significant challenges to the overall system designers and operators.

The creation of the SHM process then can be thought of as a HM-specific view of the systems engineering process, with its own lifecycle stages mapped to those of the systems engineering process. The SHM lifecycle stages are shown in Fig. 2.4. Note that the SHM systems engineering process stages exhibit the notion of iteration and feedback between stages and reference a function common to each stage labeled "Monitor and Control." This is because the SHM development process is highly distributed; elements of an integrated SHM solution are provided from the distributed elements of multi-disciplinary, multi-organizational teams. Suppliers and systems integrators must work together to achieve the most affordable and safe SHM solution possible for the system under development. Therefore, one of the primary functions of any SHM development team is to monitor these disparate process inputs, in order to coordinate and control the timing and quality of the various work products across both internal and multi-organizational design teams.

The mapping of the SHM process to the core systems engineering process also implies that there is a rough correlation in time between the two lifecycle views of system development. For example, when the primary system (or delivered product) is being manufactured, the SHM system may be in the design synthesis and integration stage. These mappings are given in Table 2.1, and we describe the HM systems engineering lifecycle phases in the following sections.

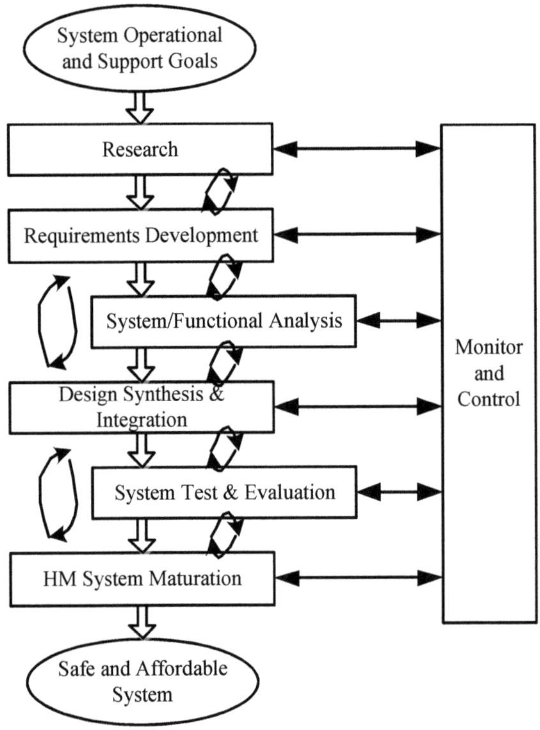

Fig. 2.4 The components of the SHM systems engineering life cycle (Stephen et al. 2011)

Table 2.1 Mapping of the SHM systems engineering process stages to core systems engineering process stages

Core systems engineering stages	Health management lifecycle stages
Preconcept	Research
Concept	Requirements development
Development	System/functional analysis
Manufacturing	Design synthesis and integration
Utilization	System test and evaluation
Support	Health management system maturation
Retirement	N/A

2.4 SHM Lifecycle Stages

2.4.1 Research Stage

The primary activity in the research stage is the identification, selection, and refinement of technologies or methods to meet customer operational needs. This is generally applied scientific research, based on more generic basic research executed by the participating organizations and the supporting academic community. This includes research in topic areas leading to product discriminators—those system features that are novel or so advanced compared to the competition as to provide some competitive advantage—as well as research performed to advance the maturity of technologies for use in the product line under development. A number of pure research centers, academic institutions, nonprofit centers, system providers, and integrators may be engaged in separate or coordinated research activities.

The initial selection of technologies and processes can have a significant effect on overall system dependability: "Numerous retrospective studies indicate that uncertainties often constitute a central consideration in the performance of engineering systems". Uncertainty can be managed in part by the selection of proven, well-developed technologies that are understood and have historical performance data available that allows the system creator to assess their dependability. Where newer technologies are anticipated for inclusion in the system proposal, careful planning of trade studies and other technology integration activities supporting risk reduction is essential. While it is not always the case that experimental or new technologies have greater chances of failure, they are less understood and inject a greater degree of uncertainty into the system creation process. The system goals and objectives will determine whether the development or use of new technologies is required. They may also place limits on the selection of existing technologies.

In addition to research into the dependability of advanced product features, this phase also is characterized by the development of advanced HM capabilities aimed at providing improved support for operational and safety goals compared to current product generations.

2.4.2 Requirements Development Stage

The purpose of the SHM requirements development stage is to define a complete set of system requirements that can be effectively used to manage the HM development process and assure that the end product will satisfy all customer needs and expectations.

From the system developer (or integrator) point of view, the primary activity in this stage is the requirements analysis leading to the development of a system concepts. These requirements define what the system must do to meet customer needs; the analysts must also consider requirements for how the building, testing, and operation of the system will be conducted.

Some of these activities relate to the management and administration of the system creation process. Budgets and schedules are developed. The management team is created, usually consisting of a project manager, systems engineers, technical experts, and administrative personnel. Legal and reporting requirements are determined. The effectiveness of these organizational and management structures has a profound influence on the dependability of the system being created. Shenhar and Bonen state: "both project management and systems engineering practice differ with each specific system [being created] and that management attitudes must be adapted to the proper system type" (1997).

This is also the time to integrate SHM into the total system design. Including consideration of SHM methods in all phases of the system life cycle and defining requirements for their implementation result in increased system dependability. It is during the requirements development phase that system development process leaders must arrive at a consensus which balances initial system delivery costs with overall system lifecycle costs.

Many, but not all, elements of SHM are developed by multi-organizational teams. As a result, much of the HM requirements definition occurs within the framework of the initial contractual process; that is, operational requirements for the system are defined as part of the published request for proposal (RFP) and are the basis for supplier selection and subsequent contracting. During this phase, supplier requirements may be further tailored as necessary to capture system technical details and programmatic constraints. The RFP typically includes formal procurement documents such as performance specifications, statements of work, supplier data list, and bidder's instructions. This information is distributed to prospective suppliers, who then submit proposals. During the source selection process, these items may be further tailored to each potential supplier based on how they propose to do business.

The goal here is to try and arrive at a clear understanding of the requirements for the multi-organizational system development team to provide the most cost-effective solution supporting systems operational and safety requirements. To that end, the procuring organization typically identifies and performs operational and support system trade studies to determine the system architectural and development features that have the greatest impact on these goals. The initial SHM concept may include elements detailing the architecture of distributed HM system elements (within either the platform boundaries, or extending to off-platform components), hardware and software configuration identification and management plans (an understanding of specific system configuration is crucial to effective HM), and HM system interfaces and data collection mechanisms—again, both on- and off-platform.

Definition of a detailed concept supported by architectural trades as just described will support the development of a cross-functional and interorganizational program management plan to support the execution of an effective SHM program and help initiate the development of another critical program element, the risk management plan.

SHM program risks can originate from numerous areas; a few notable sources to initiate the development of a SHM risk management plan could include the HM performance of similar systems in relevant field environments, knowledge of technical performance or business practices of potential system development partners, and anticipated issues with proposed customer HM requirements specifications. Customer requirements are allocated to subsystems and incorporated into the RFPs for each specific subsystem component procurement document. Specific targets for failure prediction, failure detection, and fault isolation, as well as criteria such as false alarms (FAs) and cannot duplicates (CNDs), are allocated and distributed to potential partners, as well.

Anticipation and worst-case analyses of systems performance in anticipated field conditions are a critical part of the requirements analysis process, and as these are performed, mechanisms for collecting relevant field data should be postulated and work initiated to ensure that such mechanisms are deployed to adequately support the HM systems maturation stage (as is discussed in that upcoming section of this chapter).

Another important part of initial communication with potential system development partners is to make expectations known for program information exchange, analytical tool and metric considerations, and the relationship of the delivery of data elements, analyses, and metric pass/fail criteria and expectations to proposed program schedules.

Ultimately, the goal of the requirements development stage is to determine the most affordable manner in which each product development partner can contribute to an acceptable overall technical systems design and then to enact an effective organizational infrastructure to manage the product development process. A satisfactory conclusion to this stage includes a well-formed design team with a clear understanding of mutual goals, the establishment of program management criteria as well as documented processes and tools to help ensure successful systems deployment and satisfaction of customer goals and requirements.

2.4.3 System/Functional Analysis

The purpose of the SHM system/functional analysis stage is to develop a preliminary functional design that allows all SHM program requirements to be satisfied. To do this, a system functional analysis is performed, in which the system is decomposed into the relevant top-level functions and lower-level subfunctions required to meet system performance goals. Alternative mechanisms to perform these functions are assigned and assessed by design teams, and means to assess or potentially guarantee the performance of system functions are postulated and analyzed. This is a matter of assessing the inherent system dependability from a standpoint of system reliability and developing operational health assessment and failure mitigation strategies that support contractual and operational goals. During the systems engineering development stage, requirements developed during the conceptual stage are translated into conceptual product architectures, and alternative designs for specific and tangible elements that will execute the system functions are postulated. These activities may vary significantly based on the type of system being created. Aerospace systems, software, and nuclear power plants, for example, have different approaches and methodologies for system design. The common factor is the goal of designing the product to meet the established systems operational requirements. From the SHM standpoint, however, during the corresponding system/functional analysis stage, engineers and system analysts perform detailed modeling and analyses that support the system designers' requirements and concepts through an integrated approach to system condition monitoring, failure prediction, detection and isolation, and correlation of system health effects across hierarchical system boundaries.

From a top-level functional perspective, SHM design is derived from an assessment of the needed dependability of each function in the system decomposition, and then a decision as to whether failure prevention provides sufficient reliability, or whether active operational failure mitigation methods will be required to achieve the needed reliability of each function. Put another way, every function in the system must be considered from the perspective of its failure and the consequent effect of that failure on the system. Completeness of the SHM design derives from complete coverage of SHM failure preventions or operational failure mitigations across all branches of the "function tree decomposition." Note that failure mitigation may include the possibility of doing nothing at the time of failure, if the failure is not safety or mission-critical, until a later time when proper maintenance can address the failure.

Proposed failure prevention and mitigation design mechanisms are allocated for each function and analyzed using historical and analytical techniques to determine whether the needed reliability can be achieved for that function. If failure mitigation is selected, then the proposed failure mitigation design mechanism must be able to detect the actual or potential loss of functionality and respond quickly enough so that it successfully completes before the critical failure effects that it is attempting to mitigate and propagate to cause functional failure. If the mechanism does not operate quickly enough, then it is usually true that the mitigation mechanisms must be driven further down the function tree into lower-level components, which are closer to the originating fault, thus detecting the problem faster and providing more time for the response actions to complete. System reliability is estimated by statistical summation of all component reliabilities to determine whether the system's overall dependability goals are achieved. If the system reliability does not meet these goals, then the SHM design must be improved in one or more components: The system's operational concept must be changed, or the system's dependability goals must be relaxed, or a combination of these actions must be taken to satisfy the design goals. System dependability estimates must include estimates of the potential failure of the SHM design mechanisms, which prominently include interactions between SHM detection and response mechanisms with each other, with the mission sequences, and with the system's control system. Specific activities in the SHM system/functional analysis stage include allocation of HM requirements to responsible subsystem design teams, discussions of architectural issues with strategic and technology partners, and trade studies to optimize architectural partitioning decisions (e.g., on-platform vs. off-platform, distributed vs. centralized, diagnostic vs. prognostic approaches). SHM approaches are selected from best-of-class technology resources that will satisfy operational and technological health requirements and then optimized on cost, safety, reliability, and diagnostic characteristics using a variety of engineering analysis tools.

System partners work with system integrators to ensure that distributed development schedules are all integrated into program management plans and schedules, and specific trade studies supporting HM development activities around the evolving SHM architecture are agreed to and planning initiated. Plans, schedules, and trade studies all support the requirements and functional areas covered in the requirements development stage.

Preliminary reliability, safety, and diagnostic analyses are performed using models derived from initial system design data, anticipated system operational usage, and system support specification. System failure modes are analyzed for their probability of occurrence and potential detectability. Failures with relatively low probability of occurrence or minor consequential effects are considered with respect to the resources required to detect them. Specific cost–benefit studies may be initiated in this SHM development stage—as initial product design begins to evolve from a functional architecture into a physical manifestation, evaluations of the cost-effectiveness of predictive versus diagnostic or scheduled maintenance solutions, as well as on-platform versus off-platform trades— and can be evaluated. These trade studies, besides targeting individual system drivers, can be evaluated as a whole in order to determine the most cost-effective overall approach to the overall SHM solution. In general, the cost ramifications of each design decision are considered against the projected lifecycle cost targets, and the implementation risks are evaluated against systems risk management plans initiated in the previous stage.

At the conclusion of this stage, the preliminary SHM design approach and proposed concrete implementation should be specified and supported by detailed functional analytical studies as system design becomes concrete.

2.4.4 Design, Synthesis, and Integration

The purpose of the design synthesis and integration stage is to develop and integrate a detailed design solution that meets all SHM program requirements. Implementation of selected SHM approaches is initiated, and analytical models are further refined with details of selected approaches and design knowledge as system and subsystem designs mature.

Complex systems are often characterized by two notable characteristics: (1) incorporation of an increasing number of functionalities that increase the integration of the number of parts and components (multi-component) as well as services and (2) incorporation of a number of maturing technologies (Dosi et al. 2003). Verification of the capability of these maturing technologies (providing product differentiation or other advanced capabilities to support system design goals) to support the requirements for which they are intended is performed in this stage. Partner design instantiations are verified by analysis and benchmarking against the agreed-to system performance metrics, and interfaces are evaluated for compatibility.

This stage initiates the process of verification and validation (V&V) of the SHM system. Verification is the process of ensuring that the system has metal of the developed requirements. Validation is the process of ensuring the correct system to meet customer needs has been constructed. Verification activity in the design synthesis and integration stage is strongly connected to the requirements development and system analysis stages, on one the hand, and with the validation activities in the system test and evaluation stage on the other hand. During the requirements phase, system objectives and goals are translated into requirements. These are further elaborated via the design process to greater levels of detail in the system. These requirements determine system specifications and impact the selection of materials and components; they further provide a basis for additional development of requirements and specifications and criteria for the verification of system performance and dependability.

Requirements verification is usually based on analytical approaches begun in the previous stage. Trade studies of the proposed SHM approaches are finalized and analyzed to assess the potential effectiveness of selected SHM approaches and system design decisions. Detailed reliability, safety, failure detection, fault isolation, and verticality verification analyses are performed based on specific schematic diagram and part information to support the understanding of the chosen SHM approach and ongoing assessment and management of requirements compliance.

It is usually not possible to have a "perfect" approach to detection and remediation of all failure modes as they exist in the initial design concept. At this point, in the design process, the areas that may be deficient are addressed through alternative approaches to system design, modification of the operations concept, or supplemental support procedures that may be outside the scope of initial design considerations. The coordination of this effort can be challenging because of the number of partners that may contribute to the development and production of any complex system. The close relationships of these partners in the design and production process are reflected in the coupling of the information flow and analysis required to support verification. At the completion of the design synthesis and integration stage, the SHM detailed design approach should be integrated with overall system design in a way that satisfies SHM requirements and is fully supported by interorganizational hardware design and analysis data.

Throughout the functional analysis and design synthesis stages, trade studies are continually performed that evaluate the efficacy of the proposed failure detection, isolation, prediction, and mitigation approaches with alternatives. This process continues until a satisfactory SHM approach has evolved that meets system design goals.

2.4.5 System Test and Evaluation

The purpose of the system test and evaluation phase is to qualify the implemented SHM solutions for delivery to the customer as part of the overall system. Activities in this stage will:

(1) Verify that the detailed design solution, which was previously verified by analysis to assure compliance with specification requirements, will actually achieve all SHM requirements upon delivery. As is the case in fundamentally sound software testing, system requirements are mapped to a collection of tests that will formally verify those requirements (answering the question "is the system built right?"). This can be perceived as a "bottom-up" approach that will provide traceability of the system design and development work performed to the specified requirements.

(2) Validation methods in this stage use demonstrable measures of reliability, availability, and dependability in conjunction with detailed system simulations of SHM system performance to determine whether the system is capable of achieving its goals as expressed in the system concepts—a "top-down" approach to functional verification that thoroughly exercises safety- and performance-critical HM functionality. This will provide a level of confidence that the system design is both correct and will satisfy customer goals and expectations. So, while verification determines whether the system has been built right, validation determines whether the right system has been built.

SHM verification and validation (V&V) activities include fault insertion, qualification, integration, and operational testing. Failure detection and fault isolation predictions and methodology are verified, and susceptibility to unexplained anomalies (UAs), false detections or false alarms (FAs), and cannot duplicates (CNDs) is assessed and their risk mitigated through various means. It is also possible at this point to identify and develop supplemental test and other support system elements.

The central model to the implementation of the SHM failure mitigation approach —the failure detection and fault isolation model(s)—can be verified analytically, but V&V of the SHM system requires that a fault insertion approach be employed, as diagnostic software is opportunistic (i.e., it only performs its intended job in the presence of failures). The implemented SHM approach can therefore be validated using one or more of the following approaches, depending on the criticality of the application and the corresponding stringency of customer requirements:

(1) Detailed simulation models of the subject system can be built and exercised in conjunction with the implemented SHM approach. System component faults can be inserted and the performance of the SHM system validated based on the response of the simulation models. This would require detailed validation of the simulation model itself.

(2) The system could be validated in a laboratory setting, where an actual system is placed in a simulated representative environment on a test bench, and then, system component faults can be inserted and SHM system response in the presence of failures can be verified. The simulation environment will again need to be verified to some degree, but this "hardware in the loop" approach does have the advantage of exercising the actual system.
(3) Finally, failures could be inserted in an actual operational environment (e.g., flight test of an aircraft system) and the SHM system then validated under actual operating conditions.

In practice, a mix of the three methods just described is typically employed based on factors such as availability of system hardware, operational parent systems or laboratory hardware, and cost considerations. Regardless of the methods used, as deficiencies are discovered, corrective actions are implemented and validated before the demonstration is considered successfully completed.

FAs and CNDs are undesired design characteristics that cannot truly be predicted or systematically tested, so there is no true validation possible. However, there is risk reduction activities that can be performed in conjunction with other program validation efforts to reduce the likelihood that FA/CND programs will occur during system operation. The only way to observe these system anomalies is by exercising the system in conditions as close to operating conditions as possible. There is a great deal of uncertainty associated with the design, development, and deployment of a HM system, and often, its performance cannot be accurately predicted due to emergent "metasystem" behaviors deriving from the interaction of the engineered system and its operating environment (as discussed early in this chapter). The more the system is exercised, and particularly if it can be exposed to operational conditions that may exceed the envelope considered during initial design, the better the chance one has to observe and then correct the root cause of the issue prior to actual deployment.

As the validation phase draws to a close, any system failures identified as sufficiently probable and consequential to warrant inclusion in the SHM detection, prediction, isolation, and mitigation strategy that remain undetectable should be addressed by an alternate means in accordance with customer requirements or other remedial action (redesign, additional testing, support system workarounds, etc.). Any additional support system requirements should be documented as part of the evolving requirements for the support infrastructure. Any potential impacts to customer requirements must be coordinated prior to delivering hardware and software to the customer. At the conclusion of the system test and validation stage, the SHM system is a fully qualified, production-ready integrated hardware and software design.

2.4.6 HM System Maturation

The purpose of the system maturation process is to effectively measure actual SHM system performance and to identify/implement corrective actions as required to satisfy all customer needs and expectations. The system maturation stage actually overlaps with system test and evaluation and continues throughout product deployment (corresponding to the core systems engineering utilization and support stages). The maturation process in brief consists of these major activities: (1) collect system operational (including performance and maintenance) data; (2) identify anomalous or unwanted SHM performance issues; (3) perform root cause analysis; (4) identify potential corrective actions; and (5) implement identified changes within formal closed-loop corrective action processes. Corrective actions may include physical system design changes, SHM system design changes, additional supplemental tests, or other support system element or process changes. Another way of viewing the closed-loop corrective action process is that it is just the iteration in the evolution of the SHM system within the product life cycle.

The operation phase of the systems engineering life cycle is where the system actually performs its intended functions. Operations typically involve environmental and human–machine interactions that can have a significant effect on system dependability. Interaction of the system with its operational environment has been, until deployment, a matter of engineering conjecture—actual interactions with the environment may provide significantly different outcomes than those anticipated in system design. This is because the development of effective SHM solutions requires prediction of complex systemic interactions and the effect of presupposed external stimuli. It is nearly always the case that unforeseen emergent behaviors (those that result from unpredicted system interactions) of fielded systems within their operational context create deviations from anticipated SHM system performance. Similarly, the operational infrastructure (including human, facility, and supply chain resources), details of process definitions, extent of operator training, and the functionality of human–machine interfaces are all critical influences on system dependability and may often not be accurately assessed until the system has been deployed within the support infrastructure, and interactions between these elements can be observed.

Initial test and maintenance solutions that are deployed to support new complex systems are therefore generally imperfect (by definition) and are initially liable to contribute substantially to system ownership costs. This suggests a need for processes and tools to (1) monitor the effectiveness of produce HM solutions in their application domains, (2) collect data that validate and document system performance, and (3) pinpoint and analyze relevant patterns that can help mitigate the issues that arise. The ability to mature the effectiveness of fielded system test, diagnostic, and maintenance procedures is a critical factor in an overall system operational and support posture. The process of identifying and implementing corrective actions as required to satisfy customer SHM requirements is known as the health management maturation cycle. A maturation cycle should primarily be initiated as a function of system supportability performance monitoring; however, customer requests and other internal investigations can also trigger a cycle.

Another important consideration is the feedback connection between the requirements, design synthesis/integration, and maturation stages. System requirements are the drivers that guide the designer and ensure that the system does what it is intended to do. The operation of a system in its fielded environment provides the ultimate integration testing (or validation) of system requirements. Observation of unanticipated behaviors can often trigger a new cycle through the SHM system life cycle (or its corresponding product system engineering life cycle).

If the developers who are writing requirements in the planning phase do not understand what the actual operational conditions are, then they cannot write good requirements.

2.5 A Systems-Based Methodology for PHM/CBM Design

A systems-based methodologies have a direct impact on the design of PHM/CBM systems: a formal framework to conduct trade studies that are intended to compare alternative options for the selection of components, sensors, and algorithms and to assist in the selection of "best" alternative technologies according to specified requirements. Failure modes and effects criticality analysis (FMECA) forms the foundation for good PHM/CBM design. In concert with reliability-centered maintenance (RCM), a FMECA study decides on the severity of candidate failure modes, their frequency of occurrence, and their testability. For each failure mode, it considers fault symptoms and the required sensor suite to monitor their behavioral patterns. In advanced versions, FMECA studies also may list the candidate diagnostic and prognostic algorithms that are best suited to address the identified failure modes. New fault data may be required in most cases that are essential for training and validating diagnostic and prognostic routines if historical data collected through on-system monitoring or test bench testing are not sufficient or nonexistent. Means, therefore, must be sought to devise and design a test plan, execute it, and assess the statistical significance of the collected data. Technical and economic performance metrics must be defined to guide the design process and evaluate the effectiveness and performance of the overall PHM/CBM system and its individual modules.

Figure 2.5 depicts the main modules of an integrated approach to PHM/CBM system design with the systems-based components of the architecture described. The schematic indicates feedback loops that are intended to optimize the approach and complete the data collection and analysis steps that are essential inputs to the development of the fault diagnostic and prognostic algorithms.

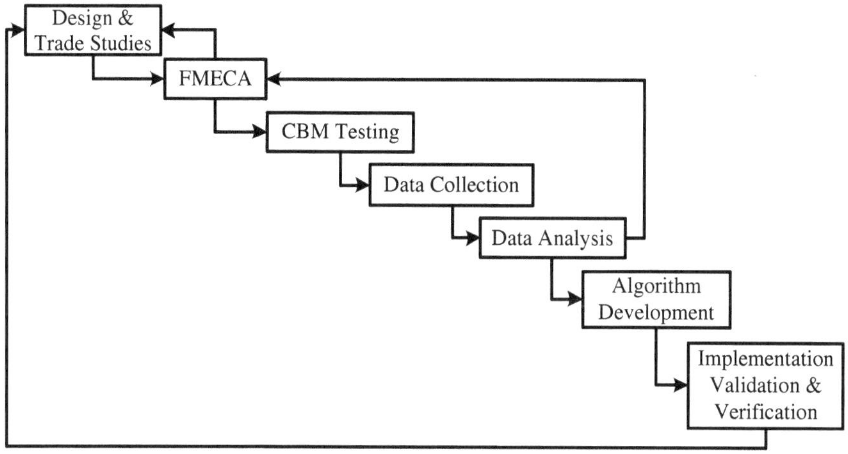

Fig. 2.5 An integrated approach to PHM/CBM design (Vachtsevanos et al. 2006)

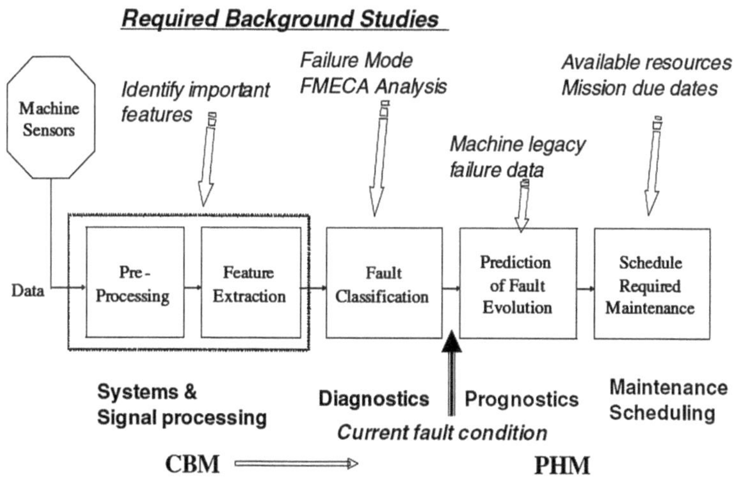

Fig. 2.6 The CBM/PHM cycle (Vachtsevanos et al. 2006)

In Fig. 2.6, one can identify a preliminary off-line phase and an online implementation phase of PHM/CBM. The online phase includes obtaining vehicle/system data from sensors, signal preprocessing, extracting the features that are the most useful for determining the current status or fault condition of the system, fault detection and classification, prediction of fault evolution, and scheduling of required maintenance.

2.6 A Proposed PHM Design Approach for Rotary Machinery Systems

To conduct step-by-step design and deployment of a PHM system, 5S approach is adopted to convert multivariate data to abstract prognostics information, utilizing different computing tools for different steps (Lee et al. 2013). 5S, as shown in Fig. 2.7, stands for Streamline, Smart Processing, Synchronize & See, Standardize, and Sustain.

The *first* "S," Streamline, focuses on identifying critical components and prioritizing data to ensure the accuracy of the second "S," which is Smart Processing. Identifying the critical components for which the prognostics should be performed is the first key step of smart processing by determining which components' degradation or failure has the most significant impact on a system in terms of performance and/or cost of downtime. In real-world applications, data collected from multiple sensors are not necessarily in a readily usable form due to issues such as missing data, redundant data, noise, or even sensor degradation problems. Therefore, it is necessary to sort, filter, and/or prioritize the raw data before processing it.

The *second* "S," Smart Processing, focuses on utilizing computing tools to convert data into information for different purposes, such as health degradation evaluation, performance trend prediction, and potential failure diagnosis. Currently, most manufacturing, mining, farming, and service machines (e.g., elevators) are actually quite "smart" on their own; many sophisticated sensors and computerized components are capable of delivering data concerning status and performance. In many situations, a large amount of data is available, but it is often not known which prognostics technologies should be applied. A systematic methodology for the design of a PHM system should include a means of selecting and combining a set of data-to-information conversion tools to convert machine data into performance-related information to provide real-time health indicators/indices for decision makers to effectively understand the current performance and make maintenance decisions before potential failures occur. This would prevent waste in terms of time, spare parts, and personnel and ensures the maximum uptime of equipment, resulting in significant cost-savings.

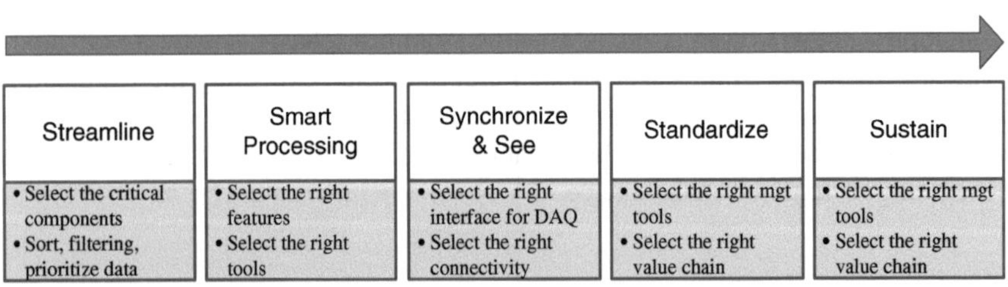

Fig. 2.7 5S approach (Lee et al. 2013)

Synchronize & See is the *third* "S" of the 5S methodology. It integrates the results of the first two S's (Streamline and Smart Processing) to enable the selection of the right hardware solutions and software platforms to most effectively facilitate data-to-information conversion and information transmission. Advanced technologies, such as embedded agents and tether-free communication, are considered to realize prognostics information transparency between manufacturing operations, maintenance practitioners, suppliers, and customers. Prognostics information is demonstrated using information visualization tools. These tools allow decision makers to use decision support tools, based on the delivered information, to assess and predict the performance of machines in order to make the right maintenance decisions before failures can occur. Prognostics information can be further integrated into an enterprise asset management system, which can greatly improve productivity and asset utilization by providing a direct link between machine status and support availability.

The *fourth* "S," Standardize, has great impacts for enterprises, especially in terms of deploying large-scale information technology applications. The interface for acquiring prognostics information from the Synchronize & See stage and importing the information into enterprise business systems, such as supply chain management (SCM) and enterprise resource planning (ERP) systems, needs to be constructed. The implementation of those applications can benefit from a standardized open architecture, information sharing interface, and plant operation flow, which brings cost-effective information integration between different systems that can aid in realizing the implementation of e-manufacturing.

The *fifth* "S," Sustain, aims to technically enable a sustainable closed-loop product life cycle. To accomplish this, management tools need to be selected and value chains need to be defined. Product information, such as product usage profiles, historical data, and middle-of-life (MOL) and end-of-life (EOL) service data, can be provided as feedback to designers and lifecycle management systems.

References

Campbell G, Johnson S, Obleski M (1992) Rocket Engine condition monitoring system. Final report—SHM design methodology prepared for Martin Marietta Space Launch Systems Company, Denver, CO

Dosi G, Hobday M, Marengo L et al (2003) The economics of system integration: toward an evolutionary interpretation. The usiness of systems integration. Oxford University Press, Oxford, pp 95–113

Lee J et al (2013) Prognostics and health management design for rotary machinery systems-reviews, methodology and applications. Mech Syst Signal Process

Stephen et al (2011) System health management with aerospace application. Wiley, pp 100–109. ISBN: 978-7-118-09350-6

Vachtsevanos G, Lewis F, Roemer M et al (2006) Intelligent fault diagnosis and prognosis for engineering systems. Wiley. ISBN: 978-0-471-72999-0

Chapter 3
Overview of Data-Driven PHM

3.1 Introduction

In this book, technical approaches of data-driven PHM are introduced. The related methodology and frameworks are explained. The CBM can be regarded as a technical architecture and engineering strategy for data-driven PHM. Therefore, here we unify CBM and data-driven PHM and give general terms and methodology.

Fault is an abnormal condition or defect at the component, equipment, or subsystem level which may lead to a failure. A system fault occurs when the condition of any system of its components or their assembly is degraded or exhibits an abnormal behavior. This may lead to failure. According to US Federal Standard 1037C, the term *fault* has the following meanings:

- An accidental condition that causes a functional unit to fail to perform its required function.
- A defect that causes a reproducible or catastrophic malfunction. A malfunction is considered reproducible if it occurs consistently under the same circumstances.

Failure in general refers to the state or condition of not meeting a desirable or intended objective. The failure of a component/subsystem occurs when one or more of the principle functions are no longer available. This generally happens when one or more of its components are in a fault condition.

Condition monitoring is defined as a means to prevent catastrophic failure of critical component/subsystem and as a maintenance scheduling tool that uses various analysis data to determine the need for corrective maintenance actions (Davies 1998). The parameters to monitor should be characteristics that will indicate an assets condition. The parameters to monitor should be selected by the ones that in normal mode remain stable but in abnormal or unhealthy mode will indicate some

G. Niu, *Data-Driven Technology for Engineering Systems Health Management*,
DOI 10.1007/978-981-10-2032-2_3

sort of a trend, e.g., increased vibration levels, increased noise, or decreased pressure. (Yam et al. 2001).

Failure analysis is the process of collecting and analyzing data to determine the cause of a failure and how to prevent it from recurring. It is an important discipline in many branches of manufacturing industry, where it is a vital tool used in the development of new products and for the improvement of existing products.

Fault diagnosis is detecting, isolating, and identifying and impending incipient failure condition, while the affected components (subsystem, system) are still operational even though at a degraded mode.

Failure diagnosis is detecting, isolating, and identifying a component (subsystem, system) that has ceased to operate.

Prognosis (*prognostics*) is the estimation of time to failure and risk for one or more existing and future failure modes. An alternative to definition of prognosis is a point estimate of the remaining life of an equipment (or component) based on one or more condition or performance signals observed at some point during its life.

The terms fault detection, isolation, and identification are topically employed to convey the following meaning:

- *Fault (failure) detection.* An abnormal operating condition is detected and reported.
- *Fault (failure) isolation.* Determining which component is failing or has failed.
- *Fault (failure) identification.* Estimating the nature and extent of the fault or failure.

Although the terms *fault isolation* and *fault detection* are sometimes used synonymously, fault detection means determining that a problem has occurred, whereas fault isolation pinpoints the exact cause and location.

As illustration, in order to estimate the remaining useful lifetime or time to failure of a failing component, that is, to conduct fault prognosis and maximize uptime through CBM. We seek to determine accurately and without false alarm impending or incipient failure condition, that is faults. Thus, fault diagnosis is aimed at determining accurately without false alarm impending or incipient failure conditions. In this case, engineer's objective is to achieve the possible performance of a given diagnostic routine by minimizing the false positive or false negative while reducing the time delay between the initiation and detection/isolation of a fault even.

In relation with CBM, the last step of a CBM program is maintenance decision making. Sufficient and efficient decision support would be crucial to maintain personnel's decisions on taking maintenance actions. Techniques for maintenance decision support in a CBM program can be divided into two main categories: diagnosis and prognosis. As mentioned earlier, fault diagnosis focuses on detection, isolation, and identification of faults when they occur. Prognosis, however, attempts to predict faults or failures before they occur. Obviously, prognosis is superior to diagnosis in the sense that prognostics can prevent faults or failures, and if impossible, be ready (with prepared spare parts and planned human resources) for the problems, and thus save extra unplanned maintenance cost. Nevertheless, prognostics cannot completely replace diagnosis since in practice there are always some faults and failures which are not predictable. Besides, prognostics, like any other prediction techniques, cannot be 100 % sure to predict faults and failures. In the case of unsuccessful prediction, diagnosis can be a complementary tool for providing maintenance decision support. In addition, diagnosis is also helpful to improving prognostics in the way that diagnosis information can be useful for preparing more accurate event data and hence building better CBM model for prognostics.

Furthermore, diagnosis information can be used as useful feedback information for system redesign. Jardine (2002) reviewed and compared several commonly used CBM decision strategies such as trend analysis that is rooted in SPC, expert systems, and neural networks. Wang and Sharp (2002) discussed decision aspect of CBM and reviewed the recent development in modeling CBM decision support.

3.2 PHM Technical Approaches

Prognostics is an engineering discipline focused on predicting the time at which a system or a component will no longer perform its intended function (Vachtsevanos 2006). This lack of performance is most often a failure beyond which the system can no longer be used to meet desired performance. The predicted time then becomes the remaining useful life (RUL), which is an important concept in decision making for contingency mitigation. Prognostics predicts the future performance of a component by assessing the extent of deviation or degradation of a system from its expected normal operating conditions (Pecht 2008). The science of prognostics is based on the analysis of failure modes, detection of early signs of wear and aging, and fault conditions. An effective prognostics solution is implemented when there is sound knowledge of the failure mechanisms that are likely to cause the degradations leading to eventual failures in the system. It is, therefore, necessary to have initial information on the possible failures (including the site, mode, cause, and mechanism) in a product. Such knowledge is important to identify the system parameters that are to be monitored. Potential uses for prognostics are in condition-based maintenance. The discipline that links studies of failure mechanisms to system life cycle management is often referred to as prognostics and health management (PHM), sometimes also system health management (SHM) or—in transportation applications—vehicle health management (VHM), or engine health management (EHM). Technical approaches to building models in prognostics can be categorized broadly into data-driven approaches, model-based approaches, and hybrid approaches as follows:

Data-driven approaches are appropriate when the understanding of first principles of system operation is not comprehensive or when the system is sufficiently complex such that developing an accurate model is prohibitively expensive (Liu and Wang 2009). Therefore, the principal advantages to data-driven approaches are that they can often be deployed quicker and cheaper compared to other approaches, and that they can provide system-wide coverage (e.g., physics-based models, which can be quite narrow in scope). The main disadvantage is that data-driven approaches may have wider confidence intervals than other approaches and that they require a substantial amount of data for training. Data-driven approaches can be further subcategorized into *fleet-based statistics* and *sensor-based conditioning*. In addition, data-driven techniques also subsume cycle counting techniques that may include domain knowledge.

As mentioned, a principal bottleneck is the difficulty in obtaining run-to-failure data, in particular for new systems, since running systems to failure can be a lengthy and rather costly process. When future usage is not the same as in the past (as with most non-stationary systems), collecting data that includes all possible future usages (both load and environmental conditions) becomes often nearly impossible. Even where data exist, the efficacy of data-driven approaches is not only dependent on the quantity but also on the quality of system operational data. These data sources may include temperature, pressure, oil debris, currents, voltages, power, vibration and acoustic signal, spectrometric data as well as calibration and calorimetric data. Features must be extracted from (more often than not) noisy, high-dimensional data (Mosallam et al. 2013).

Model-based approaches attempt to incorporate physical understanding (physical models) of the system into the estimation of RUL. Modeling physics can be accomplished at different levels, for example, micro and macro levels. At the micro level (also called material level), physical models are embodied by series of dynamic equations that define relationships, at a given time or load cycle, between damage (or degradation) of a system/component and environmental and operational conditions under which the system/component are operated. The *micro-level models* are often referred as damage propagation model. Yu and Harris's fatigue life model for ball bearings, which relates the fatigue life of a bearing to the induced stress, (Yu and Harris 2001) Paris and Erdogan's crack growth model, (Paris and Erdogan 1963) and stochastic defect propagation model [citation needed] are other examples of micro-level models. Since measurements of critical damage properties (such as stress or strain of a mechanical component) are rarely available, sensed system parameters have to be used to infer the stress/strain values. Micro-level models need to account in the uncertainty management of the assumptions and simplifications, which may pose significant limitations of that approach.

Macro-level models are the mathematical model at system level, which defines the relationship among system input variables, system state variables, and system measures variables/outputs where the model is often a somewhat simplified representation of the system, for example, a lumped parameter model. The trade-off is increased coverage with possibly reducing accuracy of a particular degradation mode. When this trade-off is permissible, faster prototyping may be the result. However, when systems are complex (e.g., a gas turbine engine), even a macro-level model may be a rather time-consuming and labor-intensive process. As a result, macro-level models may not be available in detail for all subsystems. The resulting simplifications need to be accounted for by the uncertainty management.

Hybrid approaches attempt to leverage the strength from both data-driven approaches as well as model-based approaches (Pecht 2010; Liu et al. 2012). In reality, it is rare that the fielded approaches are completely either purely data-driven or purely model-based. More often than not, model-based approaches include some aspects of data-driven approaches and data-driven approaches glean available information from models. An example for the former would be where model parameters are tuned using field data. An example for the latter is when the set point, bias, or normalization factor for a data-driven approach is given by models. Hybrid approaches can be categorized broadly into two categories, preestimate fusion and post-estimate fusion.

3.3 Data-Driven PHM/CBM System Architecture

In order for a system to achieve full potential as a data-driven PHM/CBM system, it needs to be constructed by a number of different functional capabilities. The open system architecture for condition-based maintenance organization (OSA-CBM) has specified an open standard proposal on how a CBM system should be designed technically. The OSA-CBM is an industry consortium that includes industrial, commercial, and military participants. The open, non-proprietary, standard proposal was developed in order to create a free market for CBM components, where users of CBM technology will be able to choose CBM components from different manufactures. The organization has divided a CBM system into seven different technical modules (Thurston 2001) (see Fig. 3.1) including prognosis function, which can be regarded as an infrastructure of data-driven PHM. The standard proposal covers more than the technical design of CBM systems, e.g., means of communication within the system; this chapter, though, will solely focus on the architecture design.

Layer 1 Sensor module: The sensor module provides the CBM system with digitized sensor or transducer data.

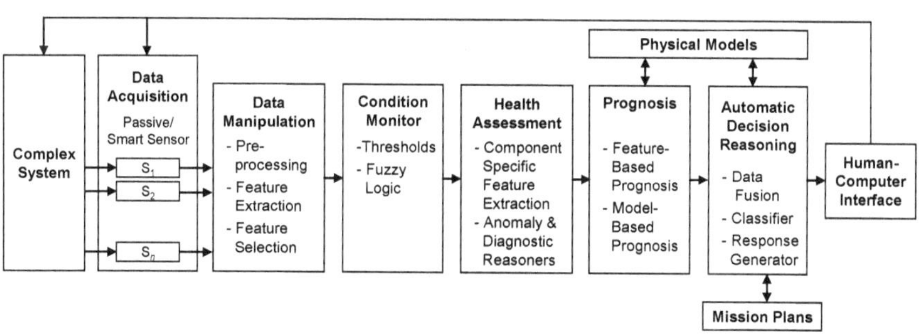

Fig. 3.1 The seven modules in the OSA-CBM architecture (Lebold et al. 2003)

Layer 2 Signal processing: The signal processing module receives signals and data from the sensor module or other signal processing modules. The output from the signal processing module includes digitally filtered sensor data, frequency spectra, virtual sensor signals, and other CBM features.
Layer 3 Condition monitor: The condition monitor receives data from the sensor modules, the signal processing modules, and other condition monitors. Its primary focus is to compare data with expected values. The condition monitor should also be able to generate alerts based on preset operational limits.
Layer 4 Health assessment: The health assessment module receives data from different condition monitors or from other health assessment modules. The primary focus of the health assessment module is to prescribe if the health of the monitored component, subsystem, or system has degraded. The health assessment module should be able to generate diagnostic records and propose fault possibilities. The diagnosing should be based upon trends in the health history, operational status and loading, and maintenance history.
Layer 5 Prognosis: The prognosis module should have the possibility to take account data from all the prior layers. The primary focus of the prognosis module is to calculate the future health of an asset, with account taken to the future usage profiles. The module should report the future health status of a specified time or the RUL.
Layer 6 Decision support: The decision support module receives data from the health assessment module and the prognosis module. Its primary focus is to generate recommended actions and alternatives. The actions can be related to maintenance or how to run the asset until the current mission is completed without occurrence of breakdown.
Layer 7 Presentation: The presentation module should present data from all the previous modules. The most important layers to present would be the data from the health assessment, prognosis, and decision support modules as well as alerts generated from the condition monitors. The ability to look even further down in the layer should be a possibility. The presentation module could be built into a regular machine interface.

3.4 Role of Condition Monitoring, Fault Diagnosis, and Prognosis

Data-driven PHM/CBM seeks to implement a policy wherein maintenance management decisions are based on the identification of the current condition of monitored machinery. The implementation of efficient maintenance management strategies based on CBM presupposes that adequate condition monitoring, as well as system fault diagnosis and prognosis are in place (Jardine 1973).

Reliable prognosis itself relies on condition monitoring and fault diagnosis, as prediction of the future state of system implies that the current state of system is known. Therefore, condition monitoring and fault diagnosis are central to the implementation of efficient maintenance management strategies (Emmanouilidis et al. 2006). In terms of condition monitoring and fault diagnosis, this means that the primary use is to predict unplanned equipment failures and assess equipment condition. Through the use of condition monitoring and fault diagnosis technologies, the following benefits can be achieved:

- Improving equipment reliability through the effective prediction (and then avoidance) of equipment failures.
- Minimizing downtime through the integrated planning and scheduling of repairs indicated.
- Maximizing component life by avoiding the conditions that reduce equipment life (for example, by ensuring ongoing precision alignment, minimal lubricant contamination).
- Reducing excessive electric power consumption caused by inefficient machinery performance.
- Reducing the number of dissatisfied customers or lost customers and rework due to poor quality—with less than optimal machine performance.
- Improving operator safety through reducing the potential for destructive failure which could cause personal injury or death.
- Minimizing maintenance costs.

3.5 Fault Diagnosis Framework

A general fault diagnosis can be performed using residual signal, and this technique is depicted in Fig. 3.2. Residual signal is a signal that represents a deviation from standard operating conditions which is generated by comparing, for example, a model output with the actual system output. Based on this signal, one makes decision about the operating condition of the machinery.

The other techniques for fault detection are model-based and data-driven, as shown in Fig. 3.3. Model-based technique relies on an accurate dynamic model of the system and is capable for detecting even unanticipated faults. It takes advantage of the actual system and model outputs to generate a discrepancy or residual, as it is known, between two outputs that are indicative of a potential fault condition. In this method, bank of filter can be used to pinpoint the identity of the actual fault component.

Fig. 3.2 General fault diagnosis

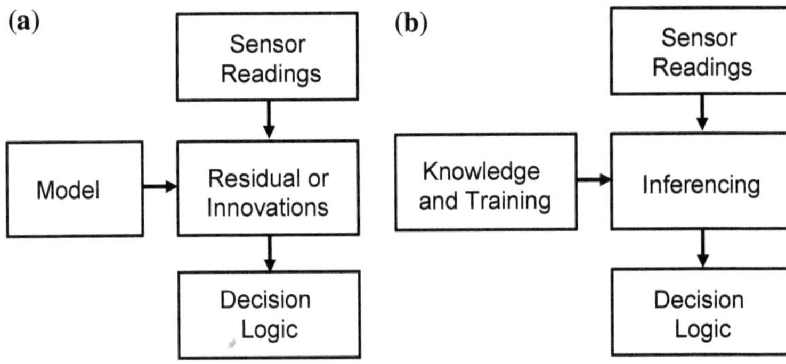

Fig. 3.3 Model-based and data-driven fault diagnosis. **a** Model-based, **b** Data-driven

On the other hand, data-driven technique often addresses only anticipated fault conditions, where a fault model now is a construct or a collection of constructs such as neural network, expert systems, fuzzy logic, support vector machines (SVMs), etc. that must be trained first with known prototype fault pattern and then employed online to detect and determine the faulty component's identity.

In this book, we will present feature-based fault diagnosis technique using fault representation. A general procedure for online fault diagnosis is shown in Fig. 3.4. The major task in this method is data collection through sensor outputs and building of a feature representation or feature vector that contains enough information about the current machine operating condition to allow for fault identification and classification. The information contained in the feature vectors can be obtained by three ways:

- Model-based method (e.g., identification of physical model parameters using, for example, Kalman filter or recursive least squares)
- Data-driven method (e.g., vibration and current signal statistical moments, vibration and current frequency spectrum information)
- Statistical regression and clustering technique on existing historical legacy data.

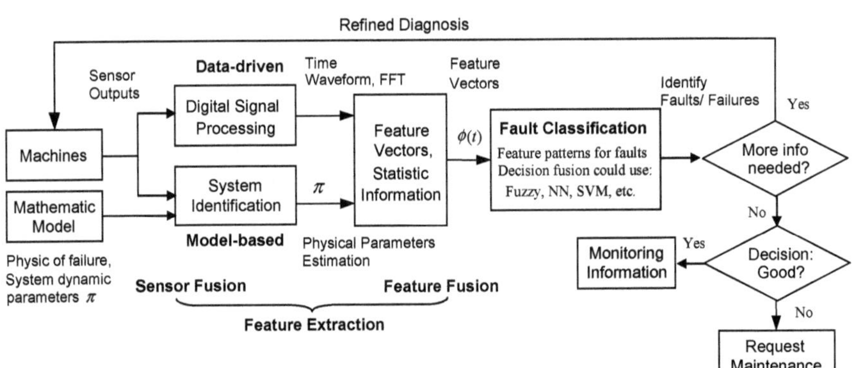

Fig. 3.4 Online fault diagnosis system

Once the feature vectors have been obtained, it is used as input to a fault classification that contains various types of decision-making algorithms. If more information is needed before a decision can be reached, probing signals can be sent back to the machine (e.g., inject test vibration signal into a machinery mount to test for loose fittings).

3.6 Problems During Implementation

While importance of condition monitoring and fault diagnosis is evident, reliable automated identification of system condition is not straightforward, as system malfunction demonstrates itself with different signal patterns, depending on the particular kind of problem encountered. These patterns of behavior can vary considerably even for the same fault types on similar system. During maintenance implementation, the problems encountered are listed:

(1) No associates training on maintenance know-how for top persons which results in lack of management support
(2) Lack of operator's comprehension
(3) Lack of resources
(4) Lack of involvement of production associates
(5) Lack of long-term vision; quick return expected by management
(6) Lack of sustained momentum.

People management level is the key to success or failure. They have to be well trained. From the training, they can learn about the benefits and effects of the maintenance program in the company. They know that management commitment for maintenance implementation is the fundamental precept. Also they can realize that maintenance is not a onetime activity for continuous improvement. It is a long term and never ending improvement activity. In this way, issues (1), (4), (5), and (6) can be resolved. Maintenance is carried out in cooperation with operators, maintenance technicians, and engineers. The advantage of having such an operator involved is that the operator knows the environment and conditions better than anyone. Thus, brief introductory training for all is necessary. Through training, issue (2) can be overcome, and they can understand the actual meaning of maintenance implementation as well as their role and responsibility. Issue (3) is a vital factor that constrains maintenance performance since current maintenance is still far away from full automation. Due to the lack of manpower and information resources, the diagnosis and repair on failed equipment usually cannot be performed immediately, hence lead to long downtime of equipment and causes a significant production loss. Advanced techniques provide a solution to this issue. Recently, artificial intelligence techniques have been widely applied for equipment degradation assessment, statistic failure analysis, prognosis, and intelligent diagnosis. For instance, expert systems, artificial neural networks (ANNs), fuzzy logic systems, and genetic algorithms have been employed to assist the diagnosis and condition monitoring task to correctly interpret the fault data. The application of AI techniques makes maintenance process intelligent, improves accuracy, and decreases people involvement. Effective maintenance case management also can relieve this pressure. Case-based reasoning should be the best choice for this task, including retain, retrieve, reuse, and revise.

Although condition monitoring, fault diagnosis, and prognosis can assist in improving maintenance during planned maintenance phase, most of final results are failure due to lack of long-term vision-quick return expected by management, lack of sustained momentum, and lack of resource, etc. Among above mentioned reasons, lack of manpower is one of the main factors.

Condition monitoring, fault diagnosis, and prognosis require experienced staff whose role is to diagnose abnormal plant conditions and advise operational personnel on the remedial actions required. This means that accurate diagnosis and prognosis of equipment failure is largely dependent on the skill and experience of individual staff. Failure to do so can lead to incorrect diagnosis and prognosis being made. The skill of specialists is based on the combination of quantitative and qualitative information regarding a problem, in addition to their extensive experience in dealing with similar problems.

Unfortunately, such specialists are relatively rare in the field. Even if specialists are available, the technical information needed by the engineers is not always to hand or received in the first instance. This is because the information is distributed centrally, but it is the responsibility of the distributor or subsidiary to relay it, and there are difficulties in remembering and applying this amount of knowledge for experts. Also, there are a number of problems associated with traditional method of condition monitoring and diagnosis as follows:

- The number of specialists skilled in the monitoring analysis, diagnosis, and prognosis of plant problems has been reduced due to the rationalization of operational functions within the industry. Also, it is sometimes practically infeasible for experts to be available to monitor the plant on a 24-h basis.
- Specialists might require additional sources of effective guidance (e.g., they may need to collaborate with other specialists, or use simulation, modeling, or statistical tools), particularly when under severe time pressure, in order to support operational decision making.
- Although modern monitoring and control systems provide plant operator with immediate access to a range of raw data, only domain specialists with clear diagnostic or prognostic knowledge are capable of providing qualitative interpretation of acquired data, an ability which will be lost when the specialists retire.

Many intelligent diagnosis and prognostics tools have been developed with successful results in the identification of faults that occur during operation. However, experience in the design, implementation, and operation of intelligent systems for fault diagnosis and prognosis in engineering applications has shown that the scope, functionality, and process of constructing such systems could be significantly enhanced through the integration of a range of appropriate techniques.

3.7 Related Techniques

In order to solve above stated problems, many techniques have been developed for last two decades. The most common and widely used tool in modern industry is spectral analysis of the vibration signals. Over the last 20 years, the connections between certain spectral properties and the fundamental nature of various vibration problems and their origins have been perfected and widely used in the industrial fields.

The fault diagnosis system should perform two tasks: fault detection and fault diagnosis. The former is to determine that a fault has occurred in the system. To achieve this goal, all the available information from the system should be collected and processed to detect any change from nominal behavior of the process. The latter is devoted to locate the fault category or the fault source. In rotating machinery, the root cause of faults is often from rolling element bearings, gears, and so on. One way to increase operational reliability of machine is to monitor and diagnose incipient faults in these mechanical elements. Generally, vibration signals can be used to detect the incipient fault of the machine components and reduce the possibility of catastrophic damage and the downtime, through the online monitoring and diagnosis system.

Recently, numerous *machine learning* techniques have been applied to the areas of fault diagnosis and prognosis, such as ANNs, genetic algorithm, fuzzy reasoning, decision tree, case-based reasoning, et al. Artificial intelligence techniques have been employed to assist the diagnosis and condition monitoring task to correctly interpret the fault data, which makes monitoring and diagnosis process intelligent, improves accuracy, and decreases people involvement. Machine learning refers to a system capable of the autonomous acquisition and integration of knowledge. This capacity to learn from experience, analytical observation, and other means, results in a system that can continuously self-improve and thereby offer increased efficiency and effectiveness. Learning, like intelligence, covers such a broad range of processes that it is difficult to define precisely. A dictionary definition includes phrases such as *to gain knowledge, or understanding of, or skill in, by study, instruction, or experience*, and *modification of a behavioral tendency by experience*. Certainly, many techniques in machine learning derive from the efforts of psychologists to make more precise their theories of animal and human learning through computational models. It seems likely also that the concepts and techniques being explored by researchers in machine learning may illuminate certain aspects of biological learning.

As regards machines, very broadly, that a machine learns whenever it changes its structure, program, or data based on its inputs or in response to external information in such a manner that its expected future performance improves. Some of these changes, such as the addition of a record to a database, fall comfortably within the province of other disciplines and are not necessarily better understood for being called learning. Machine learning usually refers to the changes in systems that perform tasks associated with artificial intelligence. Such tasks involve recognition, diagnosis, planning, prediction, etc. (Nilsson 1996).

Pattern recognition is the research area that studies the operation and design of systems that recognize patterns in data. It encloses subdisciplines like discriminant analysis, feature extraction, error estimation, cluster analysis (together sometimes called statistical pattern recognition), grammatical inference and parsing (sometimes called syntactical pattern recognition). Important application areas are image analysis, character recognition, speech analysis, man and machine diagnostics, person identification, and industrial inspection. A branch of artificial intelligence concerned with the classification or description of observations. Pattern recognition aims to classify data (patterns) based on either a priori knowledge or on statistical information extracted from the patterns. The patterns to be classified are usually groups of measurements or observations, defining points in an appropriate multi-dimensional space.

Some intelligent learning algorithms, such as ANNs and SVMs have been successfully applied to automated detection, diagnosis, and prognosis of system conditions. The main difference between ANNs and SVMs is in their risk minimization. For SVMs, structured risk minimization principle is used to minimize an upper bound based on an expected risk. In ANNs, traditional empirical risk minimization is used to minimize the error in training of data. The difference in risk minimization leads to a better generalization performance for SVMs than ANNs. However, in real-world applications, the classification result of the practically implemented SVMs is often far from the theoretically expected level, because their implementations are based on the approximated algorithms due to the high complexity of time and space.

Data mining, known as *knowledge discovery* in databases, is the process of discovering interesting patterns in databases that are meaningful in decision making and is also an application area that can provide significant competitive advantage to an organization. According to a general model (Fig. 3.5), in the first step for data

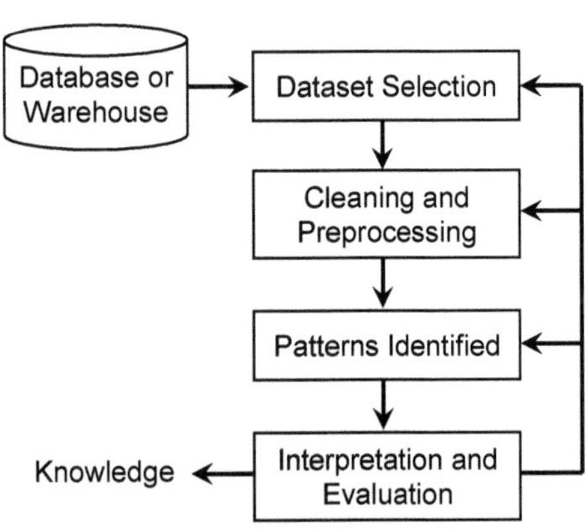

Fig. 3.5 Data mining phases

mining the main data sets that may be retrieved from operational databases or data warehouse are selected. The selected data set then undergoes cleaning and pre-processing for removing discrepancies and inconsistencies to improve its quality (Huang et al. 2007).

Next, the data set is analyzed to identify patterns that represent relationship among data by applying algorithms, such as neural networks, decision trees, and so on. Then the patterns are validated with new data set. A pattern that satisfies these conditions becomes organizational knowledge. The steps in the mining process are iterative until meaningful knowledge is extracted.

With the development of computer and information technologies, traditional data-based condition monitoring, fault diagnosis, and prognosis are gradually replaced by feature-based due to fast transfer, small storage space, and high accuracy. Here, feature means some value that can represent machines conditions. Features also can be considered as data compression. Referring to features, feature extraction and feature selection technologies are indispensable which have gained much attention recently.

References

Davies A (ed) (1998) Handbook of condition monitoring. Chapman & Hall, Great Britain, Cornwall

Emmanouilidis C, Jantunen E, MacIntyre J (2006) Flexible software for condition monitoring, incorporating novelty detection and diagnostics. Comput Ind 57:516–527

Huang MJ, Chen MY, Lee SC (2007) Integrating data mining with case-based reasoning for chronic diseases prognosis and diagnosis. Expert Syst Appl 32:856–867

Jardine AKS (1973) Maintenance replacement and reliability. Pitman, Wiley, London

Jardine AKS (2002) Optimizing condition based maintenance decisions. In: Proceeding of the annual reliability and maintainability symposium, pp 90–97

Lebold M, Reichard K, Boylan D (2003) Utilizing DCOM in an open system architecture framework for machinery monitoring and diagnostics. Aerosp Conf Proc 3:1227–1236

Liu J, Wang G (2009) A multi-step predictor with a variable input pattern for system state forecasting. Mech Syst Signal Process 23(5):1586–1599

Liu J, Wang M, Yang Y (2012) A data-model-fusion prognostic framework for dynamic system state forecasting. Eng Appl Artif Intell 25(4):814–823

Mosallam A, Medjaher K, Zerhouni N (2013) Nonparametric time series modelling for industrial prognostics and health management. Int J Adv Manuf Technol 69(5):1685–1699

Nilsson NJ (1996) Introduction of machine learning. Draft of incomplete notes. http://ai.stanford.edu/~nilsson

Paris PC, Erdogan F (1963) A critical analysis of crack propagation laws. ASME J Basic Eng 85:528–534

Pecht MG (2008) Prognostics and health management of electronics, Wiley. ISBN 978-0-470-27802-4

Pecht M (2010) A prognostics and health management roadmap for information and electronics-rich systems. Microelectron Reliab 50(3):317–323

Rockwell International Corporation (2004) Reaping the rewards of strategic maintenance. GMSG10-WP001A-EN-D

Smith AM (1993) Reliability-centered maintenance. McGraw-Hill, New York

Thurston MG (2001) An open standard for web-based condition-based maintenance systems. In: Proceedings of IEEE systems readiness technology conference, pp 401–415

Vachtsevanos G, Lewis F, Roemer M, Hess A, Wu B (2006) Intelligent fault diagnosis and prognosis for engineering systems, Wiley. ISBN 0-471-72999-X

Wang W, Sharp J (2002) Modeling condition-based maintenance decision support. Condition Monitoring: Engineering the Practice, Bury St Edmunds, pp 79–98

Wireman T (1990) World class maintenance management. Industrial Press, New York

Yam RCM, Tse PW, Li L, Tu P (2001) Intelligent predictive decision support system for condition-based maintenance. Int J Adv Manuf Technol 17(5):383–391

Yu WK, Harris TA (2001) A new stress-based fatigue life model for ball bearings. Tribol Trans 44 (1):11–18

Chapter 4
Data Acquisition and Preprocessing

4.1 Introduction

The process of condition monitoring, fault diagnosis, and prognosis can be summarized as follows: data acquisition, data processing, data analysis, and decision making. Here, data represent system condition, so it can be called condition-based (data-driven) systems health management. The main problems of such a system are data transfer and data storage. For example, when we monitor a huge generator condition, we should install several of sensors to assure diagnosis reliability. Such many sensors result in a lot of data coming out.

With the globalization and fast growth of the Internet technologies and computer and information technologies, online or continuous condition monitoring has gained much attention. Data transfer and data storage problem become more serious. If transferring plenty raw data directly, longtime delays due to heavy traffic may be experienced, which results in the lost of monitoring and diagnosis in time.

Data represented as features provide a best solution for this problem that greatly reduces the requirement of transfer number and save storage space. That data are compressed as features from many domains with keeping data information as high as possible. Relative feature construction techniques come out, such as feature representation, feature extraction, and feature selection.

The typical feature-based condition monitoring and fault diagnosis structure are illustrated in Fig. 4.1 that can be summarized as follows:

- Data are acquired online after preprocessing. The data can be vibration, voltage, current, acoustic emission, sound signals, etc. Corresponding to the object, different preprocessing approaches are used, such as filtering (high-, low-, and band-pass), wavelet transform, and averaging.
- Features are calculated from various domains: time, frequency, cepstrum, and wavelet domain. In this way, the information of raw data is kept at best to meet different analysis methods in future. Furthermore, the transfer and storage problems of data can be solved.

G. Niu, *Data-Driven Technology for Engineering Systems Health Management*,
DOI 10.1007/978-981-10-2032-2_4

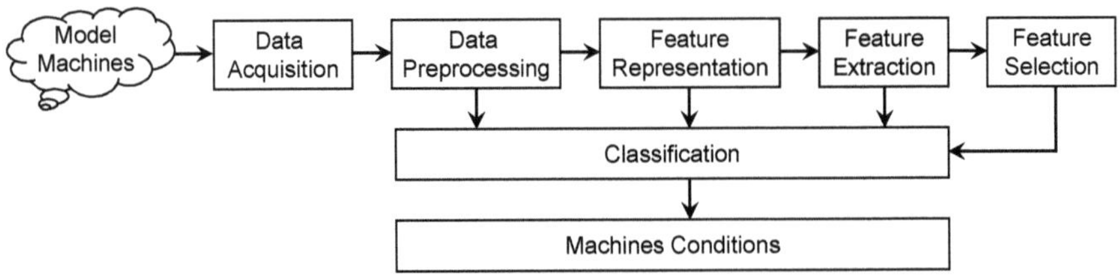

Fig. 4.1 Feature-based condition monitoring and fault diagnosis system

- Many calculation parameters and many domains result in many features. All of them are not useful for condition analysis; sometimes, some of them even can increase analysis difficulty and degrade accuracy. Reducing feature dimensionality is necessary, which removes excessive and garbage features.
- According to monitoring object, the features that can significantly represent equipment performance are selected.
- The selected features are sent to condition monitoring system and fault diagnosis system to get equipment condition.

4.2 Data Acquisition

Data acquisition is a process of collecting and storing useful data (information) from targeted engineering assets for the purpose of condition monitoring and diagnosis. This process is an essential step in implementing for machinery fault diagnosis and prognosis.

Data collected in a condition-based maintenance program can be categorized into two main types: event data and condition monitoring data. *Event data* include the information on what happened (e.g., installation, breakdown, and overhaul, and what the causes were) and/or what was done (e.g., minor repair, preventive maintenance, and oil change) to the targeted asset (Jardine et al. 2006). *Condition monitoring data* are the measurements related to the health condition/state of the engineering asset. Condition monitoring data are very versatile. It can be vibration, acoustic, oil analysis, temperature, pressure, moisture, humidity, weather or environment data, etc. Various sensors, such as accelerometer, proximity probe, ultrasonic sensors, acoustic emission sensors, current probe, and flux sensors, have been designed to collect different types of data.

Taking data acquisition of rotating machinery as example, the vibration signal is usually selected and will be reviewed in detail. The acquisition of proper vibration data is the key to effective machine monitoring, fault diagnosis, and prognosis. Quality data acquisition requires planning involving the machine, the nature of expected vibration data, available instrumentation, and the purpose of the testing. The topic presented in this section includes the selection of the measures (displacement, velocity, or acceleration), transducers, mounting technique, and location. The process of data acquisition including acquisition times and sample size must be considered well before storage process due to digitization of measured data. It is important to show the proper displays that are useful for analysis and evaluation.

4.2.1 Selecting a Proper Measure

A measure is defined as a unit or standard of measurement that provides a means for evaluating data. In vibration analysis, three measures are available: *displacement*, *velocity*, and *acceleration*. The measure is selected based on the frequency content of vibration present, type of machine, type of the analysis to be conducted, and information sought.

Absolute displacement is used for low frequency of structural vibration, 0–20 Hz. This is typically measured with a double-integrated accelerometer. *Relative shaft displacement*, which is measured with a proximity probe, shows the extent of bearing clearance taken up by vibration and is used over a wide frequency range.

Velocity is the best measure. Velocity is defined as a time rate of change of displacement which is dependent upon both frequency and displacement, and also, it is related to fatigue. It is considered to be a good measure in the span of 10 Hz–1 kHz for general condition monitoring of machinery because a single value of root mean square (RMS) or peak velocity can be used in rough assessment of condition without the need to consider frequency. This is typically measured with a single-integrated accelerometer.

Acceleration is the measure used above 1 kHz. It relates to force and is used for such high-frequency vibration as gear mesh and rolling element bearing defects. Acceleration and velocity are absolute measures taken on the bearing housing or as close to it as possible. Relative displacement between the machine housing and the rotor is typically measured by a permanently mounted proximity probe.

The summary of measures in vibration analysis is presented in Table 4.1, and the default of frequency spans of data acquisition is shown in Table 4.2.

Table 4.1 Machine vibration measure

Measure	Frequency span	Physical parameters	Application
Relative displacement	0–1 kHz	Stress/motion	Relative motion in bearing/casing
Absolute displacement	0–20 Hz	Stress/motion	Structural motion
Velocity	10 Hz–1 kHz	Energy/fatigue	General machine condition, medium-frequency vibration
Acceleration	>1 kHz	Force	General machine condition, medium-/high-frequency vibration

Table 4.2 Default of frequency spans for data acquisition

Component	Frequency span
Shaft vibration	10 × rotating frequency
Gearbox	3 × gear mesh frequency
Rolling element bearing	10 × ball frequency of inner race
Pumps	3 × vane passing frequency
Motors/generators	6 × line frequency
Fans/blowers	3 × blade-passing frequency
Sleeve bearings	10 × rotating frequency

Example 4.1 Select measure for an induction motor 2500 RPM that uses ball bearing (SKF 230–600) with $N = 29$ rolling elements.

Solution: Because the highest rolling element bearing is the ball frequency of inner race (BPFI), it can be estimated as

$$\begin{aligned}\mathrm{BPFI} &= 0.6 \times \mathrm{RPM} \times N \\ &= 0.6 \times 2500 \times 29 = 43{,}500\,\mathrm{CPM} = 725\,\mathrm{Hz}\end{aligned}$$

where 0.6 is estimated value of BPFI (see Eshleman 1999).

Therefore, the frequency span is 7,250 Hz (see Table 4.2 for BPFI). This frequency is within the acceleration range (see Table 4.1).

Often, the selection of the parameter to be measured is predetermined by related standards or by specifications. When this is not the case, it is helpful to apply the considerations given in Table 4.3 (Harris 1997).

4.2.2 *Vibration Transducers*

A transducers or sensor is a device that receives a signal and responds with an electrical signal. A sensor or other technical measurement devices have some advantages to human inspection: They are reliable and precise, they can measure in unhealthy and hazardous conditions, they work fast, they work continuously, and they can perform measurements to a relatively low cost.

The information of vibration is acquired by transducers positioned at optimal location on the machine system. Transducers convert mechanical vibrations to electronic signals that are conditioned and processed by digital signal analyzer instrument. Transducer selection is based on the sensitivity, size required, selected measure, frequency response, and machine design and speed. The response of any transducers determines how well the instruments respond to a stimulus (voltage or vibration) at a given frequency. Vibration analysts want a flat frequency response at all frequencies.

Figure 4.2 shows the frequency response of a velocity transducer that is not flat. At the lower frequencies, it rolls off, and it responds less to the same strength signal

Table 4.3 A Guide for the selection of the parameter to be measured

Measure	Consideration
Acceleration	• Suitable for high frequencies where acceleration measurements provide the highest signal outputs • Used where forces, loads, and stresses must be analyzed; force is known proportional to acceleration • Used where a transducer of small size and small mass is required, since the accelerometers usually are somewhat smaller than velocity or displacement pickup
Velocity	• Used where vibration measurements are to be correlated with acoustic measurements since the sound pressure is proportional to the velocity of the vibrating surface • Used at the intermediate frequencies where displacement measurement yields transducer output which may be too small to measure conveniently • Used extensively in measurements of machinery where the velocity spectrum usually is more uniform than either the displacement or acceleration spectra
Displacement	• Used where amplitude of displacement is particularly important, e.g., where vibrating parts must not touch or where displacement beyond a given value results in equipment damage • Used where the magnitude of displacement may be an indication of stresses to be analyzed • Used at low frequencies where output of the accelerometers or velocity pickups may be too small for useful measurement • Used to measure relative motion between rotating bodies and structures of machine

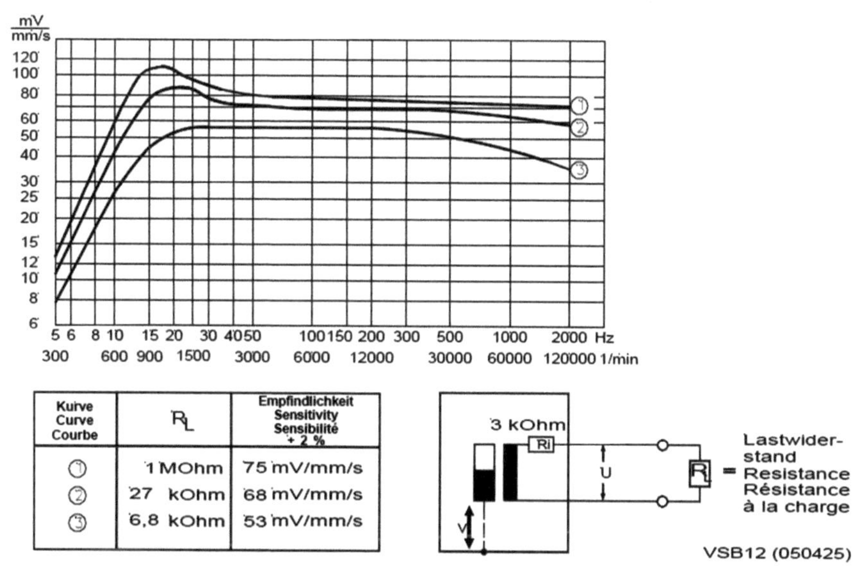

Fig. 4.2 An example of frequency response of velocity transducer (*B&K Vibro*)

than it does at frequencies greater than 30 Hz. The sensitivity of the transducer is depend on its voltage output for given vibration input, for example, 200 mV/mil, 120 mV/mm/s, 500 mV/(in/s), and 100 mV/g. The higher the voltage output per engineering unit, the more the sensitive transducers.

(1) *Proximity probes*

The proximity probe or non-contacting displacement transducer which is in a general class of transducers measures the static and dynamic displacement of a shaft relative to the bearing housing (Fig. 4.3). Several position sensing techniques are used by non-contact displacement transducers, including capacitance, optical, laser, and inductive (eddy current) transducers. The most widely used non-contacting probes on rotating machinery are the eddy current type, commonly known as proximity probes. Proximity probes are used because they are durable, and they can directly measure shaft to bearing clearance.

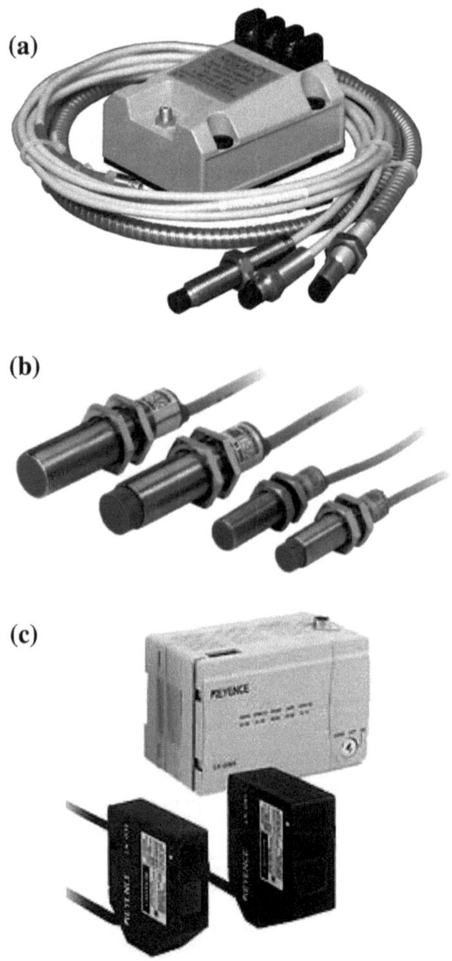

Fig. 4.3 Proximity probe. **a** Eddy current type (*Bently Navada*), **b** inductive type (*Rhein Taco*), and **c** laser type (*Keyence*)

Usually, it is permanently mounted on many machines for condition monitoring and fault diagnosis. The eddy current proximity probes consist of two parts, the probe and the oscillator–demodulator (Fig. 4.3). The high-frequency oscillator is used to induce an eddy current on the shaft's surface without actually touching it. The probes can sense the gap between the probe tip and the conductive surface of the rotating shaft. As the shaft moves relative to the sensor, more or less energy goes into the eddy current. These eddy current energy changes modulate the amplitude of the oscillator voltage. This signal is demodulated, providing an output voltage proportional to the change in gap. As the rotating surface moves away from the probe tip, less eddy current energy is lost and the voltage output of the demodulator becomes more negative. As the gap gets smaller, more eddy current is lost in the shaft surface and the output becomes less negative. The indicated gap change causes the voltage to change about the probe's set point.

The probe is coil of wire surrounded by a non-conductive plastic or ceramic material contained in a threaded body. An oscillator–demodulator is required to excite the probe that is about 1.5 MHz. The resulting magnetic field radiates from the tip of probe. When a shaft is brought close to the probe, eddy currents are induced that extract energy from the field and decrease its amplitude. This decrease in amplitude provides an AC signal directly proportional to vibration. The DC voltage from oscillator–demodulator varies in proportion to distance between the probe and the conducting material. The sensitivity of the probe generally 200 mV/mil (8 mV/μm) with a gap ranges from 0 to 80 mils. The oscillator–demodulator requires a supply of 24 V DC. The probe must be shielded and grounded.

(2) *Velocity transducers*

The velocity pickup (Fig. 4.4) is a very popular transducer for condition monitoring the vibration of rotating machinery. This transducer installs easily on machines and generally costs less than other sensors. The transducer is ideal for general purpose machine applications. Velocity transducers have been used on rotating machines for a very long time and are still utilized for a variety of applications today. Velocity transducers are available in many different physical configurations and output sensitivities.

When a coil of wire is moved through a magnetic field, a voltage is induced across the end wires of the coil. The induced voltage is caused by the transferring of

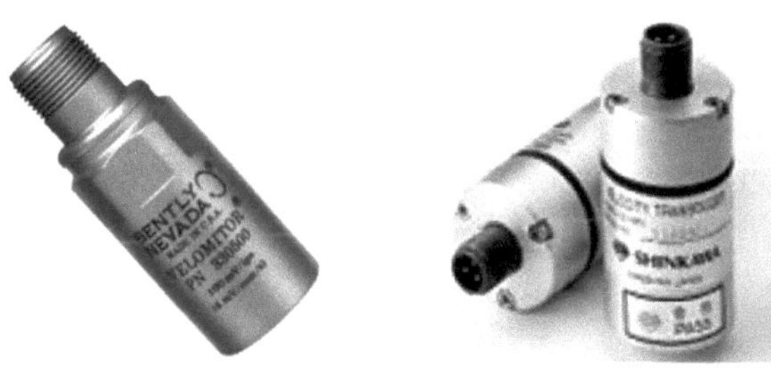

Fig. 4.4 Velocity transducer (*Courtesy of Bently Nevada, Shinkawa*)

energy from the flux field of the magnet to the wire coil. As the coil is forced through the magnetic field by vibratory motion, a voltage signal representing the vibration is produced. The self-generated signal can be directly passed to an oscilloscope or signal analyzer.

Velocity transducers will have different frequency responses depending on the manufacturer. However, most transducers have a frequency response range in the order of 10 Hz–2 kHz. This is an important consideration when selecting a velocity pickup for a rotating machine application. The transducer's frequency response must be within the expected vibration frequencies of the machine. The transducer is self-excited that requires no power supply and consists of a permanent magnet mounted on springs encased in a cylindrical coil of wires. A typical velocity transducer generates 500 mV/(in/s) except at frequencies below 10 Hz which is the natural frequency of the active element. The reduction of output below 10 Hz requires that a frequency-dependent compensation factor be applied to the amplitude of the signal. The measured phase also changes with frequency at frequencies below 10 Hz. The velocity transducer can be used to evaluate vibration velocity in order to assess machine condition when the frequency range of concern is within the flat frequency response (10 Hz–2 kHz) of the transducer. Velocity transducer can be used to measure shaft vibration with a fish tail, a simple wooden device that attaches to the transducer. A vee notch permits the fish tail to ride on the rotating shaft. Keys and other variations of the shaft surface pose to safety hazards.

(3) *Accelerometers*

Accelerometers have been a popular choice for rotating machinery vibration monitoring. They are a rugged, compact, lightweight transducer with a wide frequency response range. Accelerometers have been used extensively in many machinery monitoring applications. This transducer is typically attached to the outer surface of machinery. Generally, this machinery will have parts that generate high-frequency signals, such as rolling element bearings or gear sets. Accelerometers are used to measure vibration levels on casing and bearing housings. This will provide continuous or periodic sensing of absolute case motion (vibration relative to free space) in terms of acceleration. Accelerometers are inertial measurement devices that convert mechanical motion to an electrical signal. This signal is proportional to the vibration's acceleration using the piezoelectric principle. Inertial measurement devices measure motion relative to a mass. This follows Newton's third law of motion: Body acting on another will result in an equal action on the first.

An accelerometer (Fig. 4.5) consists of a small mass mounted on the piezoelectric crystal that produces an electrical output proportional to the acceleration when the force is applied to the form of a vibrating mass. Force transducers such as modal hammer (or impact hammer) and force gauges (Fig. 4.6) also contain a pie-piezoelectric crystal, but the output of crystal is proportional to the force applied.

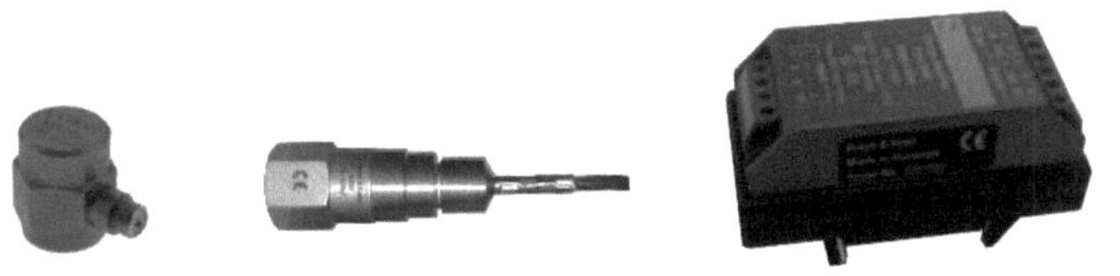

Fig. 4.5 Accelerometers and amplifier

Fig. 4.6 Impulse hammer

The piezoelectric crystal generates a high-impedance signal that must be modified by the charge of voltage conversion to low impedance. The size of accelerometer is proportional to its sensitivity. Small accelerometers may have sensitivity of 5 mV/g (1 g = 9.806 m/s^2) and a flat frequency response to 25 kHz. The vibration analyst must be aware of the properties of each accelerometer being used.

If vibration velocity is desired, the signal is usually integrated before it is recorded or analyzed; an analog integrator/power supply is needed. This device has an own frequency response characteristics and rolls off at low frequencies. Accelerometers are recommended for permanent seismic monitoring because of their extended life and because of their cross-sensitivity that is low. Cross-sensitivity means that the transducer generates a signal in x direction from vibration in y direction. However, cable noise, transmission distance, and temperature sensitivity of the accelerometer must be carefully evaluated. Usually, an excellent guidance is available from vendors for accelerometer use.

(4) *Key-phasor*

The key-phasor provides a once-per-revolution reference signal for angular measurements on rotating shaft. The purpose of a key-phasor is to relate the rotating reference frame of the shaft to the stationary reference frame of the vibration or displacement sensors. A key-phasor sensor has similar output voltage characteristics as a proximity probe shown in Fig. 4.7. The voltage swing goes from −2 to −5 V range for almost the complete revolution to −16 to −22 V as the slot passes under the probe.

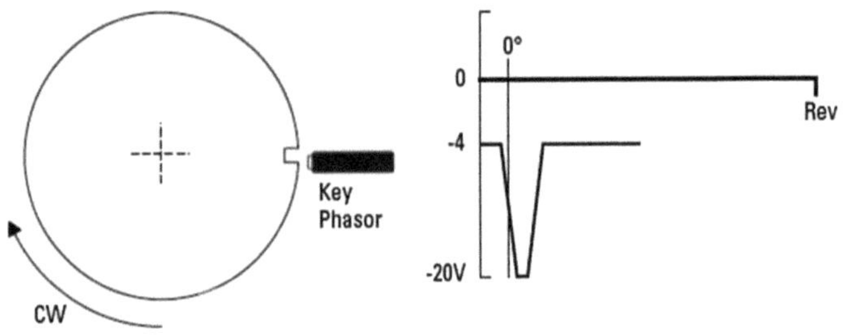

Fig. 4.7 A typical key-phasor setup with a key slot

The ideal key-phasor signals would have a very fast and large change in voltage as a high or low spot edge passes under the probe. The optimum setup for an eddy current phase sensor is to adjust it for a maximum voltage swing. In the case where a slot (keyway) is the reference transition, the probe sees a small gap for almost the complete revolution. The key or slot should be at least two probe tip diameters in width.

The optical pickup is most often used to obtain the once-per-revolution reference signal required to measure the phase angle between a piece of reflecting tape on a shaft and a once-per-revolution vibration peak. When energized by the light pulses from the reflecting tape, the pickup sends a voltage pulse to the analyzer. The analyzer can compare the timing of the tape (shaft reference pulse) to the other events, i.e., other marks on the shaft, vibration peaks, or its own reading to determine the shaft speed.

Optical pickups can also used to observe the time elapsed between equally spaced marks on a rotating shaft when torsional vibration measurements are made. The optical system includes pickup mounted adjacent to the shaft, reflective tape on the shaft, and a power amplifier.

The rapid technology developments within the sensor industry have pressured both the prices and sizes of sensors. MEMS (micro-electromechanical systems) and smart sensors have made it into the market of condition-based maintenance as the lowest level of systems health management. Sensors that have decreased in size can perform more tasks than conventional sensors.

4.2.3 Transducer Selection

Important considerations in transducer selection include frequency response, signal-to-noise ratio (SNR), the sensitivity of the transducer, and the strength of the signal being measured. The frequency range of the transducer must be compatible with the frequency generated by the mechanical components of the machine. Otherwise, another transducer must be selected, and the signal must be converted to the proper measures. For example, if the velocity measure is desired at frequency above 2 kHz, an accelerometer integrated to velocity should be selected to obtain the signal. If the time waveform of the velocity measure is desired, the signal must be acquired from a velocity pickup or analog integrated signal from an accelerometer, either within or external to the data collector.

A single transducer, usually an accelerometer because of its small size and frequency characteristics, is provided with most electronic data collector. The frequency response characteristic of the unit must be assessed so that user will not try to detect vibration to which the collector does not respond. For example, if a typical collector with an accelerometer is set up to respond to frequencies up to 8 kHz and a gearbox has gear mesh at 10 kHz, the signal will be dropped out. Acceleration is measured, and most collectors provide readouts in acceleration or velocity. The parameter selected depends on the criteria selection.

4.2.4 Transducer Mounting

Proximity probes are seldom mounted on machines in the ideal horizontal and vertical orientation due to bearing split lines and interference of oil lines. As a result, X-Y probe pairs are often encountered that are rotated from the ideal horizontal and vertical position (Fig. 4.8). The probe pairs shown are rotated +45° from horizontal and vertical. On many machinery trains, several different probe pair orientation may be encountered.

If you want to compare orbits along a machinery train with different probe orientations, you must rotate the orbits back to a common horizontal and vertical position as shown in Fig. 4.9 (Eisenmann 1997). The rotation correction is very critical when the measurement results are to be communicated clearly to others and when multi-plane balancing is done.

The method used to mount a vibration transducer can affect the frequency response curve because the natural frequency of an accelerometer can decrease, depending on the mounting method used, i.e., handheld or fixed method such as

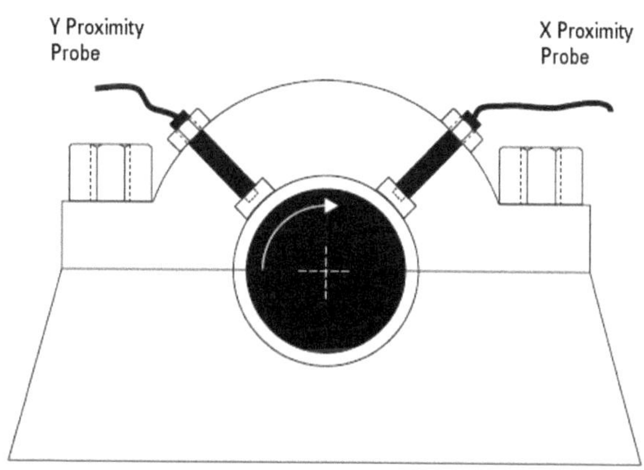

Fig. 4.8 Mounting method of proximity probes (*Agilent Technologies*)

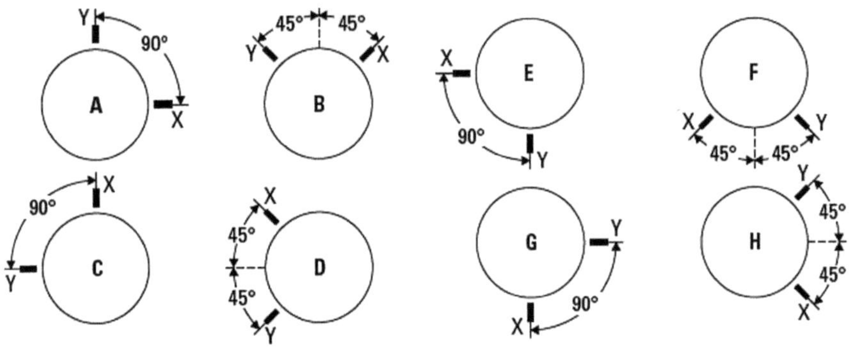

Fig. 4.9 Proximity probe pair angular locations and identification

magnetic, adhesive, and threaded stud (Fig. 4.10). Therefore, it is important to ensure that the frequency response is adequate before measurements are taken (Figs. 4.11 and 4.12).

Each of the above methods of mounting has its advantages and disadvantages. The appropriate choice for a given measurement problem depends on a number of factors, including the following:

- Effect of the mounting on the useful frequency range of the transducer;
- Effect of mass loading of the transducer mounting on the test surface;
- Maximum level of vibration the mounting can withstand;
- Measurement accuracy;
- Repeatability of measurements;
- Stability of the mounting with time;

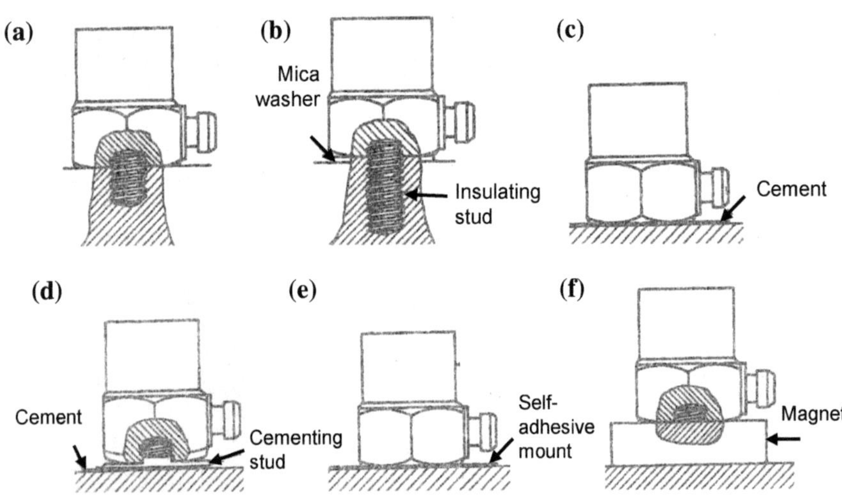

Fig. 4.10 Mounting technique of accelerometer (Harris 1997)

- Requirement that the test surface not be damaged by screw holes;
- Requirement for electrical insulation of the transducer;
- Time required for preparing mounting;
- Time required for removing mounting;
- Difficulty in cleaning the transducer after removal from test transducer;
- Difficulty in cleaning test surface after transducer removed;
- Skill required for preparing mounting;
- Cost of mounting; and
- Environmental problem (dirt, dust, oil, moisture).

(1) *Stud Mounting*

A typical stud-mounted transducer is shown in Fig. 4.10a. The transducer is fixed to the test surface by means of a threaded metal screw. One method of insulating the stud-mounted transducer from the test surface is shown in Fig. 4.10b. The metal stud is replaced with one which is fabricated of insulating material, and a mica washer is inserted between the transducer and the test surface. Other manufacturers employ a threaded, insulated stud with a flange made of the same material. The flange, midway along the length of the stud, serves as the base for the accelerometer.

Where stud mounting is practical, and it is the best type to use for the following reasons:

- It provides the highest resonance frequency (up to 20 kHz) of any mounting techniques.
- It permits measurements at very high vibration levels without the loosening of the transducer from the test surface.
- It does not reduce the maximum permissible operating temperature at which measurement position can be made.
- It permits accurate and reproducible results since the measurement position can always be duplicated.

(2) *Cement mounting*

The cement mounting of a transducer provides excellent frequency response, as shown in Fig. 4.10c for three conditions: accelerometer cemented directly of the surface, accelerometer cemented with a soft adhesive (not recommended), and accelerometer with a cementing stud which is cemented to the surface with hard cement.

This type of mounting may be used at high levels of vibration if the cementing surfaces are carefully prepared. The maximum temperature at which measurements can be made is limited by the physical characteristic of the cement employed usually about 80 °C, although some cements such as 3 M Cyanolyte 303 have an upper limit as high as 200 °C. This type of mounting has good stability with the time.

(3) *Wax Mounting*

Beeswax or a petroleum-based petrowax may be used to attach a transducer to a flat test surface. If the bonding layers are thin (no greater than 0.2 mm), it is possible to obtain a resonance frequency almost as high as for the stud mounting, but if the test surface is not smooth, a thicker layer of wax is required and the natural frequency will be reduced. If the mating surface is clean and free from moisture, the transducer can be mounted fairly and easily. The transducer can be removed rapidly with naphtha-type solvent. Disadvantages include the possibility of disattachment of the transducer at high vibration levels, a temperature limitation because of the relatively low melting point of wax, and poor longtime stability of the mounting. The maximum temperature at which measurements can be made with this mounting technique is usually about 40 °C.

(4) *Adhesive mounting*

An adhesive film may be used to mount a small transducer on a flat, clean test surface by means of the double-sided adhesive tape. This technique is shown in Fig. 4.10e and is rapid and easy to apply. Furthermore, such a mounting has the advantage of providing electrical insulation between the transducer and test surface and does not require the drilling of a hole in the test surface. It is particularly applicable for use with a transducer having no tapped hole in its base. The temperature of the test surface is usually below 95 °C.

(5) *Magnetic mounting*

The use of magnetic mounting is shown in Fig. 4.10f, a permanent magnet attaches the transducer to the test surface, which must be ferromagnetic, flat, free from dirt particles, and reasonably smooth. The transducer can be attached to the test surface easily and moved quickly from one measurement point to another. For example, in condition monitoring system, it can be used to determine a suitable measurement location for a transducer to be mounted permanently on large rotating machine. In a heavy machine, the use of magnetic mounting does not give the effect of added mass on test surface, but in the other problems, the additional mass loading on the test surface may make the use of magnetic mounting unacceptable. Furthermore, if the acceleration is sufficiently high, as the impact testing, the magnet may become loosened momentarily. This can result inaccurate reading and possibly a slight position change in the position of the transducer, which would also change the reading. The maximum temperature at which measurement can be made with this mounting technique is usually about 150 °C (Fig. 4.11).

(6) *Handheld mounting*

A transducer which is held against the test surface by hand provides the poorest performance of any of the techniques described here, but it sometimes can be useful

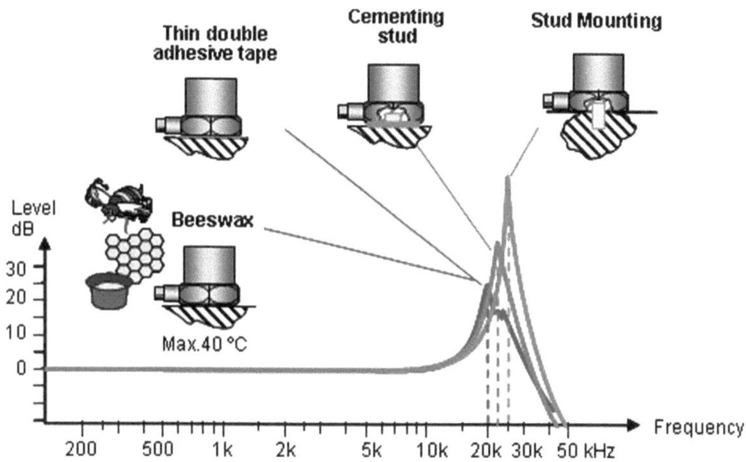

Fig. 4.11 Frequency response of fixed mounting (*B&K Vibro*)

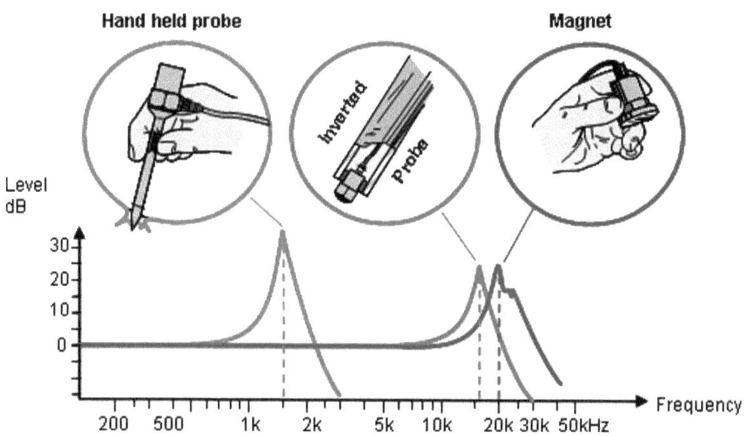

Fig. 4.12 Frequency response of handheld mounting (*B&K Vibro*)

in making a rapid survey of a test surface because the measurement location can be changed more rapidly. The frequency response is highly restricted about 20–1,500 Hz; furthermore, this technique should not be employed for accelerations greater than 1 g. Thus, this technique is used when measurement accuracy is not essential, e.g., in finding the nodal points on a vibration surface (Fig. 4.12).

4.2.5 Transducer Location

All vibration sensors measure motion along their major axis. This fact should be considered in choosing the number of sensors. Due to the structural asymmetry of

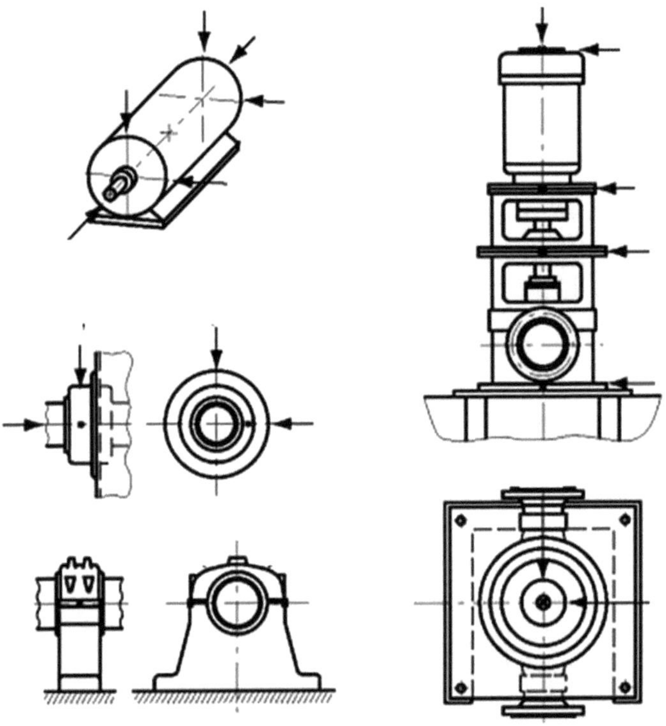

Fig. 4.13 Recommended locations for mounting transducer (*ISO 10816-1*)

machine cases, the vibration signals in the three axes of motion may differ. Where practical, a transducer should be mounted in the vertical, horizontal, and axial planes to measure vibration in the three directions. The three sensors will provide a complete definition of the vibration signature.

The key to the accurate vibration measurement is the placement of the transducer at a point that is responsive to machine condition. In any event, the transducer should be placed as close to the bearing as it is physically possible. Figure 4.13 shows the recommended locations for mounting transducer for data acquisition. The axial, horizontal, and vertical locations at the bearing centerline are shown. These locations are used to sense the vibration from radial forces such as mass unbalance. Vibrations from axially directed forces are measured in the axial direction in the load zone.

The transducer must be located as close to the bearing as possible, even though placement is restricted by such components as housings, coupling guards, and fan covers.

When comparing readings, it is essential that all readings are taken at the same location and the same plane. Even small differences in location can affect the overall readings. All vibration transducers are single-plane devices and only measure in the plane in which they are held or are mounted.

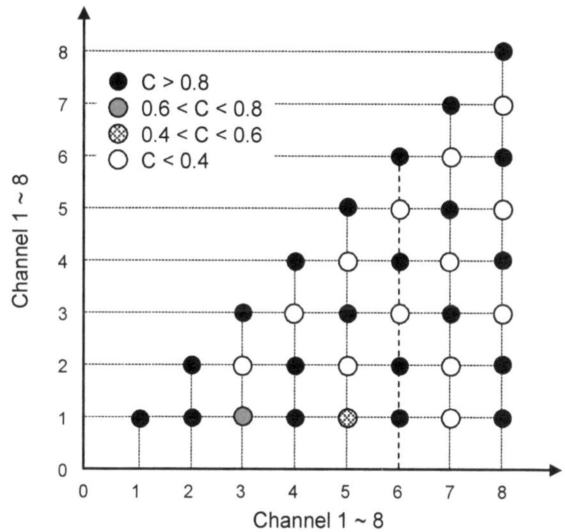

Fig. 4.14 An example of correlation matrix from channels 1–8

Clearly, we can reduce some of measuring points and still reach the target, increase the monitoring level for critical failures. Often, the sensors are systematically placed to measure at horizontal direction based on the fact that the support stiffness generally is at minimum in this direction. However, the support stiffness is not always a right ground to choose the measuring direction. The correlation matrix over the different measurement channels is shown in Fig. 4.14 (Järvinen and Miettinen 2006). The correlation is calculated by using the RMS velocity from narrow band close to the rotational frequency. By referring to the vibration features by that way, we will find the specific measuring points, in which the features are unique. It helps us to choose the right measurement channels and the minimum amount of measuring points.

4.2.6 *Frequency Span*

Spectra can be collected as part of the screening function on most data collectors. The frequency spans must properly reflect the sample. And the proper transducer must be selected. Geared units may generate must frequencies with significant harmonics that are clipped by the 2 kHz roll-off of velocity transducer. Therefore, measurement must be based on acceleration. Clipping can be also occurred when the range is less then the maximum frequency being transmitted. Table 4.2 contains recommended frequency spans for spectra taken from rotating machines for condition monitoring and analysis. The spans are based on RPM and other machine frequencies. Clipping in the frequency span is indicated by spectral energy values that are significantly lower than overall levels.

However, if the spectral span is broad, resolution can be reduced to the point that no discrete frequency information is available. If adequate resolution is not available from the default frequency spans, multiple data samples must be acquired and analyzed. An optimum configuration allows sufficient resolution to analyze the operating speed frequency and sidebands as well as the range to measure higher bearing or gear mesh frequencies. It is desirable to split the data point into two or three spans or to increase lines of resolution to obtain better resolution. Therefore, either several data acquisition cycles may be necessary at the same test point or the data collector must be capable of processing the data in several spans from a single sample.

4.2.7 Data Display

Vibration data from a machine running at a constant operating speed are generally repetitive. Small variations do occur as a result of the influenced load, temperature, and process. Environmental load conditions should be noted when the data are taken. Data are typically displayed in the spectrum, waveform, and orbit.

(1) *Time waveform*

Time waveform is a plot of vibration amplitude versus time. It reflects the physical behavior of the machine in the vibration signal. The waveform is useful in identifying unique events in a machine and the rate at which they are repeated. The length (in second) of the display of data from the time waveform depends on the information sought. It is typically related to the operating period τ of the machine; τ in second is equal to 60/RPM. The example of time waveform is presented in Fig. 4.15.

Example 4.2 Determine the frequency of waveform in Fig. 4.15 that shows the time waveform of the pump. The time spacing between the impacts is 0.0337 s.

Solution: From the time spacing information, the frequency can be determined as

$$f = \frac{1}{\tau} = \frac{1}{0.0337} = 29.67\,\text{Hz} = 1780\,\text{RPM}$$

This indicates the impact is occurring at the rotating frequency of 1 × RPM.

In most situations, the time waveform pattern is very complex as illustrated below (Fig. 4.16), and therefore, the determination of frequency components is extremely difficult using this method and is not recommended. So these time waveform data are best utilized by applying the principles of pattern recognition and if necessary calculating the frequency components of the major events in the waveform pattern.

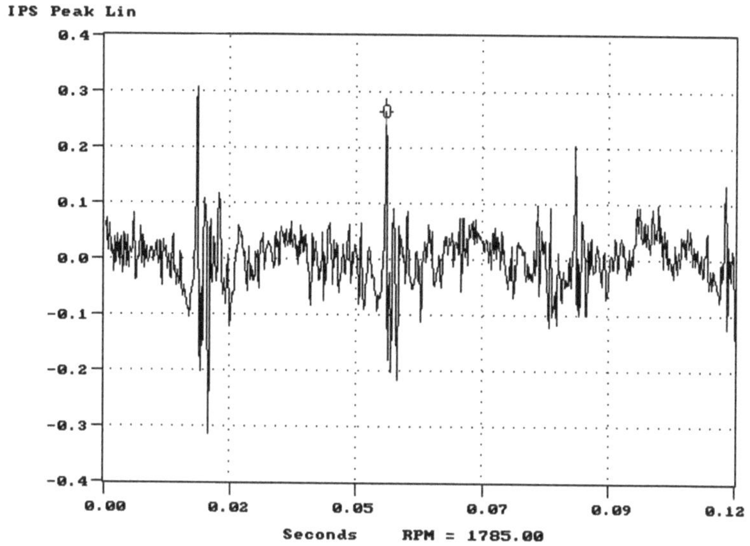

Fig. 4.15 Time waveform of pump

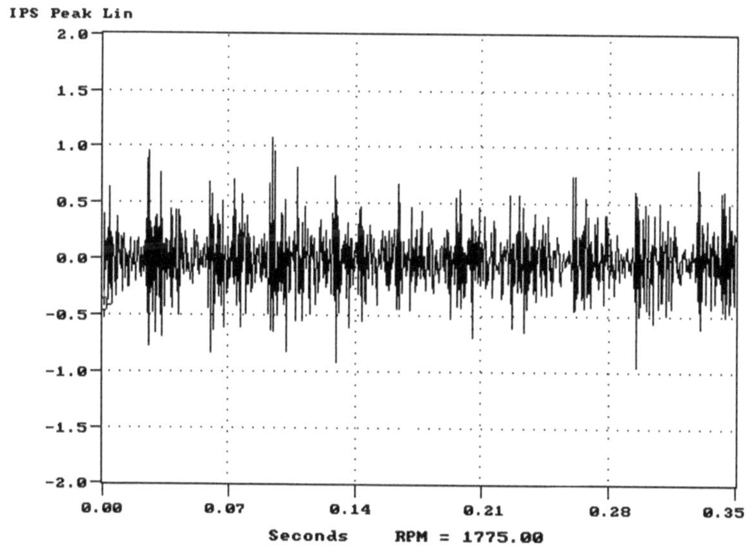

Fig. 4.16 A complex time waveform of pump

Time waveform can be used effectively to enhance spectral information in the following applications:

- Low-speed applications (less than 100 RPM),
- Indication of true amplitude in situations where impacts occur such as assessment of rolling element bearing defect severity,

- Gears,
- Sleeve bearing machines with X-Y probes (2 channel orbit analysis), and
- Looseness, rubs, beats.

(2) *Spectrum*

The setup of spectrum (Fig. 4.17) is determined by the frequency span of the data in order that all information will be obtained. Resolution, dynamic range, and amplitude accuracy are determined by the setup of the fast Fourier transform (FFT) analyzer (Fig. 4.18). In 1965, Cooley and Tukey devised a FFT algorithm, making the calculation of the frequency spectra of a signal far more efficient and rapid than that was previously possible in a digital computer. The FFT spectrum analyzer is an electronic device that is capable of taking the time waveform of a given signal and converting it into its frequency components. This analyzer is capable of rapidly calculating the Fourier transform of a vibration signal from a machine.

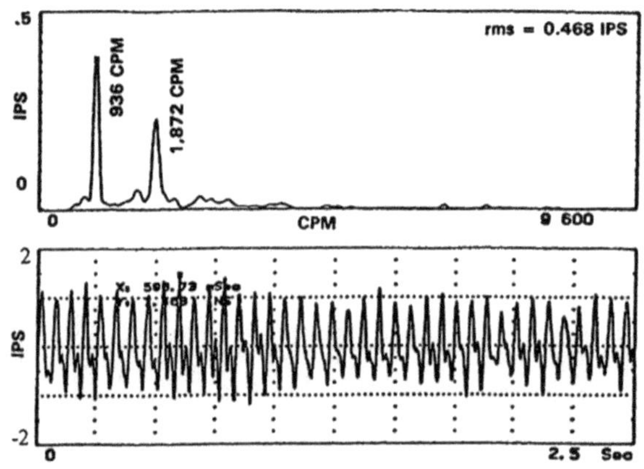

Fig. 4.17 Spectrum and time waveform

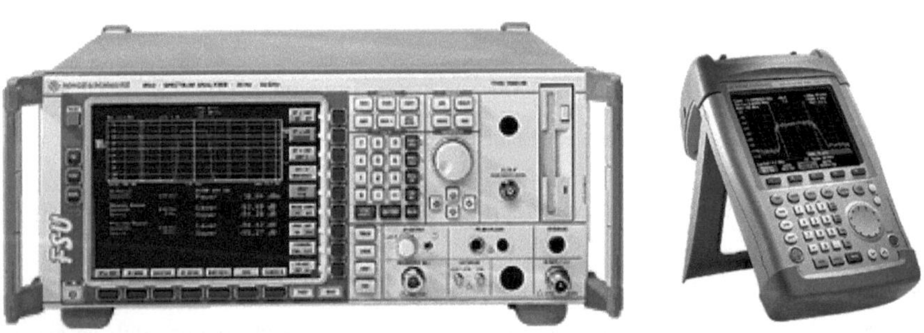

Fig. 4.18 Spectrum analyzer (*Courtesy of Rohde and Schwarz*)

In Fig. 4.17, the response at operating speed (1X) of the shaft to vibration is being analyzed. A 10X frequency span was thus selected. These data were processed in a fixed 400 line analyzer. Thus, no variability is allowed on resolution except for the window. A flattop window was used so that the correct amplitude would be obtained. In such a situation, if the frequency span and better resolution are needed, two or more spectra should be processed in difference frequency spans. With a data collector, the analyst has the option of increasing the lines of resolution instead of taking more spectra.

(3) *Orbit*

Orbits, or *X* versus *Y* displays of shaft position change versus angle of rotation, were first measured by placing a needle point on a turbine shaft end and then observing the motion with a microscope. This discovery from the 1930s was replaced long ago by pairs of proximity probes, digital vector filters, and an oscilloscope (Fig. 4.18).

Orbit displays give a two-dimensional visual picture of the motion of a rotating shaft. Orbits can be contaminated by shaft-related *noise*. Removing the noise using slow roll compensation techniques greatly improves the quality of orbit data.

Figure 4.19 shows a two-dimensional display of the vibration at point on a machine. This figure displays whirl orbit of the rotor system with squeeze fluid damper using magnetic fluid with different current (A). These orbits are collected from proximity probes, which show the physical motion of the shaft with respect to bearing. Orbits are useful in showing the motion of pedestals, piping, or any structure when a better visualization of the vibrating objects is desired (Fig. 4.20).

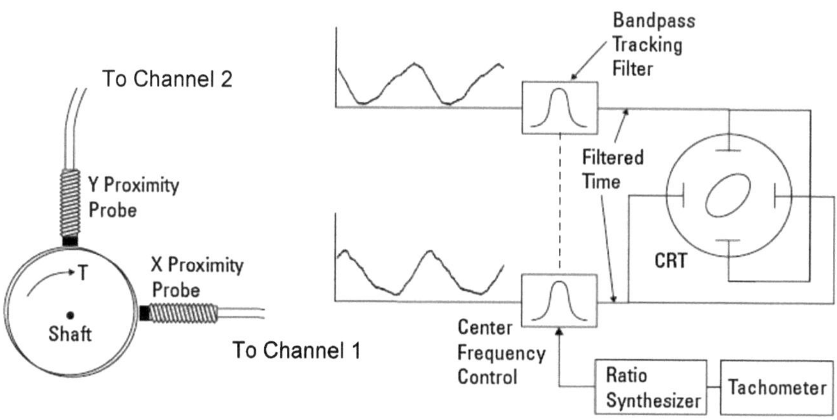

Fig. 4.19 A block diagram to measure and display an orbit (*Agilent Technologies*)

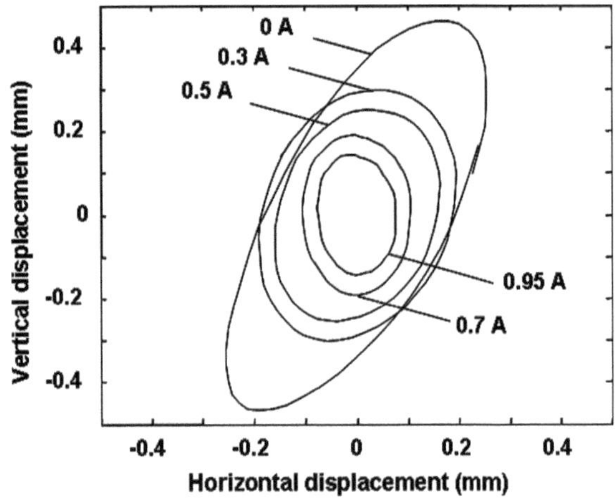

Fig. 4.20 An example of orbital display

4.3 Data Processing

Data processing for waveform data is also called *signal processing*. Various signal processing techniques have been developed to analyze and interpret waveform data to extract useful information for further diagnosis and prognosis purpose. The purpose of signal processing in diagnosis applications and condition-based maintenance is as follows:

- Remove distortions and restore the signal to its original shape,
- Remove sensor data that is not relevant for diagnosis or predictions, and
- Transform the signal to make relevant features more explicit (may be hidden in the signal, FFT analysis is an example of such a transformation).

Distortions in sensor data may be caused by an imperfect:

- Sensor,
- Media (metal, water, air, etc.) in which the signal travels before reaching the sensor, and
- Media from the sensor to an analog/digital converter.

Signal processing may also manipulate the signal that some characteristics enabling prognosis are more visible (for an analysis program or a human). Creating a feature vector from a signal is an abstraction of the signal, preserving the features used in diagnosis and prognosis.

The first step of data processing is *data cleaning* (*denosing*). This is an important step since data always contain errors. Data cleaning ensures or at least increases the chance that clean (error-free) data are used for further analysis and modeling. Without the data cleaning step, one may get into the so-called *garbage in garbage out* situation (Jardine et al. 2006). Data errors may be caused by sensor faults. In general, there is no simple way to clean data. Sometimes, it requires manual examination of data.

The next step of data processing is data analysis. A variety of models, algorithms, and tools are available in the literature to analyze data for better understanding and interpretation of data. The methods used for data analysis depend mainly on the types of data collected.

In this section, signal preprocessing techniques for waveform data are reviewed and then data analysis techniques for other types of data are discussed.

Data preprocessing describes any type of processing performed on raw data to enhance the data's reliability and, thereby, to improve the accuracy of the signal analysis. Commonly used as a preliminary data practice, data preprocessing transforms the data into a format that will be more easily and effectively processed for the purpose of the user, for example, in a neural network. There are a number of different tools and methods used for preprocessing which are as follows (Bengtsson et al. 2004):

- Sampling which selects a representative subset from a large population of data,
- Transformation which manipulates raw data to produce a single input,
- Denoising which removes noise from data,
- Normalization which organizes data for more efficient access, and
- Feature extraction which pulls out specified data that are significant in some particular context.

In this section, we mentioned data preprocessing just means denoising. In other words, we just focus on how to increase signal-to-noise ratio through highlighting the signals interested, removing, or reducing noise influence. About sampling, transformation, and normalization, they are described with other modules.

(1) *Wavelet transform*

When current signals show non-stationary or transient characteristics, the conventional Fourier transform technique is not suitable. The analysis of non-stationary signals can be performed using time–frequency techniques (short-time Fourier transform) or time-scale techniques (wavelet transform).

The wavelet transform is a remarkably powerful and general method for its many distinct merits. Newland's work (Newland 1994) made the wavelet popular in engineering applications, especially for vibration analysis; and later on, the wavelet prevailed in the machine condition monitoring and fault diagnosis. Wavelet was applied to analyze the gear vibration signals to detect incipient failure. The cracks in rotor system or in structures were other important objects for the application of the wavelets (Adewusi and Al-Bedoor 2001; Queck et al. 2001; Jones 2001; Zou et al. 2002). The wavelet in the denoising is also used widely for signal preprocessing in the monitoring and fault diagnostic field (Menon et al. 2000).

The wavelet transform decomposes a concerned signal into a linear combination of a time-scale unit. It analyzes original signals and organizes them into several signal components according to the translation of the mother wavelet (or wavelet basic function), which changes the scale and shows the transition of each frequency component (Daubechies 1992).

Several families of wavelets that have proven to be especially useful are included in wavelet toolbox of MATLAB. The families of wavelets are Haar ("Haar" or "db1"), Daubechies ("db1"–"db10"), Biorthogonal ("bior"), Coiflets ("coif"), Symlets ("sym"), Morlet ("morl"), Mexican Hat ("mexh"), and Meyer ("meyer"). *Continuous wavelet transform (CWT)*: The continuous wavelet transform is the integration with respect to the total time of the product of the target signal $f(t)$ and the mother wavelet $\psi_{a,b}$. Using the mathematical expression, the continuous wavelet transform of time function $f(t)$ can be written as follows:

$$\mathrm{CWT}(a,b) = \int_{-\infty}^{\infty} f(t)\psi_{a,b}\mathrm{d}t, \quad \psi_{a,b} = \frac{1}{\sqrt{a}}\psi\left(\frac{t-b}{a}\right) \tag{4.1}$$

where a, b, and $\psi_{a,b}$ indicate the scale, translation parameters, and the mother wavelet, respectively.

The transform result represents the correlation between the signal and the transform of the mother wavelet being scaled and translated. If the signal and mother wavelet are similar, the transform result will have large values. This means that lead and delay are translation, while the scale is an expansion and compression. As the low scale is a compressing wavelet, it becomes a rapid changing signal; that is, it improves the sensitivity in the time domain for high-frequency signals and in the frequency domains for low-frequency signals. This makes it possible to perform a multi-resolution analysis.

Example 4.3 Using CWT for denoising of a noisy sinusoidal signal is described as follows:

Solution: From the MATLAB, type:

```
≫ load noissin;
≫ whos
   Name        Size        Bytes Class
   noissin     1×1000      8000 double array
   Grand total is 1000 elements using 8000 bytes
≫ c = cwt (noissin, 1:48,'db4');
≫ c = cwt (noissin, 1:48,'db4','plot');
≫ colormap(pink);
≫ c = cwt(noissin, 2:2:128,'db4','plot'); % De-noising using scale from 2 to 128
```

After denoising, the plot gives a clearer picture of denoising using CWT (Figs. 4.21, 4.22, and 4.23).

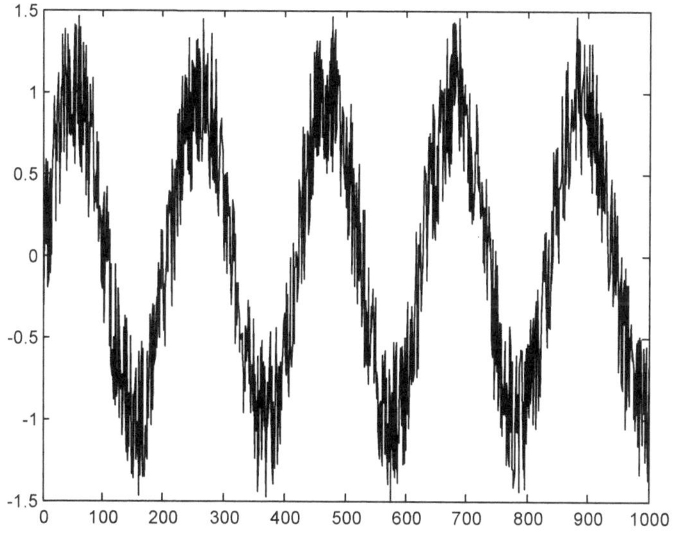

Fig. 4.21 A noisy sinusoidal signal

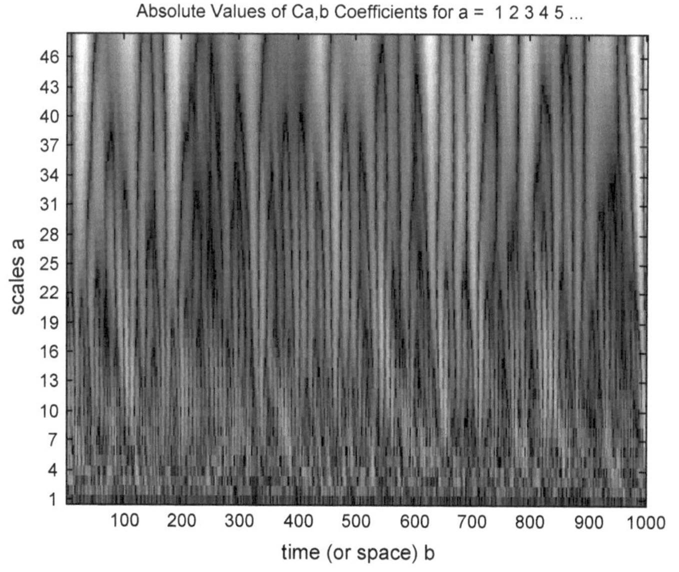

Fig. 4.22 CWT for a noisy sinusoidal signal

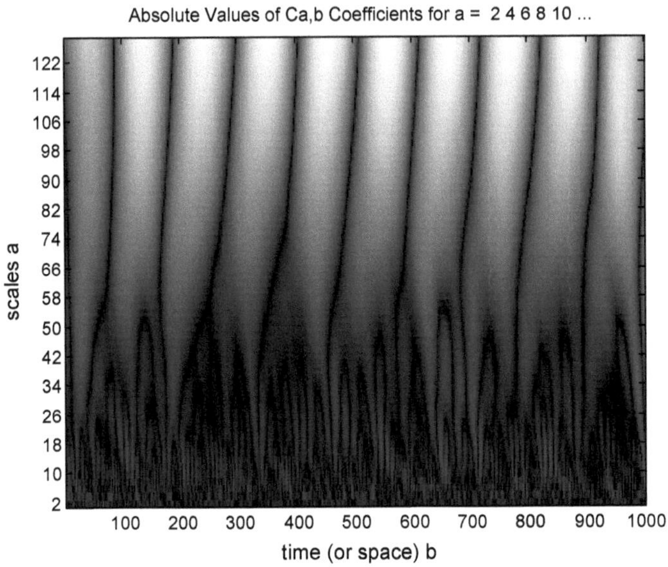

Fig. 4.23 CWT for a noisy sinusoidal signal after denoising

Discrete wavelet transform (DWT): At CWT, it takes a long time due to calculating the wavelet coefficient at all scales, and it produces a lot of data. To overcome such a disadvantage, we can reduce the elapsed time and keep accuracy by performing the wavelet transform at a scale of power 2.

At the discrete wavelet transform, the scale function $\phi(t)$ and wavelet function $\psi(t)$ are defined as follows:

$$\phi(t) = \sum_k c_k \phi(2t - k), \quad \psi(t) = \sum_k (-1)^k c_k \phi(2t + k - N + 1) \qquad (4.2)$$

where N is the number of data, which is a power of 2, and c_k is the wavelet coefficient. The input signal that passed through the high-pass filter becomes the wavelet coefficient, and the input signal, which passed through the low-pass filter, moves to next DWT transform loop. The DWT is carried out by the repetition of this process.

Example 2.4 Using DWT for denoising of a noisy signal of leleccum, this signal is particularly interesting because of the noise introduced when a defect developed in the monitoring equipment as the measurements were being made. Wavelet analysis effectively removes the noise.

Solution: From the MATLAB, type:

```
>> load leleccum;
>> plot (leleccum);
>> s = leleccum(1 : 3920); % setting variables
>> ls = length (s);
>> [C, L] = wavedec(s, 3, 'db1'); % Multi level decomposition using db1
>> % Constructing Approximations and Details
>> cA3 = appcoef(C, L, 'db1', 3);
>> cD3 = detcoef(C, L, 3);
>> cD2 = detcoef(C, L, 2);
>> cD1 = detcoef(C, L, 1);
>> % Signal reconstructions
>> A3 = wrcoef('a', C, L, 'db1', 3);
>> D1 = wrcoef('d', C, L, 'db1', 1);
>> D2 = wrcoef('d', C, L, 'db1', 2);
>> D3 = wrcoef('d', C, L, 'db1', 3);
>> subplot(2, 2, 1); plot(A3); title('Approximation A3');
>> subplot(2, 2, 2); plot(D1); title('Detail D1');
>> subplot(2, 2, 3); plot(D2); title('Detail D2');
>> subplot(2, 2, 4); plot(D3); title('Detail D3');
```

Denoised leleccum signal is shown in Fig. 4.26 of level 3 approximation (Figs. 4.24, and 4.25).

```
>> subplot(2, 1, 1); plot(s); title('Original');
>> subplot(2, 1, 2); plot(A3); title('Level 3 Approximation');
```

Using wavelet to remove noise from a signal requires identifying which components contain the noise and then reconstruct the signal without those components. In this example, we note that successive approximations become less noisy as more and more high-frequency information is filtered out of the signal. The level 3 approximation, A3, is quite clean as comparison between it and original signal shows.

There are many basic wavelet functions; they have their own characteristics. For different conditions, different basic wavelet functions are used. For an example, when the signal is composed of impulses, such as the bearing fault signal and gear signal, the Morlet wavelet should be used due to his wave feature similar with impulse (Lin and Qu 2000).

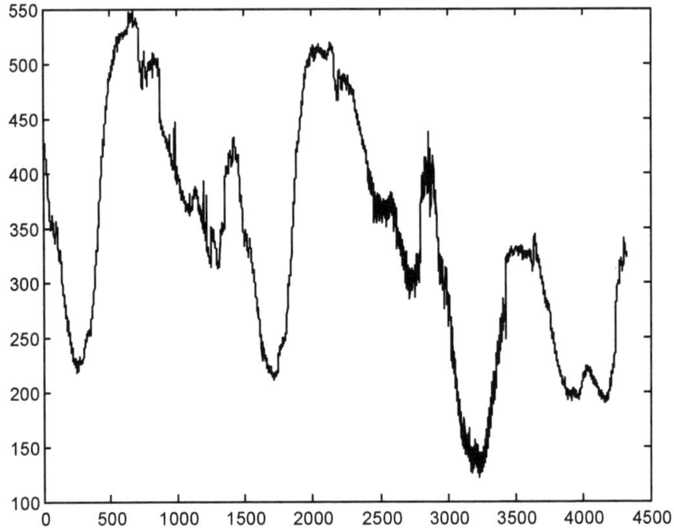

Fig. 4.24 Signal leleccum

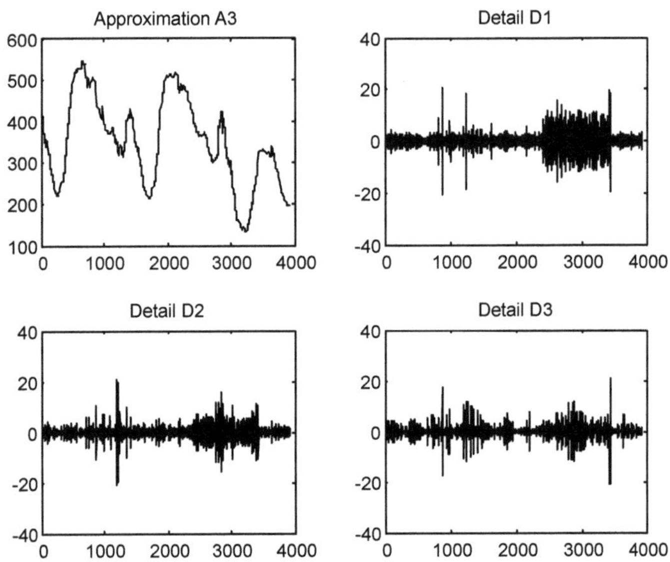

Fig. 4.25 Three-level decompositions using "db1"

The fast Fourier transform (FFT) and the discrete wavelet transform (DWT) are both linear operations that generate a data structure that contains $\log_2^n$ segments of various lengths, usually filling and transforming it into a different data vector of length 2^n. The mathematical properties of the matrices involved in the transforms

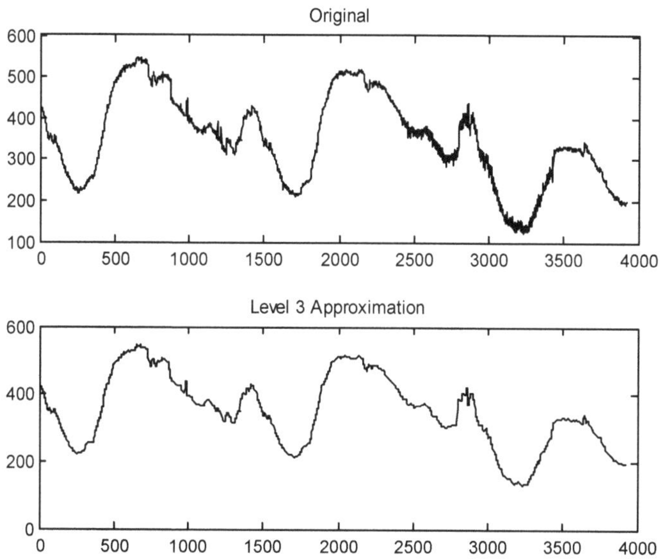

Fig. 4.26 Original signal and denoised signal of leleccum

are similar as well. The inverse transform matrix for both the FFT and the DWT is the transpose of the original. As a result, both transforms can be viewed as a rotation in function space to a different domain. For the FFT, this new domain contains basis functions that are sines and cosines. For the wavelet transform, this new domain contains more complicated basis functions called wavelets, mother wavelets, or analyzing wavelets.

Both transforms have another similarity. The basis functions are localized in frequency, making mathematical tools such as power spectra (how much power is contained in a frequency interval) and scalegrams (to be defined later) useful at picking out frequencies and calculating power distributions. The most interesting dissimilarity between these two kinds of transforms is that individual wavelet functions are localized in space. Fourier sine and cosine functions are not. This localization feature, along with wavelets' localization of frequency, makes many functions and operators to use wavelets' *sparse* when transformed into the wavelet domain. This sparseness, in turn, results in a number of useful applications such as data compression, detecting features in images, and removing noise from time series. One way to see the time–frequency resolution differences between the Fourier transform and the wavelet transform is to look at the basis function coverage of the time–frequency plane. Figure 4.27 shows a windowed Fourier transform, where the window is simply a square wave. The square wave window truncates the sine or cosine function to fit a window of a particular width. Because a single window is used for all frequencies in the WFT, the resolution of the analysis is the same at all locations in the time–frequency plane.

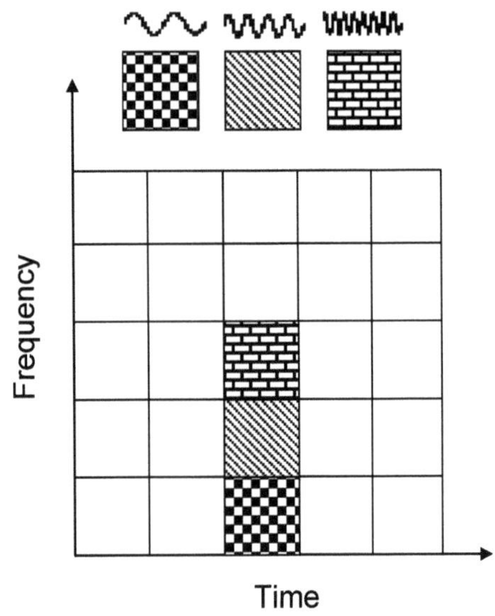

Fig. 4.27 Fourier basis functions, time–frequency tiles, and coverage of the time–frequency plane

An advantage of wavelet transforms is that the windows *vary*. In order to isolate signal discontinuities, one would like to have some very short basis functions. At the same time, in order to obtain detailed frequency analysis, one would like to have some very long basis functions. A way to achieve this is to have short high-frequency basis functions and long low-frequency ones. This happy medium is exactly what you get with wavelet transforms. Figure 4.28 shows the coverage in the time–frequency plane with one wavelet function, the Daubechies wavelet.

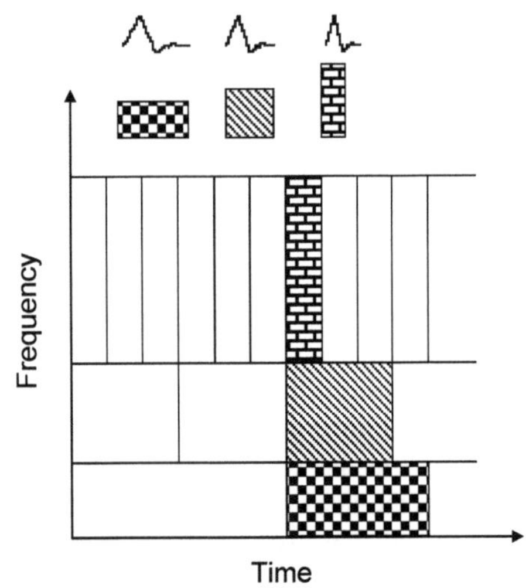

Fig. 4.28 Daubechies wavelet basis functions, time–frequency tiles, and coverage of the time–frequency plane

One thing to remember is that wavelet transforms do not have a single set of basis functions like the Fourier transform, which utilizes just the sine and cosine functions. Instead, wavelet transforms have an infinite set of possible basis functions. Thus, wavelet analysis provides immediate access to information that can be obscured by other time–frequency methods such as Fourier analysis.

(2) *Averaging*

Averaging can be divided into two types: One is *synchronous averaging* (sometimes also called *time synchronous averaging*), and the other is *spectrum averaging*. From the view of process domain, they are done in time domain and frequency domain, respectively. Synchronous averaging is very useful in reducing both non-synchronous components and non-coherent components in the measurement, or in reducing the effect of other interfering signals such as components from another nearby machine, which requires a tachometer to synchronize each *snapshot* of the signal to the running speed of the machine. This is a fundamental process to many shaft diagnosis, gear diagnosis, and balance algorithms. A common rule of thumb is that the amount of attenuation is inversely proportional to $\sqrt{N}$ where N is the number of averages. This rule is highly representative for the non-coherent components but is not representative for the non-synchronous components (Hochmann and Sadok 2004).

Synchronous averaging is a fundamentally different process than the usual spectrum averaging that is generally done in FFT analysis. It is used to greatly reduce the effects of unwanted noise in the measurement. The waveform itself is averaged in a time buffer before the FFT is calculated, and the sampling of the signal is initiated by a trigger pulse input to the analyzer. If the trigger pulse is synchronized with the repetition rate of the signal in question, the averaging process will gradually eliminate the random noise because it is not synchronized with the trigger. However, the signal that is synchronous with the trigger will be emphasized as shown in Fig. 4.29.

For doing time-domain averaging on the vibration signal from a real machine, the averaged time record gradually accumulates those portions of the signal that are synchronized with the trigger, and other parts of the signal, such as noise and any other components such as other rotating parts of the machine, are effectively averaged out. This is the only type of averaging that actually does reduce noise.

An important application of time synchronous averaging is in the waveform analysis of machine vibration, especially in the case of gear drives. In this case, the trigger is derived from a tachometer that provides one pulse per revolution of a gear

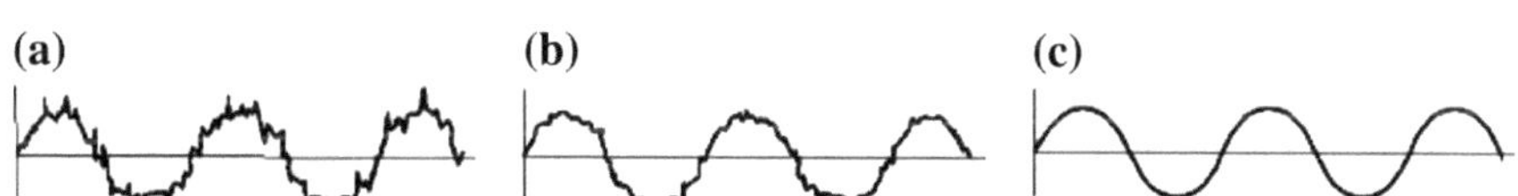

Fig. 4.29 Time synchronous averaging. **a** 1st average, **b** 10th average, **c** 100th average

in a machine. This way, the time samples are synchronized in that they all begin at the same exact point in the angular position of the gear.

A signal, which is synchronous with the rotation of the shaft of the gear being monitored, is measured to obtain a synchronizing reference that is synchronous with the rotation of the shaft. Hence, the synchronizing reference signal remains synchronous with the excitation when the excitation frequency or shaft speed changes. However, the response is not synchronous with the reference signal, owing to the phase change imposed by the transmission path when the rotational speed of the machine changes. It is only practically feasible to measure a reference signal which is synchronous with the excitation and shaft rotation. The concern about phase shifting becomes apparent when rotation domain averaging (RDA) is performed. In the RDA process, a constant number of samples are interpolated from the measured vibration signal for predetermined angles of shaft rotation. The angle of shaft rotation is calculated from the reference signal, which is synchronous with the shaft rotation. If the rotational speed of a gear being monitored changes, the RDA process will ensure that the order content in the signal remains correct, but the phase shift will be neglected.

Another type of averaging that is important in machinery monitoring is frequency averaging. In performing spectrum analysis, regardless of how it is done, some form of time averaging must be done to accurately determine the level of the signal at each frequency. In vibration analysis, the most important type of averaging employed is linear spectrum averaging, where a series of individual spectra are added together and the sum is divided by the number of spectra.

Averaging is very important when performing spectrum analysis of any signal that changes with time, and this is usually the case with vibration signals of machinery. Linear averaging smoothes out the spectrum of the random noise in a spectrum making the discrete frequency components easier to see, but it does not actually reduce the noise level. Unlike time synchronous averaging, spectrum averaging does not reduce noise. Instead, it finds the average magnitude at each frequency, where a series of individual spectra are added together and the sum is divided by the number of spectra.

Vibration response signals containing periodic components such as obtained from rotating equipment and gears can be effectively analyzed in the frequency domain. The averaging of spectra wherein the Fourier transform operation is synchronized with a particular periodic component will reinforce that frequency component and all of its harmonics, while all non-synchronous data will tend to be non-reinforced, thereby enhancing the synchronous spectral components. Features such as gear teeth and turbine blade-passing frequencies, and their related sidebands, can be extracted as signatures for baseline comparison and fault evaluation.

(3) *Enveloping*

Enveloping is a method for intensifying the repetitive components of a dynamic signal to provide an early warning of deteriorating mechanical condition. Enveloping is a tool that can give more information about the life and health of important plant assets. Common applications are concerned with rolling element bearings and gear mesh fault analyses in gearboxes. Usually, envelop analysis first filters out the rotational components with low frequency from the complex signal. The high-frequency repetitive components are enhanced and converted down to the fault spectrum range while machine noise is reduced by a significant signal-to-noise factor.

Enveloping is a multiple-step process that extracts signals of interest from an overall vibration signal (Fig. 4.30). In a rolling element bearing, the interaction between bearing elements and defects excites a structural resonance in the bearing support structure. A seismic transducer measures the vibration, and this signal is band-pass filtered to keep only signal components around the resonance frequency. The filtered signal is rectified and then enveloped, which removes the structural resonance frequency and preserves the defect impact frequency. A low-pass filter then eliminates some of the extraneous high-frequency components, and a spectrum is generated. Frequency components are correlated with physical bearing parameters, and a trend of the spectra can show progression of defects (Weller 2004).

Enveloped acceleration is an especially valuable parameter to trend, as the progression of machine condition can be evaluated. Enveloping can reveal faults in their earliest stages of development, before they are detectable by other machinery vibration measurements. Without an early fault detection technique like enveloping, personnel must wait until the latter stages of failure; when overall vibration increases, lubricants become contaminated, and temperatures rise. By this time, the remaining usable life of the failing machine elements could be very short and the damage more extensive than if the fault had been detected earlier. The enveloping technique enables the detection and analysis of low-level, repetitive vibrations by extracting them from the overall machinery vibration signal. It is important to note that the successful application and interpretation of enveloping data require

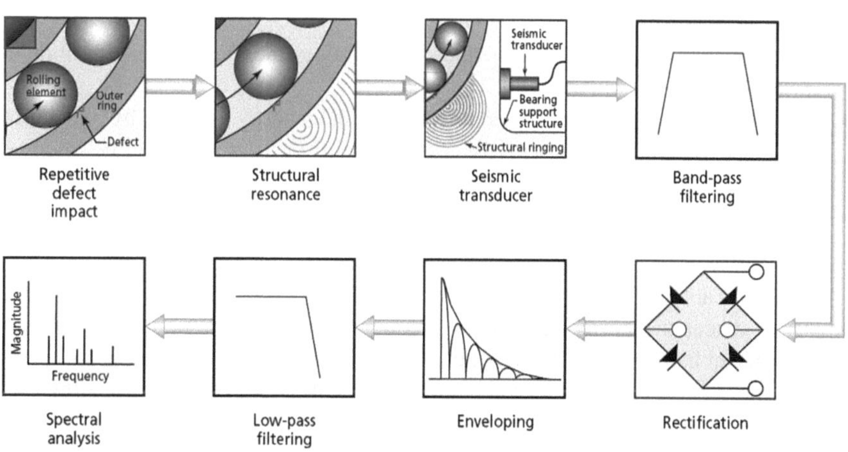

Fig. 4.30 Typical steps in the implementation of enveloping (Weller 2004)

experience. Enveloping is just one tool in the analyst's toolbox, and it is best used as one of a number of techniques for complete monitoring of a machine.

(4) *Cepstrum*

The power cepstrum introduced by Bogert et al. (1963) was used first for the detection or the suppression of echoes. It has also been used in vibratory rotating machine diagnosis, because the presence of faults induces in signals some recurrent patterns (echoes).

Cepstrum is the name given to a range of techniques all involving functions which can be considered as a *spectrum of a logarithmic spectrum or the forward Fourier transform of logarithm of the spectrum of the signal.*

Cepstrum analysis is the name given to a calculation technique involving a function that can be described as a "spectrum of a logarithmic spectrum" or a backward transform to the time domain. Because it is basically a "spectrum of a spectrum," the name was derived by reversing part of the word spectrum, and a number of terms are commonly used for the parameters of a cepstrum, namely

- Quefrency instead of frequency
- Rahmonics instead of harmonics, and
- Gamnitude instead of magnitude.

The real cepstrum of a signal z is defined as follows:

$$C_r(z) = \mathrm{real}(F^{-1}\log(|F(z)|)) \tag{4.3}$$

where F denotes Fourier transform and F^{-1} inverse Fourier transform.

The following behaviors have already been observed and have been used in diagnosis (Randall 1980):

- The cepstrum includes as many sets of decreasing positive peaks as there are rotating members in the system. The spacing of these sets corresponds to each rotation period.
- The existence of a comb peak is due exclusively to the presence of a periodicity in the signal.
- The amplitude of a comb increases when the energy emitted by the corresponding member increases, for example, in the case where the member develops a defect.
- The sum of the first peaks of every comb is constant; therefore, when the amplitude of one comb increases, it is to the detriment of others.

The application of the power cepstrum to machine fault extraction is based on its ability to detect the periodicity in the frequency spectrum, e.g., families of uniformly spaced harmonics and sidebands, while binge insensitive to the transmission path of the signal from an internal source to an external measurement point. Any periodicities in a frequency spectrum will be shown as one or two specific components in the cepstrum. The value of the main cepstrum peak was shown to be an excellent trend parameters; as it represents the average over a large number of individual harmonics, fluctuations in latter (e.g., as a result of load variations) were largely averaged out in the cepstrum value, which gave a smooth trend curve with time.

The cepstrum can therefore be said to be extremely advantageous for the following two tasks in vibration monitoring and analysis (Brüel & Kjær Vibro 2006):

For fault detection,

- It is a sensitive measure of the growth of harmonic/sideband families and
- The data is reduced to a single line per family,
- It is insensitive to
  - Measurement point location,
  - Phase combination, amplitude, and frequency, and
  - Loading.

For fault diagnosis

- It is an accurate measure of spacing,
- Can be calculated from any section of a spectrum,
- Can be used for separation of different families, and
- It is sensitive to tooth and blade differences but not uniform wear.

With the appropriate application cepstrum, we can make the job of the diagnosis technician easier and more certain when it comes to a correct diagnosis of a fault that may require a shutdown of a machine (Figs. 4.31, 4.32, and 4.33).

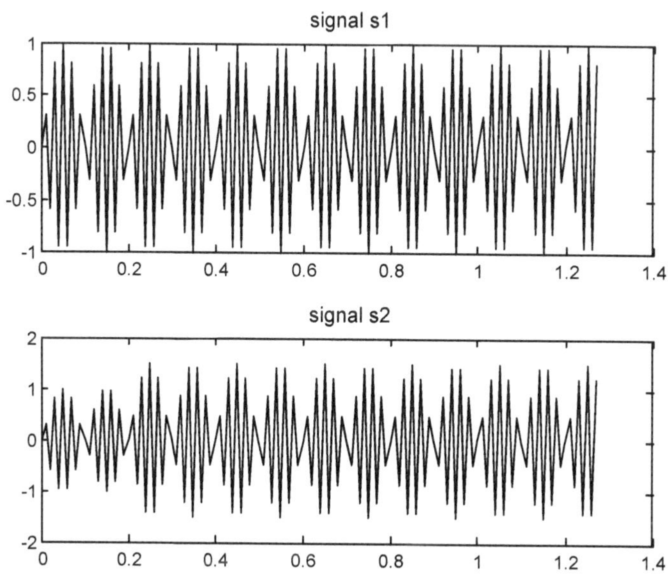

Fig. 4.31 Signals s_1 and s_2

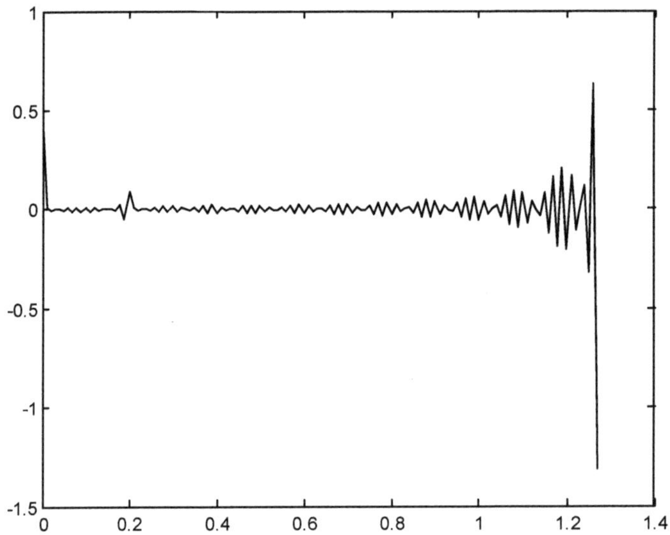

Fig. 4.32 A complex cepstrum of s_2

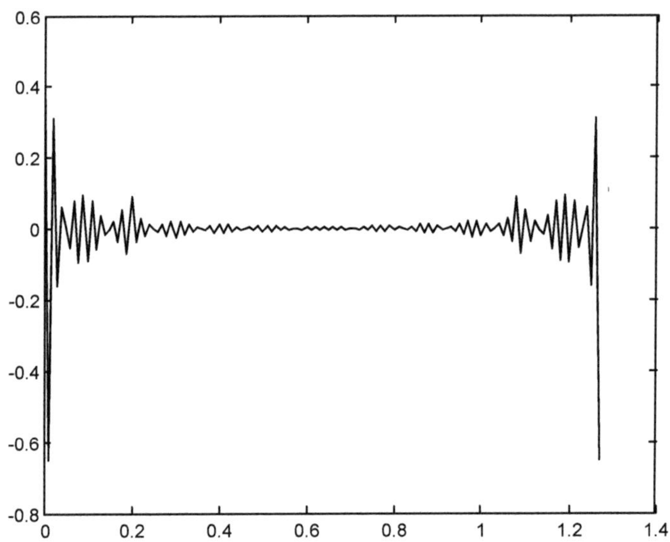

Fig. 4.33 A real cepstrum of s_2

Example 4.5 Try using cceps in an echo detection application. First, create a 45 Hz sine wave sampled at 100 Hz.

Solution: From the MATLAB, type:

```
≫ t = 0:0.01:1.27;
≫ s1 = sin(2*pi*45*t);
≫ % Add an echo of the signal, with half the amplitude,
   0.2 seconds after the beginning of the signal.
≫ s2 = s1 + 0.5*[zeros(1, 20) s1(1:108)];
≫ c = cceps(s2); % The complex cepstrum of this new signal is
≫ plot(t, c)
≫ y = rceps(s2); % The real cepstrum of s2
```

4.4 Data Analysis

Data analysis is one of the most important steps used for condition monitoring, fault diagnosis, and prognosis, whose aim is to find a simple and effective transform to the original signals. The important information (dominant features) contained in the signals can be extracted for condition monitoring and fault diagnosis.

In many pattern classification applications, data are represented by features that can be characteristic values, colors, and so on. For condition monitoring and fault diagnosis, features are some representative values, which can indicate machine conditions. For continuous, online, or remote monitoring, the way of feature is necessary and important since the transform and storage problems can be solved.

The represented features include time-domain features such as root mean square (RMS), variance, skewness, and kurtosis, frequency-domain features such as content at the feature frequency and the amplitude of FFT spectrum, and time-domain and frequency-domain features such as the statistical characteristics of short-time Fourier transform (STFT), Wigner-Viller distribution, and wavelet transform.

Transforming from data to features, feature representation module plays a very important role, which directly affects the performance of the whole system. In other words, the better the features can be tailored to reflect the requirements of the task, the better the results will be. In order to keep data information at highest level, features are calculated from time domain, frequency domain, and autoregression estimation.

4.4.1 Features in Time Domain

When carrying out vibration analysis in time domain, some simple quantities can be used such as RMS, crest factor, kurtosis, and other statistical moments, but often they do not offer enough information on the vibrations for diagnosis. A traditional tool for more enhanced signal description in time domain is time-series analysis. The time-series analysis is suitable for the description of dynamic phenomenon.

(1) *Cumulants*

The features described here are termed statistics because they are based only on the distribution of signal samples with the time series treated as a random variable. Many of these features are based on *moments* or *cumulants.* In most of the cases, the probability density function (pdf) can be decomposed into its constituent moments. If a change in condition causes a change in the probability density function of the signal, then the moments may also change, and therefore, monitoring these can provide diagnosis information.

The moment coefficients of time waveform data can be calculated using following equations:

$$m_n = E\{x^n\} = \frac{1}{N}\sum_{i=1}^{N} x_i^n \tag{4.4}$$

where $E\{\cdot\}$ represents the expected value of the function, x_i is the ith time historical data, and N is the number of data points.

The first four cumulants, mean c_1, standard deviation c_2, skewness c_3, and kurtosis c_4, can be computed from the first four moments using the following relationships:

$$\begin{aligned} c_1 &= m_1 \\ c_2 &= m_2 - m_1^2 \\ c_3 &= m_3 - 3m_2m_1 + 2m_1^3 \\ c_4 &= m_4 - 3m_2^2 - 4m_3m_1 + 12m_2m_1^2 - 6m_1^4 \end{aligned} \tag{4.5}$$

The first moment is the mean of the signal, while the second is defined as the signal's mean squared value. The moments are commonly defined about the mean so as to obtain the statistical central moment. The second central moment is called variance. The third and fourth central moments are usually normalized. As a result of the normalization, the skewness and kurtosis are obtained, respectively. The skewness measures the asymmetry of the process's pdf, and the kurtosis is a measure of the degree of flatness of the pdf distribution near its center.

In the case of a Gaussian process (i.e., a stationary signal with a Gaussian pdf), the first- and second-order statistics completely describe the properties of the signal. In particular, a Gaussian process has null skewness, while its kurtosis value is 3. It may be concluded that if a signal is non-Gaussian, then higher-order moments are needed to completely describe its properties.

In addition, non-dimensional feature parameters in time domain are more popular, such as shape factor SF and crest factor CF.

$$\mathrm{SF} = \frac{x_{\mathrm{rms}}}{x_{\mathrm{abs}}}, \quad \mathrm{CF} = \frac{x_{\mathrm{p}}}{x_{\mathrm{rms}}} \tag{4.6}$$

where x_{rms}, x_{abs}, and x_{p} are root-mean-squared value, absolute value, and peak value, respectively.

(2) *Upper and lower bound of histogram*

The histograms, which can be thought of as a discrete probability density function, are calculated in the following way. Let d be the number of divisions we wish to divide the ranges into, and let h_i with $0 \leq i < d$ be the columns of the histogram. Assume we are doing it for the time x_i only. Then,

$$h_i = \sum_{j=0}^{n} \frac{1}{n} r_i(x_i), \quad \forall i,\ 0 \leq i < d \tag{4.7}$$

$$r_i(x) = \begin{cases} 1, & \text{if } \dfrac{i(\max(x_i) - \min(x_i))}{d} \leq x < \dfrac{(i+1)(\max(x_i) - \min(x_i))}{d} \\ 0, & \text{otherwise} \end{cases} \tag{4.8}$$

The lower bound h_{L} and upper bound h_{U} of histogram are defined as follows:

$$h_{\mathrm{L}} = \max(x_i) - \Delta/2, \quad h_{\mathrm{U}} = \max(x_i) + \Delta/2 \tag{4.9}$$

where $\Delta = \{\max(x_i) - \min(x_i)\}/(n-1)$.

Effectively, this normalizes by two things: the length of the sequence and the column divisions. Since the sum term above includes a $1/n$ term, and every x_i must fall into exactly one h_i column, the net effect is that $\Sigma\, h_i = 1$ ($i = 0, \ldots, d - 1$). The column divisions are relative to the bounding box, and thus, most of h_i above will not be zero. This is desirable, since it essentially removes the issue of size of a sign and low resolution on small signs, with lots of empty columns. The alternative would be to have absolute locations, which would be nowhere near as closely correlated with the information in the sign itself.

Other approach is based on statistical properties of intensity histogram. One such measure is based on statistical moments. The expression for the nth-order moments about the mean is given by

$$\mu_n = \sum_{i=0}^{L-1} (Z_i - m)^n p(z_i) \tag{4.10}$$

where Z_i is a random variable indicating intensity, $p(z_i)$ is the histogram of the intensity levels in a region, L is the number of possible intensity levels, and m is the mean (average) intensity. We have considered similarly other five more descriptors, and their details are shown in Table 4.4.

Table 4.4 Descriptors of the intensive histogram

Moment	Expression	Measure
Mean	$m = \sum_{i=0}^{L-1} z_i p(z_i)$	A measure of average intensity
Standard deviation	$\sigma = \sqrt{\mu_2(z)} = \sqrt{\sigma^2}$	A measure of average contrast
Smoothness	$R = 1 - \frac{1}{1+\sigma^2}$	Measures the relative smoothness of the intensity in a region
Skewness	$\mu_3 = \sum_{i=0}^{L-1} (Z_i - m)^3 p(z_i)$	Measures the skewness of a histogram
Uniformity	$U = \sum_{i=0}^{L-1} p^2(z_i)$	Measures the uniformity of intensity in the histogram
Entropy	$e = -\sum_{i=0}^{L-1} p(z_i) \log_2 p(z_i)$	A measure of randomness

(3) *Entropy estimation and error*

In information theory, uncertainty can be measured by entropy. The entropy of a distribution is the amount of a randomness of that distribution. Entropy estimation is two-stage processes; first, a histogram is estimated, and thereafter, the entropy is calculated. The entropy estimation $E_s(x_i)$ and standard error $E_e(x_i)$ are defined as follows:

$$E_s(x_i) = -\sum p(x_i)\ln p(x_i), \quad E_e(x_i) = \sum p(x_i)\ln p(x_i)^2 \tag{4.11}$$

where x_i is discrete-time signals, and $p(x_i)$ is the distribution on the whole signal. Here, we estimate the entropy of vibration signal and stator current signal with using unbiased estimate approach.

(4) *Autoregression coefficients*

Since different faults display different characteristics in the time series, autoregression (AR) is used to establish a model. Then, the autoregressive coefficients are extracted as fault features. The first 8-order coefficients of AR models are selected through Burg's lattice-based method using the harmonic mean of forward and backward squared prediction errors. A common approach for modeling univariate time series is the autoregressive (AR) model as follows:

$$y_t = a_1 y_{t-1} + a_2 y_{t-2} + \cdots + a_n y_{t-n} + \varepsilon_t = \sum_{i=1}^{n} a_i y_{t-i} + \varepsilon_t \tag{4.12}$$

where a_i are the autoregression coefficients, y_t is the time series under investigation, and n is the order of the AR model (n = 8) which is generally very much less than the length of the series.

The noise term or residual ε_t is almost always assumed to be Gaussian white noise. An autoregression model is simply a linear regression of the current value of the series against one or more prior values of the series. The current term of the series can be estimated by a linear weighted sum of previous terms in the series. The weights are the autoregression coefficients.

The problem in AR analysis is to derive the *best* values for a_i given a series y_t. The majority of methods assume the series y_t is linear and stationary. By convention, the series y_t is assumed to be zero mean, if not this is simply another term a_0 in front of the summation in the equation above.

A number of techniques exist for computing AR coefficients. The main two categories are least squares and Burg method. Within each of these there are a few variants, the most common least squares method is based upon the Yule–Walker equations. MATLAB has a wide range of supported techniques; note that when comparing algorithms from different sources, there are two common variations: First is whether or not the mean is removed from the series, and the second is the sign of the coefficients returned (this depends on the definition and is fixed by simply inverting the sign of all the coefficients).

The most common method for deriving the coefficients involves multiplying the definition above by y_{t-d}, taking the expectation values, and normalizing (Box and Jenkins 1976) that gives a set of linear equations called the Yule–Walker equations that can be written in matrix form as follows:

$$\begin{bmatrix} 1 & r_1 & \cdots & r_{n-1} \\ r_1 & 1 & \cdots & r_{n-2} \\ \cdot & \cdot & \cdots & \cdot \\ r_{n-1} & r_{n-2} & \cdot & 1 \end{bmatrix} \begin{bmatrix} a_1 \\ a_2 \\ \cdot \\ a_n \end{bmatrix} = \begin{bmatrix} r_1 \\ r_2 \\ \cdot \\ r_n \end{bmatrix} \tag{4.13}$$

where r_d is the autocorrelation coefficient at delay d.

4.4.2 Features in Frequency Domain

Through the frequency-domain parameter indices, the primary diagnosis is available. In other words, the features can indicate the faults. Because these index calculations are simple and fast, they can be used in the online condition monitoring. When there are some changes on the parameters, the faults should be there. Then, precision diagnosis and prognosis can deal with the problem.

The signals of the ball bearing are composed of many stochastic elements; different faults have different spectrum in the frequency domain. But in some cases, we cannot distinguish them from the power spectrum. As mentioned above, frequency parameter indices can show the faults in the beginning of the failure. So we also can use the indices to run the condition monitoring and diagnosis.

(1) *Fourier transform*

The fast Fourier transform (FFT) which used the Fourier transform is one of the most widely used and well-established methods. The Fourier transform is a mathematical tool that allows the representation of a time signal $f(t)$ in the frequency domain as a function $F(\omega)$. In this domain, the explicit harmonic content of a signal and its distribution in terms of the frequency ω are revealed. If the temporal variable t is continuous, then the Fourier transform is continuous (continuous Fourier transform, CFT), but if t belongs to a discrete set of values, we talk about discrete Fourier transform (DFT). Its definition is given by

$$F(k) = \sum_{n=0}^{N-1} f(nT)\, e^{-j2\pi nk/N}, \quad f(nT) = \frac{1}{N}\sum_{n=0}^{N-1} F(k)\, e^{j2\pi nk/N} \tag{4.14}$$

Unfortunately, the FFT-based methods are not suitable for non-stationary signal analysis and are not able to reveal the inherent information of non-stationary signal. However, the change of the environment and the fault from the machine itself often make the output signals of the running machine to contain non-stationary components which include abundant information about machine faults.

(2) *Spectral analysis*

Spectral analysis of vibrations has been used in machine fault diagnosis for decades. It is claimed that vibration monitoring is the most reliable method of assessing the overall health of the machinery. Usually, machines have complex mechanical structures that vibrate, and coupled parts of machines transmit these vibrations. This results in machine-related frequency spectrum that characterizes healthy machine behavior. When a mechanical part of the machine either wears or breaks up, some frequency components in the spectrum will change. In fact, each fault in a machine produces vibrations with distinctive characteristics that can be measured and compared with reference ones in order to perform the fault detection and diagnosis. Figure 4.34 shows the comparison of normal condition and bearing faults (outer race and inner race) of machine taken from machine fault simulator. From this figure, we can see the differences among machine under normal condition and faulty bearing condition. These differences indicate that any degradation has occurred in the machine, and the analyst must keep attention to it and then do actions for preventing catastrophic condition.

One of the most used tools in signal analysis is the power spectrum. For a discrete-time series $x(n)$, the discrete Fourier transform (DFT) $X(f)$ of $x(n)$ is defined to be

$$X(f) = \sum_{n=-\infty}^{\infty} x(n)\, e^{-2j\pi n} \tag{4.15}$$

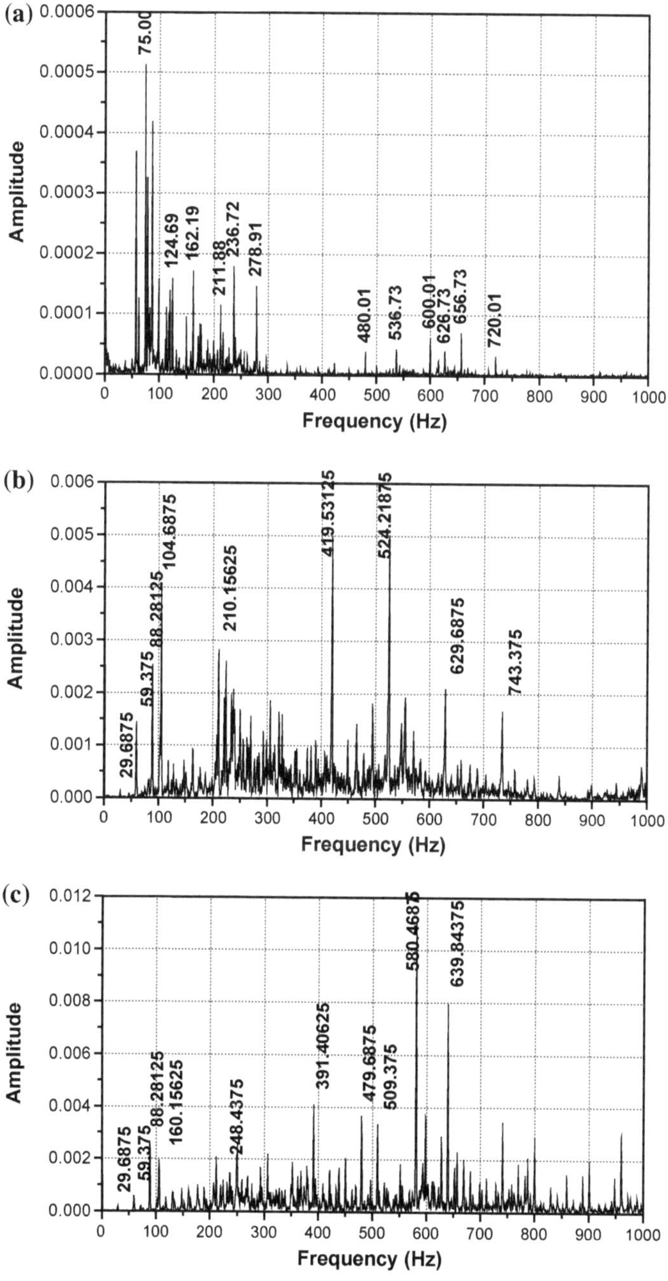

Fig. 4.34 Spectral analysis of machine with different condition. **a** Normal condition, **b** bearing outer race fault, and **c** bearing inner race fault

In the well-known second-order measure, the power spectral density (PSD) $P(f)$ of $x(n)$ can be defined in terms of $X(f)$ as follows:

$$P(f) = E[X(f)\,X^*(f)] \tag{4.16}$$

where $E[\cdot]$ is the statistical expectation, or average, and $X^*(f)$ is the complex conjugate of $X(f)$.

The PSD is a linear transform and is a function of the frequency index f. The power spectrum can be considered as the decomposition of the signal power, i.e., the signal's second moment, over frequency. The power spectrum is of limited value in analyzing signals where nonlinearities are involved. Extending these definitions to third- and fourth-order measures gives rise to the bispectrum $B(f_1, f_2)$ and trispectrum $T(f_1, f_2, f_3)$, which are defined, respectively, as

$$B(f_1,f_2) = E[X(f_1)\,X(f_2)\,X^*(f_1+f_2)] \tag{4.17}$$

$$T(f_1,f_2,f_3) = E[X(f_1)\,X(f_2)\,X(f_3)\,X^*(f_1+f_2+f_3)] \tag{4.18}$$

Unlike the PSD, the bispectrum and trispectrum are functions of more than one frequency index, and further, it may also be seen that they are complex quantities; i.e., they contain both magnitude and phase information about the original time series $x(n)$.

(3) *Frequency parameter indices*

The signal power spectrum shows the power distribution with the frequency. When the frequency elements and their power change, the position of the main spectrum peak will change. The characteristics of the frequency domain can be shown well, through frequency parameter indices as follows:

Frequency center (FC)

$$\mathrm{FC} = \frac{\int_0^\infty f\,s(f)\mathrm{d}f}{\int_0^\infty s(f)\mathrm{d}f} \tag{4.19}$$

Mean square frequency (MSF)

$$\mathrm{MSF} = \frac{\int_0^\infty f^2\,s(f)\mathrm{d}f}{\int_0^\infty s(f)\mathrm{d}f} \tag{4.20}$$

Root-mean-squared frequency (RMSF)

$$\mathrm{RMSF} = \sqrt{\mathrm{MSF}} \tag{4.21}$$

Variance frequency (VF)

$$\mathrm{VF} = \frac{\int_0^\infty (f - \mathrm{FC})^2 s(f)\mathrm{d}f}{\int_0^\infty s(f)\mathrm{d}f} \tag{4.22}$$

Root variance frequency (RVF)

$$\mathrm{RVF} = \sqrt{\mathrm{VF}} \tag{4.23}$$

where $s(f)$ is the signal power spectrum. FC, MSF, and RMSF show the position changes of main frequencies; VF and RVF describe the convergence of the spectrum power.

From the view of physics, we consider the power spectrum $s(f)$ as the mass density function of a stick in the ordinate axis. Accordingly, FC is the mass center in the abscissa. When larger the density is near the origin, the distance between FC and the origin is closer. RMSF is a rotating radial circling the stick. The relation between the distance and density is same with above-mentioned FC. Due to real calculation, the frequency spectrum needs to be discrete:

$$\mathrm{FC} = \frac{\sum_{i=2}^{N} \dot{x}_i x_i}{2\pi \sum_{i=1}^{N} x_i^2}, \quad \mathrm{MSF} = \frac{\sum_{i=2}^{N} \dot{x}_i^2}{4\pi^2 \sum_{i=1}^{N} x_i^2}, \quad \mathrm{VF} = \mathrm{MSF} - \mathrm{FC}^2 \tag{4.24}$$

where $\dot{x}_i = (x_i - x_{i-1})/\Delta$.

(4) *Higher-order spectra (HOS)*

HOS (also known as *polyspectra*) are relatively new tool in the area of signal processing. The second-order spectrum, that is the power spectrum, is very satisfactory in many vibration signals, but it is obviously of limited value in other signals where nonlinearities are involved. Nevertheless, the traditional linear spectral analysis can be employed in analyzing vibrations of nonlinear system. As a matter of fact, the second-order frequency response function and the coherence function are commonly used to detect deviations from linearity of a system. However, their computation requires two measurement sensors. Since only Gaussian process can be completely described by their second-order statistics, HOS give information about the signal's non-Gaussianity. HOS are defined in term of higher-order statistics, and they have attractive properties considering signal processing for the purpose of condition monitoring.

The k-order polyspectrum is defined as the Fourier transform of the corresponding cumulant sequence in Eq. (4.25)

$$s_2(\omega) = \sum_{k=-\infty}^{\infty} c_2(k) e^{-j2\pi\omega k} \tag{4.25}$$

$$s_3(\omega_1, \omega_2) = \sum_{k=-\infty}^{\infty} \sum_{l=-\infty}^{\infty} c_3(k, l) e^{-j2\pi\omega_1 k} e^{-j2\pi\omega_2 l} \quad (4.26)$$

$$s_4(\omega_1, \omega_2, \omega_3) = \sum_{k,l,m=-\infty}^{\infty} c_4(k) e^{-j2\pi(\omega_1 k + \omega_2 l + \omega_3 m)} \quad (4.27)$$

which are, respectively, the power spectrum (second-order spectrum), the bispectrum, and the trispectrum. Note that bispectrum is a function of two frequencies, whereas the trispectrum is a function of three frequencies. In contrast to the power spectrum which is real valued and nonnegative, bispectra and trispectra are complex valued.

For a real-valued process, symmetry properties of cumulant carry over to symmetry properties of polyspectra. The power spectrum is symmetric $s_2(\omega) = s_2(-\omega)$. The symmetry properties of the bispectrum are given by

$$\begin{aligned} s_3(\omega_1, \omega_2) &= s_3(\omega_2, \omega_1) = s_3(\omega_1, -\omega_1 - \omega_2) \\ &= s_3(-\omega_1 - \omega_2, -\omega_2) = s_3^*(-\omega_1, -\omega_2) \end{aligned} \quad (4.28)$$

while the symmetry properties of trispectrum include

$$\begin{aligned} s_4(\omega_1, \omega_2, \omega_3) &= s_4(\omega_1, \omega_3, \omega_2) = s_4(\omega_2, \omega_1, \omega_3) \\ &= s_4(-\omega_1, \omega_2 - \omega_1, \omega_3 - \omega_1) = s_3^*(-\omega_1, -\omega_2, -\omega_3) \end{aligned} \quad (4.29)$$

The bispectrum is a decomposition of the skewness of a signal over frequency and as such can detect asymmetric nonlinearities. The bicoherence is a normalization method for the bispectrum. The trispectrum is the fourth-order spectrum and is sensitive to the signal kurtosis. Therefore, it can detect symmetric nonlinearities. The normalization of the trispectrum leads to the tricoherence which is sensitive to the cubic phase coupling.

4.4.3 Features in Time–Frequency Domain

As some supplementary methods for non-stationary signal analysis, a variety of time–frequency analysis methods such as Wigner–Ville distribution (WVD) (Russell et al. 1998), the short-time Fourier transform (STFT) (Koo and Kim 2000), and instantaneous power spectrum distribution (IPS) (Cohen 1989) have been introduced for the analysis of vibration signals to extract useful diagnostic information. These methods perform a mapping of one-dimensional signal $x(t)$ to a two-dimensional function of time and frequency and therefore are able to provide true time–frequency representation for the signal. It is no doubt that the WVD has good concentration in the time–frequency plane. However, even support areas of the signal do not overlap each other; interference terms will appear on the time–frequency plane which misleads the signal analysis. In order to overcome these disadvantages, many improved methods have been proposed such as Choi–Williams distribution (CWD) and cone-shaped distribution.

These methods can exhibit local features of signals and give an account of how energy distribution over frequencies changes from one instant to the next. Of these methods, the WVD and IPS transforms are bilinear distributions, which result in interference terms when a multi-component signal is analyzed, and this might make the interpretation of distribution difficult. In contrast, the STFT and CWT perform linear decompositions of the analyzed signal and therefore do not cause any interference, which might detract from the interpretation of the targeted signal. However, the STFT employs a constant window size during the analysis and, hence, results in a constant time–frequency resolution. On the contrary, the CWT performs a decomposition of the analyzed signal into a set of waves (or wavelets), which are derived from a single wavelet, and wavelets at different frequencies are generated by introducing dilation into the analyzing wavelet. A large window is used for low-frequency estimates with poor time resolution, whereas the window automatically narrows at high frequencies, improving time resolution of the transform, but the frequency resolution deteriorates according to the uncertainty principle. Therefore, the wavelet transform provides a good compromise between localization and frequency resolution.

(1) *Short-time Fourier transform (STFT)*

Spectral representation is a stationary measure, which indicates the frequency content of the signal, but it does not reveal any time information regarding these frequency components. When the amplitude and frequency content of the signal change in time, combined time–frequency methods can be used effectively. The STFT, which maps the signal into a two-dimensional function in a time–frequency plane, can be expressed as follows (Cohen 1989; Wang and McFadden 1993):

$$\mathrm{STFT}(t,f) = \int_{t-T/2}^{t+T/2} x(\tau)w^*(\tau - t)e^{-j2\pi f\tau}\mathrm{d}\tau \tag{4.30}$$

where $x(t)$ denotes time waveform being analyzed, t is the analysis instant, f is frequency, τ is time variable, * represents complex conjugation, T is window interval, and $w(\tau)$ is the windowing function which satisfies $w(\tau) = 0$ for $|\tau| > T/2$.

Using the STFT transform, an energy density can be achieved by computing the spectrogram $S(t, f)$, which is the squared magnitude of the STFT. That is,

$$S(t,f) = |\,\mathrm{STFT}(t,f)\,|^2 \tag{4.31}$$

which represents the energy density of the original signal $x(t)$ windowed by $w(t)$.

The ordinary energy density can be obtained by setting $\|w\|_2 = 1$, where $\|\cdot\|_2$ denotes the norm for the space $L^2(R)$. Signal energy is preserved by the spectrogram and can be obtained by integrating spectrogram over the whole t–f plane. That is,

$$E = \int_{-\infty}^{\infty} |x(t)|^2 \mathrm{d}t = \int_{-\infty}^{\infty} \int_{-\infty}^{\infty} S(t,f)\mathrm{d}f\,\mathrm{d}t \tag{4.32}$$

It is shown in Eqs. (4.30) and (4.31) that the window length remains constant during the spectrogram analysis, and consequently, a constant time–frequency resolution is obtained in the whole t–f plane. However, analysis of signals with a mix of short-term impulsive events requires the use of a short window, whereas a longtime window should be employed when long-term harmonic components are analyzed.

(2) *Wavelet transform*

Over the past 10 years, wavelet theory has become one of the emerging and fast-evolving mathematical and signal processing tools. The novel concept of wavelet was first put forward definitely by Morlet in 1984. Different from the STFT, the wavelet transform can be used for multi-scale analysis of a signal through dilation and translation, so it can extract time–frequency features of a signal effectively. Therefore, the wavelet transform is more suitable for the analysis of non-stationary signals.

A time function $f(t) \in L^2$ (square integral) can be expressed in terms of some basis functions $\psi_{j,k}(t)$ and $\phi_k(t)$ by means of the following decomposition.

$$f(t) = \sum_{k=-\infty}^{\infty} c_k \varphi_k(t) + \sum_{j=0}^{\infty} \sum_{k=-\infty}^{\infty} d_{j,k} \psi_{j,k}(t) \tag{4.33}$$

where $\psi_{j,k}(t)$ and $\phi_k(t)$ are orthonormal set of basis function of L^2 which satisfy some properties and are denominated wavelet functions. The coefficients c_k and $d_{j,k}$ correspond to the discrete wavelet transform coefficients of the function $f(t)$ known as approximation and decomposition coefficients, respectively, which is defined as follows:

$$d_{j,k} = \langle f(t), \psi_{j,k}(t) \rangle = \int_{-\infty}^{\infty} f^*(x)\, \psi_{j,k}(t)\, \mathrm{d}t \tag{4.34}$$

$$c_k = \langle f(t), \phi_k(t) \rangle = \int_{-\infty}^{\infty} f^*(t)\, \phi_k(t)\, \mathrm{d}t \tag{4.35}$$

The more commonly used basis is the so-called Harr wavelets whose definition are given by

$$\psi_{j,k}(t) = 2^{j/2}\psi(2^j t - k) \quad j,k \in Z \tag{4.36}$$

$$\phi_k(t) = \phi(t - k) \tag{4.37}$$

where $\psi_{j,k}(t)$, denominated as mother wavelet, and $\phi_k(t)$, the scale function, are defined as follows:

$$\psi(t) = \phi(2t) - \phi(2t - 1) \tag{4.38}$$

$$\phi(t) = \begin{cases} 1 & \text{if } 0 \le t \le 1 \\ 0 & \text{other} \end{cases} \tag{4.39}$$

where the subindexes j and k are denominated decomposition level and the shifting time, respectively.

Table 4.5 shows a comparison of performances of continuous wavelet transform (CWT), short-time Fourier transform (STFT), Wigner–Ville distribution (WVD), Choi–Williams distribution (CWD), and cone-shaped distribution (CSD) (Peng and Chu 2004).

Table 4.5 Comparison of the performances of the different methods (Peng and Chu 2004)

Methods	Resolution	Interference term	Speed
CWT	Good frequency resolution and low time resolution for low-frequency components. Low-frequency resolution and good time resolution for high-frequency components	No	Fast
STFT	Dependant on window function, good time, or frequency resolution	No	Slower than CWT
WVD	Good time and frequency resolution	Severe interference terms	Slower than STFT
CWD	Good time and frequency resolution	Less interference than WVD	Very slow
CSD	Good time and frequency resolution	Less interference than CWD	Very slow

References

Adewusi SA, Al-Bedoor BO (2001) Wavelet analysis of vibration signals of an overhang rotor with a propagating transverse crack. J Sound Vib 246:777–793

Agilent Technologies, Rotor dynamics measurement techniques. Product Note

Bengtsson1 M, Olsson E, Funk P, Jackson M (2004) Technical design of condition based maintenance system—a case study using sound analysis and case-based reasoning. In: Maintenance and reliability conference

Brüel & Kjær (2006) Primer for cepstrum analysis—a powerful tool for simpler diagnosis of REB and gear vibrations. Appl Note 1–3

Bogert BP, Healy MJR, Tukey JW (1963) The frequency analysis of time series of echoes: cepstrum pseudo-autocovariance cross-cepstrum and saphe cracking. In: Proceedings of the symposium on time series analysis, pp 209–243

Box GEP, Jenkins GM (1976) Time series analysis: forecasting and control, 2nd edn. Holden-Day, San Francisco

Cohen L (1989) Time-frequency distributions: a review. Proc IEEE 77(7):941–981

Eisenmann R (1997) Machinery malfunction diagnosis and correction, vibration analysis and troubleshooting for the process industries. PTR Prentice Hall

Daubechies CI (1992) Ten lectures on wavelet. SIAM, Pennsylvania, USA

Eshleman RL (1999) Basic machinery vibration. VI Press Incorporated, Illinois

Harris CM (1997) Shock and vibration. McGraw-Hill, Singapore

Hochmann D, Sadok M (2004) Theory of synchronous averaging. In: IEEE aerospace conference proceedings, pp 3636–3653

Jardine AKS, Lin D, Banjevic D (2006) A review on machinery diagnostics and prognostics implementing condition-based maintenance. Mech Syst Signal Process 20(7):1483–1510

Järvinen V, Miettinen J (2006) Tools for diagnostics and prognostics of disturbances and faults in air fans. VTT Symp 243:97–114

Jones KJ (2001) Wavelet crack detection algorithm for smart structures. Proc SPIE 4328:306–313

Koo IS, Kim WW (2000) Development of reactor coolant pump vibration monitoring and diagnostic system in the nuclear power plant. ISA Trans 39:309–316

Lin J, Qu L (2000) Feature extraction based on Morlet wavelet and its application for mechanical fault diagnosis. J Sound Vib 234:135–148

Menon S, Schoes JN, Hamza R, Bush D (2000) Wavelet-based acoustic emission detection method with adaptive thresholding. Proc SPIE 3986:71–77

Newland DE (1994) Wavelet analysis of vibration. Trans ASME J Vibr Acoust 409–417

Peng ZK, Chu FL (2004) Application of the wavelet transform in machine condition monitoring and fault diagnostics: a review with bibliography. Mech Syst Signal Process 18:199–221

Queck ST, Wang Q, Zhang L, Ang KK (2001) Sensitivity analysis of crack detection in beams by wavelet technique. Int J Mech Sci 43:2899–2910

Randall RB (1980) Cepstrum analysis and gearbox fault diagnosis. Brüel Kjaer Techn Bull Edn 2:1–19

Russell PC, Cosgrave J, Tomtsis D, Vourdas A, Stergioulas L, Jones GR (1998) Extraction of information from acoustic vibration signals using Gabor transform type devices. Meas Sci Technol 9:1282–1290

Wang WJ, McFadden PD (1993) Early detection of gear failure by vibration analysis—I. Calculation of time-frequency distribution. Mech Syst Signal Process 7(3):193–2033
Weller N (2004) Acceleration enveloping-higher sensitivity, earlier detection. Orbit 10–19
Zou J, Chen J, Pu YP, Zhong P (2002) On the wavelet-time frequency analysis algorithm in identification of a crack rotor. J Strain Anal Eng Des 37:239–246

Chapter 5
Statistic Feature Extraction

5.1 Introduction

Feature extraction is a commonly used technique applied before diagnosis and prognosis when a number of measures, or features, have been taken from a set of objects in a typical statistical pattern recognition or trending reasoning task. The goal is to define a mapping from the original representation space into a new space where the classes are more easily separable. This will reduce the classifier or prediction complexity, increasing in most cases accuracy.

Accurate data-driven PHM/CBM needs multi-sensor to obtain detailed condition information, which results in plenty of raw data, thereby many features are calculated corresponding to last section to keep data information at the highest level. Too many features can cause *curse of dimensionality* and *peaking phenomenon* that greatly degrades classification accuracy. Also many features still can bring traffic jam or storage problem. So what are the curse of dimensionality and peaking phenomena and how to handle them?

Two important phenomena that can be identified are the so-called curse of dimensionality and peaking phenomenon. The performance of a data-driven PHM/CBM system depends on the interrelationship between sample size, number of features, and algorithm complexity. If one consider a very simple naive table-lookup technique consisting in partitioning the feature space into cells and associating a class label to each cell, it can be pointed out that this technique requires a number of training data points which is exponential in the feature space dimension (Bishop 1995). This phenomenon is termed the curse of dimensionality which produces as a consequence the peaking phenomenon in classifier design. This is a paradoxical effect that appears by considering the following; it is well known that the probability of misclassification of a decision rule does not increase as the number of considered features increases, as long as the class-conditional densities are known (or alternatively the number of training samples is arbitrarily large and representative of the underlying densities).

G. Niu, *Data-Driven Technology for Engineering Systems Health Management*,
DOI 10.1007/978-981-10-2032-2_5

However, it has been often noticed in practice that increasing the features to be considered by a classifier may degrade its performance if the number of training examples that are used to design the classifier is small relative to the number of features. This paradoxical behavior is termed the peaking phenomenon (Raudys and Jain 1991). The explanation stands in the following: The most commonly used parametric classifier estimates the unknown parameters and plugs them in for the true ones in the class-conditional densities. For a fixed sample size, as the number of features increases, and consequently the number of unknown parameters to be estimated from the sample, the reliability of parameter estimation decreases. As a consequence, the performance of the resulting classifier, for a fixed sample size, may degrade with an increase in the number of features.

Trunk (1979) provided a simple example to illustrate the curse of dimensionality which we reproduce below. Consider the two-class classification problem with equal prior probabilities, and a d-dimensional multivariate Gaussian distribution with the identity covariance matrix for each class. The mean vectors for the two classes have the following components

$$\mathbf{m}_1 = \left(1, \frac{1}{\sqrt{2}}, \frac{1}{\sqrt{3}}, \ldots, \frac{1}{\sqrt{d}}\right), \quad \mathbf{m}_2 = \left(-1, -\frac{1}{\sqrt{2}}, -\frac{1}{\sqrt{3}}, \ldots, -\frac{1}{\sqrt{d}}\right) \tag{5.1}$$

Note that the features are statistically independent and the discriminating power of the successive features decreases monotonically with the first feature providing the maximum discrimination between the two classes. The only parameter in the densities is the mean vector, $\mathbf{m} = \mathbf{m}_1 = -\mathbf{m}_2$.

Trunk considered the following two cases:

- The mean vector $\mathbf{m}$ is known. In this situation, we can use the optimal Bayesian decision rule (with a 0/1 loss function) to construct the decision boundary. The probability of error as a function of d can be expressed as:

$$P_e(d) = \int\limits_{\sqrt{\sum_{i=1}^{d}\left(\frac{1}{i}\right)}}^{\infty} \frac{1}{\sqrt{2\pi}} \cdot e^{-\frac{1}{2}z^2}\,\mathrm{d}z \tag{5.2}$$

  It is easy to verify that $\lim_{d\to\infty} p_e(d) = 0$. In other words, we can perfectly discriminate the two classes by arbitrarily increasing the features, d.
- The mean vector $\mathbf{m}$ is unknown and n label training samples are available. Trunk found the maximum likelihood estimate $\hat{\mathbf{m}}$ of $\mathbf{m}$ and used the plug-in decision rule (substitute $\hat{\mathbf{m}}$ for $\mathbf{m}$ in the optimal Bayes decision rule). Now the probability of error which is a function of both n and d can be written as:

$$P_e(n, d) = \int\limits_{\theta(d)}^{\infty} \frac{1}{\sqrt{2\pi}} \cdot e^{-\frac{1}{2}z^2}\,\mathrm{d}z \tag{5.3}$$

where

$$\theta(d) = \frac{\sum_{i=1}^{d}\left(\frac{1}{i}\right)}{\sqrt{\left(1+\frac{1}{n}\right)\sum_{i=1}^{d}\left(\frac{1}{i}\right)+\frac{d}{n}}} \tag{5.4}$$

Trunk showed that $\lim_{d\to\infty} p_e(n,d) = 1/2$, which implies that the probability of error approaches the maximum possible value of 0.5 for this two-class problem. This demonstrates that, unlike the first case, we cannot arbitrarily increase the number of features when the parameters of class-conditional densities are estimated from a finite number of training samples. The practical implication of the curse of dimensionality is that a system designer should try to select only a small number of salient features when confronted with a limited training set.

All of the commonly used classifiers can suffer from the curse of dimensionality. While an exact relationship between the probability of misclassification, the number of training samples, the number of features, and the true parameters of the class-conditional densities is very difficult to establish, some guidelines have been suggested regarding the ration of the sample size to dimensionality. It is generally accepted that using at least ten times as many training samples per class as the number of features ($n/d > 10$) is good practice to follow in classifier. The more complex the classifier, the greater should the ratio of sample size to dimensionality be to avoid the curse of dimensionality.

There are two methods that can realize feature dimensionality reduction: feature extraction and feature selection, see in Fig. 5.1. Methods that create new features based on transformations or combinations of the original feature set are called feature extraction. The term feature selection refers to algorithms that select the best feature subset from the all features. Often feature extraction precedes feature selection; firstly feature dimensionality is greatly reduced by extraction and then the significant features are selected from transformed features. Feature extraction leads to savings in computation time cost. Feature selection contributes to monitoring and diagnosis accuracy.

Problem with high-dimensional data, known as the curse of dimensionality in pattern recognition, implies that the number of training samples must grow exponentially with the number of features in order to learn an accurate model. Therefore,

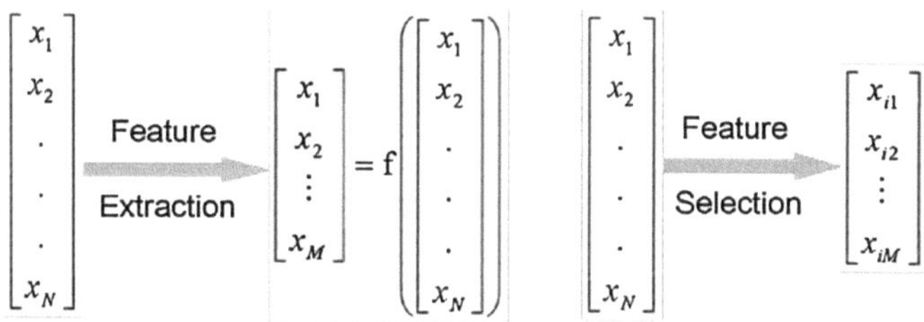

Fig. 5.1 Feature extraction and feature selection

reducing the number of features by extracting or selecting only the relevant and useful ones is desirable. There are two ways to reduce the dimensionality: feature extraction and feature selection. Feature extraction means transforming the existing features into a lower-dimensional space, and feature selection means selecting a subset of the existing features without any transformation (Han et al. 2005).

Multivariate analysis is an important branch of statistics whose purpose is to study random systems with more than one random variable of interest. Many multivariate statistical techniques had been developed and applied to fault diagnosis such as distance-based methods, principal component analysis (PCA), probability principal component analysis (PPCA), partial least squares (PLS), Fisher discriminant analysis (FDA), and discriminant partial least squares (DPLS). Given a number of data in the historical database, each associated with a different fault, the goal of fault diagnosis is equivalent to that of classification so that the out-of-control observation is assigned to the most similar or closely related fault group or class.

Feature extraction methods can be divided into linear and nonlinear, depending on the nature of the mapping function. They can also be classified as supervised or unsupervised, depending on whether the class information is taken into account or not. Most of feature extraction techniques have based on linear technique such as PCA (unsupervised), linear discriminant analysis (LDA, supervised), and independent component analysis (ICA, unsupervised). Schematically, PCA preserves as much variance of the data as possible; LDA attempts to group patterns of the same class while separating them from the other classes, and ICA obtains a new set of features by extracting the less correlated (in a broad sense) directions in the data set (Perez-Jimenez and Perez-Cortes 2006). After doing feature extraction sometimes, there is still high noise, irrelevant or redundant information in these extracted features.

On the other hand, well-known nonlinear methods are Sammon's mapping (unsupervised) (Sammon 1969), nonlinear discriminant analysis (NDA, supervised) (Mao and Jain 1995), self-organizing map (SOM, unsupervised) (Kohonen 1990), evolutionary extraction (supervised) (Liu and Motoda 1998), and so on. Sammon's mapping tries to keep the distances among the observations using hill-climbing or neural network methods; NDA obtains new features from the coefficients of the second hidden layers of a multilayer perceptron and Kohonen maps project data in an attempt to preserve the topology. Finally, evolutionary extraction uses a genetic algorithm to find combinations of original features in order to improve classifier accuracy (Perez-Jimenez and Perez-Cortes 2006).

The examples of using PCA and ICA are presented as follows. PCA with statistical process control has employed to enhance the discrimination features from the undamaged and damaged structures. Sohn et al. (2000) implemented visualization and dimension reduction for damage detection (Worden and Manson 1999). Beside of this, a number of applications of ICA have been reported in image processing (Antonini et al. 2006), biomedical signal processing (Vigario 1997), financial (Back and Weigend 1998), and medical area (Biswall and Ulmer 1999). The use of ICA in machine condition monitoring and fault detection application has reported in the field of structural damage detection (Zang et al. 2004) and submersible pump (Ypma and Pajunen 1999). However, there are still relatively few real engineering applications of ICA in machine condition monitoring and fault diagnosis.

The kernel trick is one of the crucial tricks for machine learning. Its basic idea is to project the input data into a high-dimensional implicit feature space with a nonlinear mapping, and then the data are analyzed so that nonlinear relations of the input data can be described. Recently, Bach and Jordan (2002) presented a new learning method of ICA which use contrast function based on canonical correlation in reproducing a kernel Hilbert space (RKHS). The other method proposed by Harmeling et al. (2001) uses a kernel-based blind source separation algorithm in the blind separation of nonlinearly mixed speech signal.

5.2 Basic Concepts

5.2.1 Pattern and Feature Vector

The corrected data of an object, $y(t)$, are firstly processed by a parameter extractor. Information relevant to pattern classification is extracted from $y(t)$ in the form of a p-dimensional parameter vector $\mathbf{y}$. Vector $\mathbf{y}$ is then transformed to a feature vector $\mathbf{x}$, which has a dimensionality m ($m \leq p$), by a feature extractor. Feature vector $\mathbf{x}$ is assigned to one of the K classes, $\Omega_1, \Omega_2, \ldots, \Omega_K$, by the classifier based on a certain type of classification criteria.

Pattern is a quantitative or structural description of an object or some other entity of interest. It is usually arranged in the form of a feature vector as

$$\mathbf{x} = \begin{bmatrix} x_1 \\ x_2 \\ \vdots \\ x_n \end{bmatrix}$$

where $x_1, x_2, \ldots, x_n$ are the features.

Depending on the measurements of an object, features in a pattern can be rather discrete number or real continuous values. The requirement on features is that the features can reflect the characteristics of desired objects and differ from those of other objects to the largest extent.

5.2.2 Class

Class or pattern class is a set of patterns that share some common properties. The feature vectors of the same type of objects will naturally form one set. Due to the diversity of the objects, the patterns extracted from the same type of objects are seldom identical. This can be interpreted as clusters of points in an n-dimensional space, which are called distributions of classes. Since the purpose of pattern recognition is to classify these patterns, the distributions of classes are desired to be separable and not empty. Suppose we have K classes, in a mathematical form, the requirement is

$$\Omega_k \neq \phi, \quad k = 1, 2, \ldots, K$$
$$\Omega_k \cap \Omega_l = \phi, \quad k \neq l \in \{1, 2, \ldots, K\}.$$

5.3 Parameter Evaluation Technique

To select the optimal parameters that can well represent the fault features, a feature extraction method based on the distance evaluation technique was presented by Yang et al. (2004, 2005).

Step 1: calculate the average distance of the same condition data ($d_{i,j}$) and then get the average distance of total conditions (d_{ai})
The equations can be defined as follows:

$$d_{i,j} = \frac{1}{N \times (N-1)} \sum_{m,n=1}^{N} \left| p_{i,j}(m) - p_{i,j}(n) \right|; \quad (m, n = 1, 2, \ldots, N, m \neq n) \tag{5.5}$$

where N is the number of the same condition, $P_{i,j}$ is the eigenvalue, $d_{i,j}$ is the average distance of the same condition, and i and j represent the number of parameters and conditions, respectively.

$$\underset{ai}{d} = \frac{1}{M} \sum_{j=1}^{M} d_{i,j} \tag{5.6}$$

where M is the number of different conditions.

Step 2: calculate the average distance between different condition data (d'_{ai})

$$d'_{ai} = \frac{1}{M \times (M-1)} \sum_{m,n=1}^{M} \left| p_{ai,m} - p_{ai,n} \right|; \quad (m, n = 1, 2, \ldots, M; m \neq n) \tag{5.7}$$

where d'_{ai} is the average distance of different conditions data, and $P_{ai,\,j}$ is the average value of the same condition data.

$$p_{ai,j} = \frac{1}{N}\sum_{n=1}^{N} p_{i,j}(n); \quad (n = 1, 2, \ldots, N) \tag{5.8}$$

Step 3: calculate the ratio d_{ai}/d'_{ai}

Step 4: selecting the feature parameters α_i from large value to small value. d_{ai} is the smaller, the better; on the contrary, d'_{ai} is the bigger, the better. So, bigger α_i represents the feature well.

$$a_i = d'_{ai}/d_{ai} \tag{5.9}$$

where α_i is the effectiveness factor of features.

5.4 Principal Component Analysis (PCA)

PCA is a statistical technique that linearly transforms an original set of variables into a substantially smaller set of uncorrelated variables that represents most of the information in the original set of variables (Jolliffe 1986). It can be viewed as a classical method of multivariate statistical analysis for achieving a dimensionality reduction. Because of the fact that a small set of uncorrelated variables is much easier to understand and use in further analysis than a larger set of correlated variables, this data compression technique has been widely applied to virtually every substantive area including cluster analysis, visualization of high-dimensionality data, regression, data compression, and pattern recognition.

Given a set of centered input vectors $\mathbf{x}_t$ ($t = 1, \ldots, l$ and $\Sigma \mathbf{x}_t = 0$), each of which is of m dimension $\mathbf{x}_t = [x_t(1), x_t(2), \ldots, x_t(m)]^T$ usually $m < l$, PCA linearly transforms each vector $\mathbf{x}_t$ into a new one $\mathbf{s}_t$ by

$$\mathbf{s}_t = \mathbf{U}^T \cdot \mathbf{x}_t \tag{5.10}$$

where $\mathbf{U}$ is the $m \times m$ orthogonal matrix whose ith column $\mathbf{u}_i$ is the eigenvector of the sample covariance matrix

$$\mathbf{C} = \frac{1}{l}\sum_{t=1}^{l} \mathbf{x}_t \cdot \mathbf{x}_t^T \tag{5.11}$$

In other words, PCA firstly solves the eigenvalue problem

$$\lambda_i \mathbf{u}_i = \mathbf{C} \cdot \mathbf{u}_i, \quad i = 1, \ldots, m \tag{5.12}$$

where λ_i is one of the eigenvalues of $\mathbf{C}$, $\mathbf{u}_i$ is the corresponding eigenvector. Based on the estimated $\mathbf{u}_i$, the components of $\mathbf{s}_t$ are then calculated as the orthogonal transformations of $\mathbf{x}_t$

$$\mathbf{s}_t(i) = \mathbf{u}_i^T \mathbf{x}_t, \quad i = 1, \ldots, m \tag{5.13}$$

The new components are called principal components. By using only the first several eigenvectors sorted in descending order of the eigenvalues, the number of principal components in $\mathbf{s}_t$ can be reduced. So PCA has the dimensional reduction characteristic. The principal components of PCA have the following properties: $\mathbf{s}_t(i)$ are uncorrelated and has sequentially maximum variances and the mean squared approximation error in the representation of the original inputs by the first several principal components is minimal.

Example 5.1 Calculate the covariance matrix of features vector acquired from vibration signal and determine its eigenvalue and eigenvector. Based on the eigenvalues, reduce the dimension of feature vectors.

From Fig. 5.2, the feature dimension can be reduced based on eigenvalues; for example, we can select only 5 feature vectors based on biggest eigenvalue of covariance matrix.

Example 5.2 Calculate principal component of vibration signal in Example 5.1. The principal components of signal vibration from channel 1 are presented in Fig. 5.3. Each principal component consists of 9 conditions of induction motor with 20 measurements.

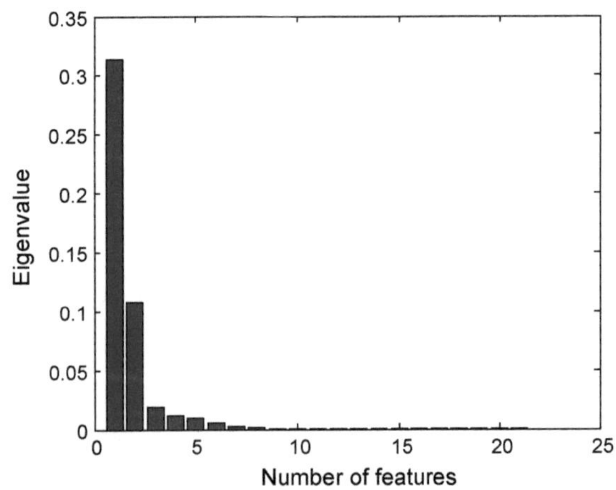

Fig. 5.2 Eigenvalue for dimensional reduction

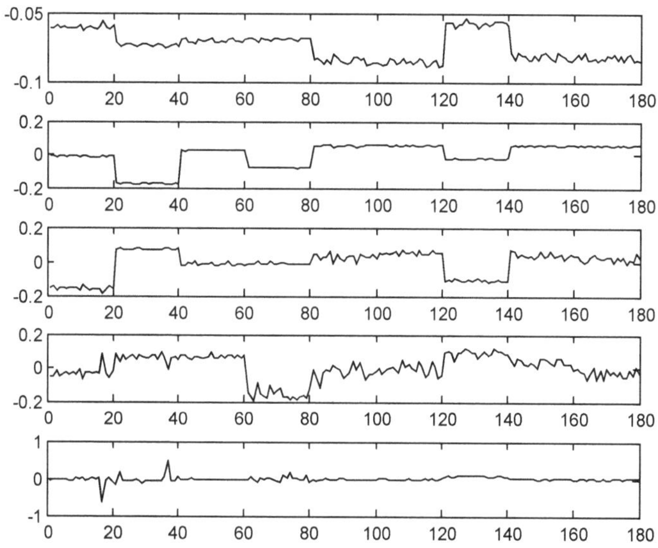

Fig. 5.3 Five principal components of vibration signal channel 1

5.5 Independent Component Analysis (ICA)

ICA is a technique that transforms multivariate random signal into a signal having components that are mutually independent in complete statistical sense. Recently, this technique has been demonstrated to be able to extract independent components from the mixed signals. Here independence means the information carried by one component cannot inferred from the others. Statistically, this means that joint probability of independent quantities is obtained as the product of the probability of each of them. A generic ICA model can be written as

$$\mathbf{x} = \mathbf{A} \cdot \mathbf{s} \tag{5.14}$$

where **A** is an unknown full-rank matrix, called the mixing matrix, and **s** is the independent component (IC) data matrix, and **x** is the measured variable data matrix. The basic problem of ICA is to estimate the independent component matrix **s** or to estimate the mixing matrix **A** from the measured data matrix **x** without any knowledge of **s** or **A**.

The ICA algorithm normally finds the independent components of a data set by minimizing or maximizing some measures of independence. Cardoso (1998) gave a review of the solution to the ICA problem using various information theoretic criteria, such as mutual information, negentropy, and maximum entropy, as well as maximum likelihood approach. The fixed-point algorithm used due to its suitability for handling raw time domain data and good convergence properties. This algorithm will now be described briefly.

The first step is to pre-whiten the measured data vector $\mathbf{x}$ by a linear transformation, to produce a vector $\tilde{\mathbf{x}}$ whose elements are mutually uncorrelated and all have unit variance. Singular value decomposition (SVD) of the covariance matrix $\mathbf{C} = E\left[\mathbf{x}\mathbf{x}^T\right]$ yields

$$\mathbf{C} = \boldsymbol{\Psi}\boldsymbol{\Sigma}\boldsymbol{\Psi}^T \tag{5.15}$$

where $\sum = \text{diag}(\sigma_1, \sigma_2, \ldots, \sigma_n)$ is a diagonal matrix of singular values and $\boldsymbol{\Psi}$ is the associated singular vector matrix. Then, the vector $\tilde{\mathbf{x}}$ can be expressed as

$$\tilde{\mathbf{x}} = \boldsymbol{\Psi}\boldsymbol{\Sigma}^{-1/2}\boldsymbol{\Psi}^T\mathbf{x} = \mathbf{Q}\mathbf{A}\mathbf{s} = \mathbf{B}\mathbf{s} \tag{5.16}$$

where $\mathbf{B}$ is an orthogonal matrix as verified by the following relation:

$$E\left[\tilde{\mathbf{x}} \cdot \tilde{\mathbf{x}}^T\right] = \mathbf{B}\,E\left[\mathbf{s} \cdot \mathbf{s}^T\right]\mathbf{B}^T = \mathbf{B}\,\mathbf{B}^T = \mathbf{I} \tag{5.17}$$

An advantage of using an SVD-based technique is the possibility of noise reduction by discarding singular values smaller than a given threshold. We have therefore reduced the problem of finding an arbitrary full-rank matrix $\mathbf{A}$ to the simpler problem of finding an orthogonal matrix $\mathbf{B}$ since $\mathbf{B}$ has fewer parameters to estimate as a result of the orthogonality constraint.

The second step is to employ the fixed-point algorithm. Define a separating matrix $\mathbf{W}$ that transforms the measured data vector $\mathbf{x}$ to a vector $\mathbf{y}$, such that all elements $\mathbf{y}_i$ are both mutually correlated and have unit variance. The fixed-point algorithm then determines $\mathbf{W}$ by maximizing the absolute value of kurtosis of $\mathbf{y}$. The vector $\mathbf{y}$ has the properties required for the independent components, thus

$$\tilde{\mathbf{s}} = \mathbf{y} = \mathbf{W}\mathbf{x} \tag{5.18}$$

From Eq. (5.16), we can estimate $\mathbf{s}$ as follows:

$$\tilde{\mathbf{s}} = \mathbf{B}^T\tilde{\mathbf{x}} = \mathbf{B}^T\mathbf{Q}\mathbf{x} \tag{5.19}$$

From Eqs. (5.18) and (5.19), the relation of $\mathbf{W}$ and $\mathbf{B}$ can be expressed as

$$\mathbf{W} = \mathbf{B}^T\mathbf{Q} \tag{5.20}$$

To calculate $\mathbf{B}$, each column vector $\mathbf{b}_i$ is initialized and then updated so that ith independent component $\mathbf{s}_i = (\mathbf{b}_i)^T\tilde{\mathbf{x}}$ may have great non-Gaussianity. Hyvärinen and Oja (2000) showed that non-Gaussian represents independence using the central limit theorem. There are two common measures of non-Gaussianity: kurtosis and negentropy. Kurtosis is sensitive to outliers. On the other hand, negentropy is based on the information theoretic quantity of (differential) entropy. Entropy is a measure of the average uncertainty in a random variable and the differential entropy H of random variable y with density $f(y)$ is defined as

$$H(y) = -\int f(y) \log f(y) \mathrm{d}y \tag{5.21}$$

A Gaussian variable has maximum entropy among all random variables with equal variance (Hyvärinen and Oja 2000). In order to obtain a measure of non-Gaussianity that is zero for a Gaussian variable, the negentropy J is defined as follows:

$$J(y) = H(y_{\mathrm{Gauss}}) - H(y) \tag{5.22}$$

where y_{Gauss} is a Gaussian random variable with the same variance as y. Negentropy is nonnegative and measures the departure of y from Gaussianity. However, estimating negentropy using Eq. (5.22) would require an estimate of the probability density function. To estimate negentropy efficiently, simpler approximations of negentropy are suggested as follows:

$$J(y) \approx [E\{G(y)\} - E\{(\upsilon)\}]^2 \tag{5.23}$$

where y is assumed to be of zero mean and unit variance, v is a Gaussian variable of zero mean and unit variance, and G is any non-quadratic function. By choosing G wisely, one obtains good approximations of negentropy. A number of functions for G are:

$$G_1(\upsilon) = \frac{1}{a_1} \log \cos h(a_1 \upsilon) \tag{5.24}$$

$$G_2(\upsilon) = \exp(-a_2 \upsilon^2/2) \tag{5.25}$$

$$G_3(\upsilon) = \upsilon^4 \tag{5.26}$$

where $1 \leq a_1 \leq 2$ and $a_2 \approx 1$. Among these three functions, G_1 is a good general-purpose contrast function and was therefore selected for use in the present study.

Based on the approximate form for the negentropy, Hyvärinen (1999) introduced a very simple and highly efficient fixed-point algorithm for ICA, calculated over sphered zero-mean vector $\tilde{\mathbf{x}}$. This algorithm calculates one column of the matrix $\mathbf{B}$ and allows the identification of one independent component; the corresponding independent component can then be found using Eq. (5.19). The algorithm is repeated to calculate each independent component.

5.6 Kernel PCA

Kernel PCA is one approach of generalizing linear PCA into nonlinear case using the kernel method. The idea of kernel PCA is to firstly map the original input vectors $\mathbf{x}_t$ into a high-dimensional feature space $\boldsymbol{\varphi}(\mathbf{x}_t)$ and then calculate the linear PCA in $\boldsymbol{\varphi}(\mathbf{x}_t)$.

Given a set of centered input vectors $\mathbf{x}_t$ (t = 1, …, l and $\Sigma\mathbf{x}_t = 0$), each of which is of m dimension $\mathbf{x}_t = (x_t(1), x_t(2), \ldots, x_t(l))^T$. By mapping $\mathbf{x}_t$ into $\boldsymbol{\varphi}(\mathbf{x}_t)$ whose dimension is assumed to be larger than the number of training samples l, kernel PCA solves the eigenvalue problem

$$\lambda_i \mathbf{u}_i = \tilde{\mathbf{C}} \mathbf{u}_i, \quad i = 1, \ldots, l \tag{5.27}$$

where $\tilde{\mathbf{C}}$ is the sample covariance matrix of $\boldsymbol{\varphi}(\mathbf{x}_t)$, λ_i is one of the nonzero eigenvalues of $\tilde{\mathbf{C}}$, and $\mathbf{u}_i$ is the corresponding eigenvectors. The $\tilde{\mathbf{C}}$ on the feature space can be constructed by

$$\tilde{\mathbf{C}} = \frac{1}{l} \sum_{t=1}^{l} \boldsymbol{\varphi}(\mathbf{x}_t) \boldsymbol{\varphi}(\mathbf{x}_t)^T \tag{5.28}$$

From Eq. (5.28), we can obtain the nonzero eigenvalues that are positive. Let us define matrix $\mathbf{Q}$ as

$$\mathbf{Q} = [\boldsymbol{\varphi}(\mathbf{x}_1), \ldots, \boldsymbol{\varphi}(\mathbf{x}_l)] \tag{5.29}$$

Then, Eq. (5.28) can be expressed by

$$\tilde{\mathbf{C}} = \frac{1}{l} \mathbf{Q}\mathbf{Q}^T \tag{5.30}$$

Moreover, we can construct a Gram matrix using Eq. (5.29) in which their element can be determined by kernel

$$\mathbf{R} = \mathbf{Q}^T \mathbf{Q} \tag{5.31}$$

$$\mathbf{R}_{ij} = \boldsymbol{\varphi}(\mathbf{x}_t)^T \boldsymbol{\varphi}(\mathbf{x}_t) = (\boldsymbol{\varphi}(\mathbf{x}_i) \cdot \boldsymbol{\varphi}(\mathbf{x}_j)) = K(\mathbf{x}_i, \mathbf{x}_j) \tag{5.32}$$

Denote $\mathbf{V} = (\gamma_1, \gamma_2, \ldots, \gamma_l)$ and $\boldsymbol{\Lambda} = \mathrm{diag}(\lambda_1, \lambda_2, \ldots, \lambda_l)$ are eigenvectors and eigenvalues of $\mathbf{R}$, respectively, we can calculate the orthonormal eigenvectors $\boldsymbol{\beta}_j$ as

$$\boldsymbol{\beta}_j = \frac{1}{\sqrt{\lambda_j}} \mathbf{Q}\, \boldsymbol{\gamma}_j, \quad j = 1, \ldots, m \tag{5.33}$$

Table 5.1 Formulation of kernel functions

Kernel	$K(\mathbf{x}, \mathbf{x}_j)$
Linear	$\mathbf{x}^T \cdot \mathbf{x}_j$
Polynomial	$(\gamma\, \mathbf{x}^T \cdot \mathbf{x}_j + r)^d$, $\gamma > 0$
Gaussian RBF	$\exp(-\lVert \mathbf{x} - \mathbf{x}_j \rVert^2 / 2\gamma^2)$
Sigmoid	$\tan h(\tau_0\, (\mathbf{x}^T \cdot \mathbf{x}_j) + \tau_1)$

Then we define matrix **B** as

$$\mathbf{B} = (\beta_1, \beta_2, \ldots, \beta_m) = \mathbf{QV\Lambda}^{-1/2} \tag{5.34}$$

The whitening matrix **P** can be derived from Eq. (5.34) and expressed by

$$\mathbf{P} = \mathbf{B}\left(\frac{1}{l}\mathbf{\Lambda}\right)^{-1/2} = \sqrt{l}\mathbf{QV\Lambda}^{-1} \tag{5.35}$$

The mapped data in feature space can be whitened by the following transformation:

$$\begin{aligned} \mathbf{r} &= \mathbf{P}^T \boldsymbol{\varphi}(\mathbf{x}) \\ &= \sqrt{l}\mathbf{\Lambda}^{-1}\mathbf{V}^T\mathbf{Q}^T\boldsymbol{\varphi}(\mathbf{x}) = \sqrt{l}\mathbf{\Lambda}^{-1}\mathbf{V}^T[K(\mathbf{x}_1, \mathbf{x}), K(\mathbf{x}_2, \mathbf{x}), \ldots, K(\mathbf{x}_l, \mathbf{x}) = \sqrt{l}\mathbf{\Lambda}^{-1}\mathbf{V}^T\mathbf{R}_x \end{aligned} \tag{5.36}$$

There are several kernel functions described in Table 5.1 which can be used to accomplish Eq. (5.32), such as linear, polynomial, and Gaussian RBF.

5.7 Kernel ICA

Practically speaking, the kernel ICA is the combination of centering and whitening process using kernel PCA as previously explanation and iterative section using ICA. The following task is to find the mixing matrix **W** in the kernel PCA-transformed space to recover independent components **s** from **r**, recall Eq. (5.18)

$$\hat{\mathbf{s}} = \mathbf{W}\,\mathbf{x} = \mathbf{W}\,\mathbf{r} \tag{5.37}$$

In summary, the nonlinear feature extraction using kernel ICA in this chapter performs two phases: whitening process using kernel PCA and ICA transformation in the kernel PCA-whitened space.

Example 5.3 Find 5 independent components of vibration signal using kernel PCA and kernel ICA with RBF kernel function (Figs. 5.4 and 5.5).

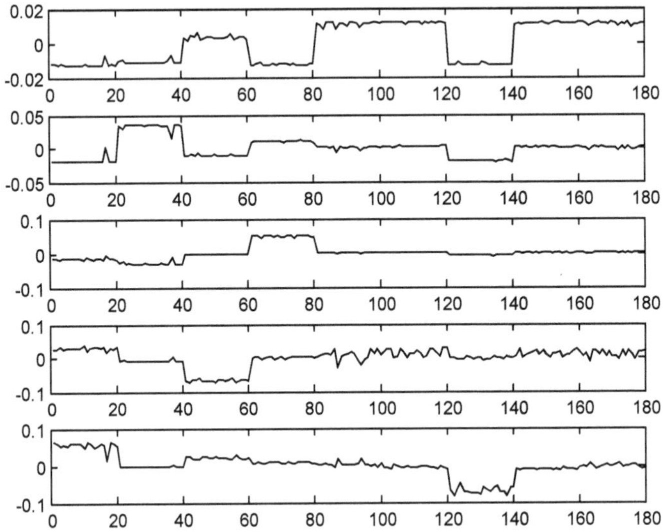

Fig. 5.4 Five principal components of vibration signal channel 1 using kernel PCA

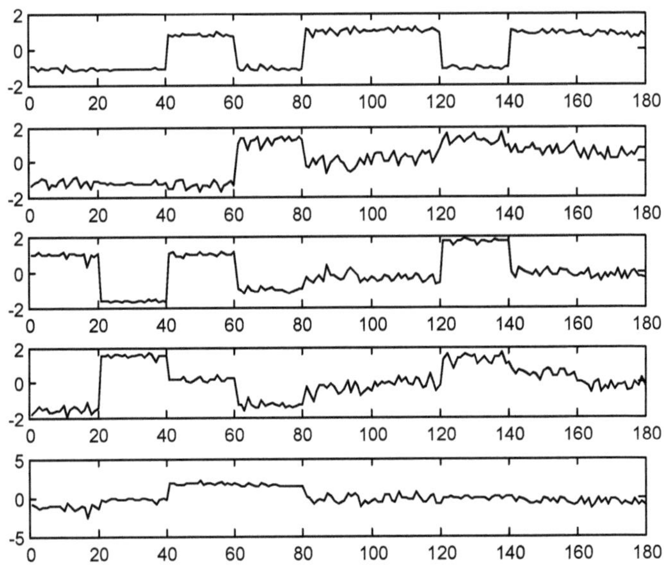

Fig. 5.5 Five independent components of vibration signal channel 1 using kernel ICA

5.8 Fisher Discriminant Analysis (FDA)

Fisher discriminant analysis (FDA) is a traditional statistical method and a linear dimensionality reduction technique, optimal in terms of maximizing the separation between several classes for feature extraction. It provides a lower-dimensional representation of data in that several groups or classes can be discriminated as clearly as possible. FDA has been shown to be the good linear technique for fault diagnosis and to outperform PCA-based diagnosis methods. A disadvantage of using *linear* FDA (LFDA) in multivariate fault diagnosis problems is that nonlinear behavior in the data cannot be represented well. Although LFDA is an effective technique for feature extraction, it is still a linear technique in nature. To overcome such a limitation, *kernel trick* has been used to develop a nonlinear kernel version of FDA, called kernel FDA (KFDA) (Baudat and Anouar 2000). The basic idea of the kernel trick is that input data are mapped into a *kernel feature space* by a nonlinear mapping function and then these mapped data are analyzed. A number of powerful kernel-based techniques have been developed, including support vector machines (Cortes and Vapnik 1995), kernel PCA (Scholkopf et al. 1998), kernel PLS (Rosipal and Trejo 2000), and kernel FDA (Baudat and Anouar 2000).

The selection of linear (LFDA) and nonlinear (KFDA) techniques for solving fault diagnosis problems depends on the characteristics of target processes of interest. In linear case, fault groups or cases in data are separated easily using linear technique. Such a linear case is the simplest problem, in which both linear and nonlinear techniques are expected to produce a good classification performance. The use of linear techniques in nonlinear case, however, may not classify most of data correctly. In this respect, a nonlinear technique such as KFDA will be useful for diagnosis because they have more nonlinear characteristics involved when compared to continuous processes. Thus, it is essential to develop a more efficient diagnosis method for engineering systems.

In this section, we describe KFDA briefly as follows. As above mentioned, KFDA is the nonlinear kernel version of linear FDA to deal with the feature extraction of nonlinear characteristics.

Let us consider a set of M observations in an n-dimensional space $\mathbf{x}_k \in \Re^n$, $k = 1, \ldots, M$. For a given nonlinear mapping Φ, the input space can be mapped into feature space F, Φ: $\Re^n \to F$, $\mathbf{x} \to \Phi(\mathbf{x})$. Note that the feature space F could have a much higher, possibly infinite, dimensionality. The objective is kernel FDA is to find certain directions in the original variable space, along which latent groups or clusters in $\Re^n$ are discriminated as clearly as possible. Kernel FDA performs LFDA in the feature space F, which is nonlinearly related to the input space $\Re^n$. As a result, kernel FDA produces a set of nonlinear discriminant vectors in the input space. The discriminant weight vector is determined by maximizing between-class scatter matrix $\mathbf{S}_b^\Phi$ while minimizing total scatter matrix $\mathbf{S}_t^\Phi$, which are defined in F as follows:

$$\mathbf{S}_b^{\Phi} = \frac{1}{M}\sum_{i=1}^{C} c_i(\mathbf{m}_i^{\Phi} - \mathbf{m}^{\Phi})(\mathbf{m}_i^{\Phi} - \mathbf{m}^{\Phi})^T \tag{5.38}$$

$$\mathbf{S}_t^{\Phi} = \frac{1}{M}\sum_{i=1}^{M} (\Phi(\mathbf{x}_i) - \mathbf{m}^{\Phi})(\Phi(\mathbf{x}_i) - \mathbf{m}^{\Phi})^T \tag{5.39}$$

where $\mathbf{m}_i^{\Phi}$ represents the mean vector of the mapped observations of class i, $\mathbf{m}^{\Phi}$ the mean vector of the mapped M observations, c_i the number of observations of class i, and C the total number of class of $\mathbf{x}_k$, $k = 1, \ldots, M$.

Similarly to the formation of LFDA, this can be done by maximizing the Fisher criterion:

$$J^{\Phi}(\mathbf{\Psi}) = \frac{\mathbf{\Psi}^T\mathbf{S}_b^{\Phi}\mathbf{\Psi}}{\mathbf{\Psi}^T\mathbf{S}_t^{\Phi}\mathbf{\Psi}}, \quad \mathbf{\Psi} \neq \mathbf{0} \tag{5.40}$$

The optimal discriminant vectors in feature space F can be obtained by solving the eigenvalue problem $\mathbf{S}_b^{\Phi}\mathbf{\Psi} = \lambda\,\mathbf{S}_t^{\Phi}\mathbf{\Psi}$ instead of Eq. (5.40). They are actually the eigenvectors of $\mathbf{S}_b^{\Phi}\mathbf{\Psi} = \lambda\,\mathbf{S}_t^{\Phi}\mathbf{\Psi}$.

The use of a kernel function $K(\mathbf{x}_i, \mathbf{x}_j)$ allows us to compute dot products in F without nonlinear mapping Φ. By replacing canonical (Euclidean) dot products $\langle\mathbf{x},\mathbf{y}\rangle$ in F by a kernel function of the form $K(\mathbf{x}_i,\mathbf{x}_j) = \langle\Phi(\mathbf{x}_i),\Phi(\mathbf{x}_j)\rangle$, we do not need to execute nonlinear mapping Φ and dot products in F, which is called the *kernel trick* (Cortes and Vapnik 1995). Consider a simple mapping Φ:

$$\mathbf{x} = (x_1, x_2) \rightarrow \Phi(\mathbf{x}) = (x_1^2, \sqrt{2}x_1x_2, x_2^2) \tag{5.41}$$

The computation of a dot product between two mappings $\Phi(\mathbf{x})$ and $\Phi(\mathbf{y})$ in F can be easily expressed in terms of a kernel function of the form $k(\mathbf{x},\mathbf{y}) = \langle\mathbf{x},\mathbf{y}\rangle^2$ as follows:

$$\langle\Phi(\mathbf{x}),\,\Phi(\mathbf{y})\rangle = \left(x_1^2, \sqrt{2}x_1x_2, x_2^2\right)^T\left(y_1^2, \sqrt{2}y_1y_2, y_2^2\right) = \left((x_1,x_2)^T(y_1,y_2)\right) = \langle\mathbf{x},\mathbf{y}\rangle^2 \tag{5.42}$$

where $K(\mathbf{x},\mathbf{y}) = \langle\mathbf{x},\mathbf{y}\rangle^2$ is the second-order polynomial kernel function.

Some of the most widely used kernel functions are described in Table 5.1.

Example 5.4 Perform kernel Fisher discriminant analysis (KFDA) and compare the feature extraction with linear Fisher discriminant analysis (FDA).

Solution: Firstly, we must calculate the kernel matrix for mapping data into feature space using kernel function, i.e., RBF kernel function. User must adjust and select proper kernel parameter to find optimal feature extraction using KFDA. Try to use different kernel parameter for comparison with Fig. 5.6 (Fig. 5.7).

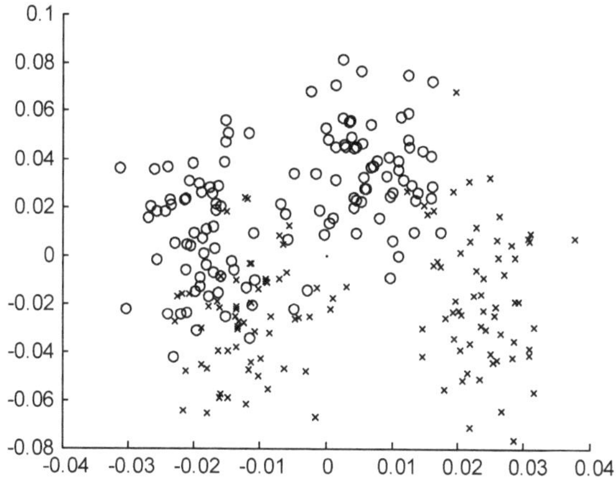

Fig. 5.6 Feature extraction using KFDA

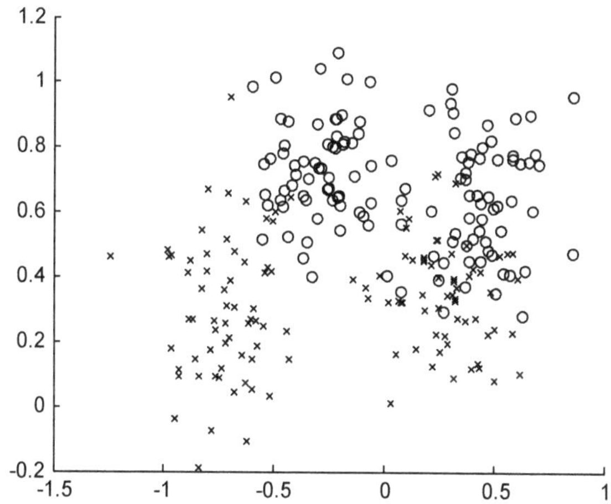

Fig. 5.7 Feature extraction using FDA

5.9 Linear Discriminant Analysis (LDA)

Although PCA finds components that are useful for representing data, there is no reason to assume that these components must be useful for discriminating between data in different classes. Where PCA seeks directions that are efficient for representation, discriminant analysis seeks directions that are efficient for discrimination.

LDA projects feature from parametric space onto feature space through a linear transformation matrix $\mathbf{T}$. Suppose the input observation vector $\mathbf{x}$ be a p-dimensional feature and $\mathbf{T}$ be a $p \times m$ $(p \geq m)$ matrix. The extracted features are $\mathbf{y} = \mathbf{T}^T\mathbf{x}$. This method maximizes the ratio of between-class variance to the within-class variance.

One of linear discriminant methods is Fisher's linear discriminant that is to well separate the classes by projecting classes' samples from p-dimensional space onto a finely orientated line. For a K-class problem, $m = \min(K - 1, p)$ different lines will be involved. Thus, the projection is from a p-dimensional space to a c-dimensional space.

Suppose we have K classes, $\mathbf{X}_1, \mathbf{X}_2, \ldots, \mathbf{X}_K$. Let the ith observation from the $\mathbf{X}_i$ be $\mathbf{x}_{ji}$, where $j = 1,\ldots, K$, $i = 1,\ldots, N_j$ and N_j is the number of observations from class j. The within-class covariance matrix $\mathbf{S}_W$ is as follows:

$$\mathbf{S}_W = \sum_{j=1}^{K} \mathbf{S}_j \tag{5.43}$$

where

$$\mathbf{S}_j = \frac{1}{N_j} \sum_{i=1}^{N_j} (\mathbf{x}_{ji} - \boldsymbol{\mu}_j)(\mathbf{x}_{ji} - \boldsymbol{\mu}_j)^T; \; \boldsymbol{\mu}_j = \frac{1}{N_j} \sum_{i=1}^{N_j} \mathbf{x}_{ji} \tag{5.44}$$

The between-class covariance matrix $\mathbf{S}_B$ is given as:

$$\mathbf{S}_B = \sum_{j=1}^{K} N_j (\boldsymbol{\mu}_j - \boldsymbol{\mu})(\boldsymbol{\mu}_j - \boldsymbol{\mu})^T \tag{5.45}$$

where

$$\boldsymbol{\mu} = \frac{1}{N} \sum_{j=1}^{K} \sum_{i=1}^{N_j} \mathbf{x}_{ji}. \tag{5.46}$$

Here we give a two-class example. In order to obtain good separation of the projected data, we really want the difference between the means to be large and the total within-class covariance to be smaller. Thus, the Fisher linear discriminant employs that linear function $\mathbf{T}^T\mathbf{x}$ for which the criterion function

$$J(\mathbf{w}) = \frac{|\tilde{\boldsymbol{\mu}}_1 - \tilde{\boldsymbol{\mu}}_2|^2}{\tilde{\mathbf{s}}_1^2 + \tilde{\mathbf{s}}_2^2} \tag{5.47}$$

To obtain $J(\cdot)$ as an explicit function of $\mathbf{w}$, after transformation the criterion function can be written as:

$$J(\mathbf{w}) = \frac{\mathbf{w}^T \mathbf{S}_B \mathbf{w}}{\mathbf{w}^T \mathbf{S}_W \mathbf{w}} \tag{5.48}$$

It is easy to show that a vector $\mathbf{w}$ that maximizes $J(\cdot)$ must satisfy

$$\mathbf{S}_B \mathbf{w} = \lambda \mathbf{S}_W \mathbf{w} \rightarrow \mathbf{S}_W^{-1} \mathbf{S}_B \mathbf{w} = \lambda \mathbf{w} \tag{5.49}$$

It is unnecessary to solve for the eigenvalues and eigenvectors of $\mathbf{S}_W^{-1}\mathbf{S}_B$ due to the fact that $\mathbf{S}_B\mathbf{w}$ is always in the direction of $\mathbf{m}_1 - \mathbf{m}_2$. Because the scale factor for $\mathbf{w}$ is immaterial, the solution for the $\mathbf{w}$ is that optimizes $J(\cdot)$:

$$\mathbf{w} = \mathbf{S}_W^{-1}(\mathbf{m}_1 - \mathbf{m}_2) \tag{5.50}$$

Example 5.5 Compute the matrices $\mathbf{S}_W$ and $\mathbf{S}_B$ of iris data (downing from https://archive.ics.uci.edu/ml/datasets/Iris).

Solution: Iris data are very famous for simulation of classification problem. This data contain 3 classes that describe iris flower: Setosa, Versicolor, and Virginia.

```
≫ data = load ('iris');
≫ [dim,num_data] = size(data.X);
   nclass = max(data.y);
≫ mean_X = mean(data.X, 2);
   Sw=zeros(dim,dim);
   Sb=zeros(dim,dim);
≫ for i = 1:nclass,
   inx_i = find(data.y==i);
   X_i = data.X(:,inx_i);
   mean_Xi = mean(X_i,2);
   Sw = Sw + cov(X_i', 1);
   Sb = Sb + length(inx_i)*(mean_Xi - mean_X)*(mean_Xi - mean_X)';
   end;
≫ Sw
   Sw =
   0.0499   0.0315   0.0361   0.0229
   0.0315   0.0704   0.0214   0.0357
   0.0361   0.0214   0.0457   0.0290
   0.0229   0.0357   0.0290   0.0790
≫ Sb
   Sb =
   4.0514    -2.2479   12.1200   14.4533
   -2.2479   2.2681    -7.3854   -8.1791
   12.1200   -7.3854   36.6850   43.3410
   14.4533   -8.1791   43.3410   51.5866
```

Example 5.6 Compute the projection matrix of $\mathbf{S_W}$ and $\mathbf{S_B}$.

Solution: The projection matrix can be calculated using singular value decomposition by svd.m that is available in MATLAB.

$$\gg [U, D, V] = svd(inv(Sw)*Sb);$$

$\gg$ U

U =

−0.3386	0.1426	−0.8329	−0.4139
−0.3536	−0.6131	0.3451	−0.6164
0.7870	0.1497	0.0031	−0.5985
0.3755	−0.7624	−0.4327	0.3008

$\gg$ D

D =

1.0e + 003*

2.1606	0	0	0
0	0.0135	0	0
0	0	0.0000	0
0	0	0	0.0000

$\gg$ V

V =

0.2064	−0.1337	0.6897	0.6811
−0.1311	−0.9068	−0.3422	0.2083
0.6281	0.2350	−0.5884	0.4517
0.7388	−0.3233	0.2469	−0.5373

Example 5.7 Perform dimensionality reduction of iris data from original 4 features dimension into 2 features using LDA algorithm as shown in Fig. 5.8.

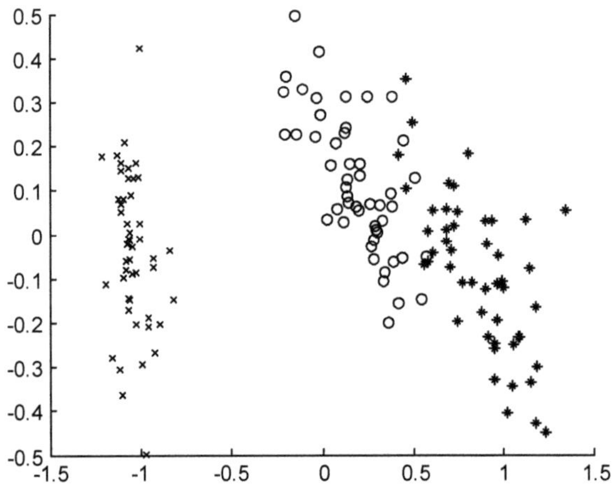

Fig. 5.8 Dimensionality reduction of iris data using LDA algorithm

5.10 Generalized Discriminant Analysis (GDA)

The generalized discriminant analysis deals with a nonlinear discriminant analysis using kernel function operator. GDA method provides a mapping of the input vectors into high-dimensional feature space.

Given a set of centered input vectors $\mathbf{x}_t$ ($t = 1, \ldots, l$ and $\Sigma \mathbf{x}_t = 0$); each of which is of m dimension $\mathbf{x}_t = (x_t(1), x_t(2), \ldots, x_t(m))^T$ usually $m < l$ in the input set X. For N is the number of classes, the cardinality of subset X_{m} is denoted by k_{m}, thus $\sum_{l=1}^{N} k_m = l$. The covariance matrix can be calculated as

$$\mathbf{C} = \frac{1}{l}\sum_{t=1}^{l} \mathbf{x}_t \cdot \mathbf{x}_t^T \tag{5.51}$$

Suppose that the space X is mapped into a Hilbert space F through a nonlinear mapping function ϕ, so that

$$\begin{aligned} \phi &: X \to F \\ x &\to \phi(x) \end{aligned} \tag{5.52}$$

The covariance matrix in the feature space F is

$$\mathbf{C} = \frac{1}{l}\sum_{i=1}^{l} \phi(x_i)\phi^T(x_i) \tag{5.53}$$

Let $\mathbf{B}$ the covariance matrix of the class centers. $\mathbf{B}$ represents the interclass inertia in the space F

$$\mathbf{B} = \frac{1}{l}\sum_{j=1}^{l} k_m \bar{\phi}_j \bar{\phi}_j^T \tag{5.54}$$

where $\bar{\phi}_j$ is the mean value of the class j

$$\bar{\phi}_j = \frac{1}{k_m}\sum_{k=1}^{k_m} \phi(x_{jk}) \tag{5.55}$$

where x_{jk} is the element k of the class j.

In the same manner, covariance matrix in Eq. (5.53) can be rewritten using the class indices

$$\mathbf{C} = \frac{1}{l}\sum_{j=1}^{l}\sum_{k=1}^{k_m} \phi(x_{jk})\phi^T(x_{jk}) \tag{5.56}$$

C represents the total inertia of the data into feature space F.

In order to simplify, when there is no ambiguity in index of x_{jk}, the class index j is omitted.

Introducing the kernel function into Eq. (5.56) yields

$$\mathbf{C} = \frac{1}{l}\sum_{j=1}^{l}\sum_{k=1}^{k_m} K(x_{jk}, x_{jk}) \tag{5.57}$$

The kernel operator K allows the construction of nonlinear separating function in the feature space F. As such for the LDA, the purpose of the GDA is to maximize the interclass inertia and minimize the intraclass inertia. This maximization is equivalent to the following eigenvalue decomposition

$$\lambda\mathbf{C}\mathbf{v} = \mathbf{B}\mathbf{v} \tag{5.58}$$

The largest eigenvalue of Eq. (5.58) gives the maximum of the following quotient of the inertia

$$\lambda = \frac{\mathbf{v}^T\mathbf{B}\mathbf{v}}{\mathbf{v}^T\mathbf{C}\mathbf{v}} \tag{5.59}$$

As the eigenvector is the linear combination of F elements, there exist coefficients α such that

$$\mathbf{v} = \sum_{p=1}^{N}\sum_{q=1}^{k_m} \alpha_{pq}\phi(x_{pq}) \tag{5.60}$$

All solutions v lie in the span on $\phi(x_{jk})$. Then, Eq. (5.59) is equal to the following quotient

$$\lambda = \frac{\boldsymbol{\alpha}^T\mathbf{KWK}\boldsymbol{\alpha}}{\boldsymbol{\alpha}^T\mathbf{KK}\boldsymbol{\alpha}} \tag{5.61}$$

where **W** is the block diagonal matrix ($l \times l$) with terms all equal to $1/k_m$; **K** is the kernel matrix defined on the class element that is composed by dot product in feature space F. Interested readers are suggested to see Boudat and Anouar (2000) for detailed derivation of equation.

Premultiplying Eq. (5.58) by $\phi^T(x_{jk})$ yields

$$\lambda\phi^T(x_{jk})\mathbf{C}\mathbf{v} = \phi^T(x_{jk})\mathbf{B}\mathbf{v} \tag{5.62}$$

We can rewrite two terms of Eq. (5.62) in a matrix form using matrices **K** and **W** which gives Eq. (5.63). Interested readers must see Boudat and Anouar (2000) for complete derivation.

Considering kernel matrix **K**, let us use its eigenvector decomposition so that

$$\mathbf{K} = \mathbf{P}\Gamma\Gamma^T \tag{5.63}$$

where $\mathbf{\Gamma}$ is the diagonal matrix of nonzero eigenvalue and **P** is the matrix of normalized eigenvectors associated with $\mathbf{\Gamma}$, thus $\mathbf{\Gamma}^{-1}$ exists. **P** is the orthonormal matrix that is

$$\mathbf{P}^T\mathbf{P} = \mathbf{I} \tag{5.64}$$

Substituting **K** into Eq. (5.61) yields

$$\lambda = \frac{(\mathbf{\Gamma P}^T\boldsymbol{\alpha})^T\mathbf{P}^T\mathbf{WP}(\mathbf{\Gamma P}^T\boldsymbol{\alpha})}{(\mathbf{\Gamma P}^T\boldsymbol{\alpha})^T\mathbf{P}^T\mathbf{P}(\mathbf{\Gamma P}^T\boldsymbol{\alpha})} \tag{5.65}$$

Let us define Eq. (5.66) such that

$$\boldsymbol{\beta} = \mathbf{\Gamma P}^T\boldsymbol{\alpha} \tag{5.66}$$

$$\lambda = \frac{\boldsymbol{\beta}^T\mathbf{P}^T\mathbf{WP}\boldsymbol{\beta}}{\boldsymbol{\beta}^T\mathbf{P}^T\mathbf{P}\boldsymbol{\beta}} \tag{5.67}$$

Therefore, we obtain

$$\lambda\mathbf{P}^T\mathbf{P}\boldsymbol{\beta} = \mathbf{P}^T\mathbf{WP}\,\boldsymbol{\beta} \tag{5.68}$$

As **P** is orthonormal, the latter equation can be simplified and gives Eq. (5.69), in which solutions are to be found by maximizing λ

$$\lambda\boldsymbol{\beta} = \mathbf{P}^T\mathbf{WP}\boldsymbol{\beta} \tag{5.69}$$

For a given $\boldsymbol{\beta}$, there exists at least one $\boldsymbol{\alpha}$ satisfying Eq. (5.66) in the form $\boldsymbol{\alpha} = \mathbf{P\Gamma}^{-1}\boldsymbol{\beta}$ that is not unique.

Thus, the first step of the system resolution consists in finding $\boldsymbol{\beta}$ according to Eq. (5.69), which corresponds to a classical eigenvector system resolution. Once $\boldsymbol{\beta}$ is calculated, the α is computed then normalized by requiring that the corresponding vectors **v** be normalized in feature space F so that

$$\mathbf{v}^T\mathbf{v} = 1 \tag{5.70}$$

Using Eq. (5.69), we obtain

$$\begin{aligned} \mathbf{v}^T\mathbf{v} &= \sum_{p=1}^{N}\sum_{q=1}^{k_m}\sum_{r=1}^{N}\sum_{s=1}^{k_n} \alpha_{pq}\alpha_{rs}\phi^T(x_{pq})\phi(x_{rs}) = 1 \\ &:= \sum_{p=1}^{N}\sum_{r=1}^{N} \alpha_p^T K_{pr}\alpha_r = 1 \\ &= \boldsymbol{\alpha}^T\mathbf{K}\boldsymbol{\alpha} = 1 \end{aligned} \tag{5.71}$$

The coefficients α are divided by $\sqrt{\boldsymbol{\alpha}^T\mathbf{K}\boldsymbol{\alpha}}$ in order to get normalized vector $\mathbf{v}$. Knowing the normalized vector $\mathbf{v}$, the projections of test point z are computed by

$$\mathbf{v}^T\phi(z) = \sum_{p=1}^{N}\sum_{q=1}^{k_m} \alpha_{pq}K(x_{pq}, z) \tag{5.72}$$

Finally, the GDA procedure can be summarized as follows:

Step 1: Compute the matrices $\mathbf{K}$ and $\mathbf{W}$.
Step 2: Decompose $\mathbf{K}$ using eigenvectors and eigenvalue.
Step 3: Compute eigenvectors $\boldsymbol{\beta}$ and eigenvalues of the system.
Step 4: Compute eigenvectors $\mathbf{v}$ using $\boldsymbol{\alpha}$ and normalize them.
Step 5: Compute projections of test points onto the eigenvectors $\mathbf{v}$.

Example 5.8 Calculate the decomposition of kernel matrix in Eq. (5.63) of iris data. Use RBF kernel function with kernel width parameter $\gamma = 1$. Then calculate block diagonal matrix $\mathbf{W}$ and finally perform dimensionality reduction from 4 features into 2 features of 3-class iris as shown in Fig. 5.9.

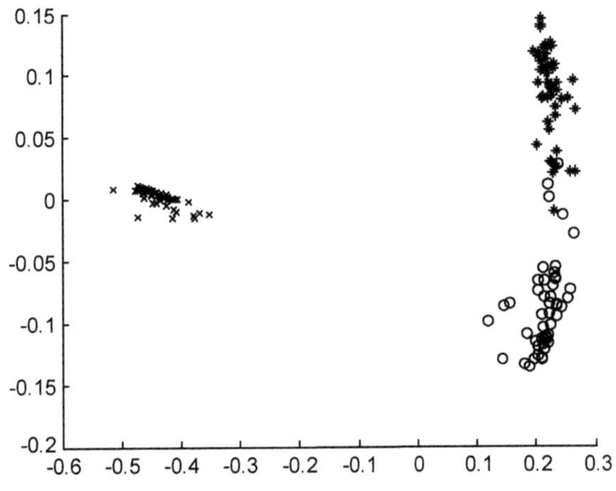

Fig. 5.9 Dimensionality reduction of iris data using GDA

Example 5.8 shows that 2-dimensional data extracted from the originally 4-dimensional iris data set using the GDA. In contrast to LDA (see Fig. 5.8 for comparison), the class clusters are more compact but at the expense of a higher risk of overfitting.

5.11 Clustering

Clustering can be considered the most important in unsupervised learning problem. So, as very other problem of this kind, it deals with finding a structure in a collection of unlabeled data. A loose definition of clustering could be the process of organizing objects into groups whose members are similar in some way. A cluster is therefore a collection of objects which are similar between them and are dissimilar to the objects belonging to other clusters. We can see this determination by a simple example in Fig. 5.10.

In Fig. 5.10, we can easily identify the 4 clusters into which the data can be divided. The similarity criterion is distance: Two or more objects belong to the same cluster if they are close according to a given distance (in this case, geometrical distance). This is so-called *distance-based clustering*. Another kind of clustering is *conceptual clustering*: Two or more objects belong to the same cluster if this one defines a concept common to all that objects. In other words, objects are grouped according to their fit to descriptive concepts, not according to simple similarity measures.

The goal of clustering is to determine the intrinsic grouping in a set of unlabeled data. But how to decide what constitutes a good clustering is relatively depended on the need of clustering. It can be said that there is no absolute best criterion which would be independent of the final aim of the clustering. Consequently, it is the user that must supply this criterion, in such way that the result of the clustering will suit their needs. For instance, we could be interested in finding representatives for homogeneous groups (data reduction), in finding natural cluster and describe their unknown properties (natural data types), in finding useful and suitable groupings (useful data classes), or in finding unusual data objects (outlier detection).

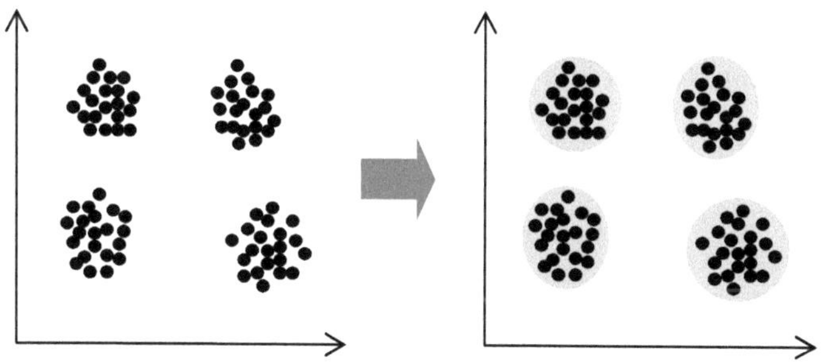

Fig. 5.10 A simple clustering

The main requirements that a clustering algorithm should satisfy are as follows:

- Scalability
- Dealing with different types of attributes
- Discovering clusters with arbitrary shape
- Minimal requirement for domain knowledge to determine the input parameters
- Ability to deal with noise and outliers
- Insensitivity to order of input records
- High dimensionality
- Interpretability and usability

When we deal with clustering, there are a number of problems may occur as follows:

- Current clustering techniques do not address all the requirements adequately (and concurrently)
- Deal with large number of dimension of data items can be problematic due to complexity
- The effectiveness of the method depends on the definition of distance (for distance-based clustering)
- If an obvious distance measure does not exist, we must define it, which not always easy, especially in multi-dimensional spaces
- The result of the clustering algorithm can be interpreted in different ways

An important component of a clustering algorithm is the distance measure between data points. This section will discuss the distance metrics that frequently used for clustering measures.

A general class of metrics for d-dimensional patterns is the *Mikowski metrics* defined by

$$L_p(\mathbf{x}, \mathbf{y}) = \left(\sum_{i=1}^{d} |x_i - y_i|^p \right)^{1/p} \tag{5.73}$$

where $\mathbf{x}$ and $\mathbf{y}$ are vectors represent clustered data points, and p is a positive integer. It is also referred to as the L_p norm. A normalized version can be defined if the measured values are normalized via division by the maximum value in the sequence.

The *Euclidean distance* is a special case where $p = 2$, it so-called the L_2 norm given by

$$L_2(\mathbf{x}, \mathbf{y}) = \left(\sum_{i=1}^{d} |x_i - y_i|^2 \right)^{1/2} \tag{5.74}$$

The *Manhattan* or *city block* distance is the other case where $p = 1$; it is known as L_1 norm

$$L_1(\mathbf{x}, \mathbf{y}) = \sum_{i=1}^{d} |x_i - y_i| \tag{5.75}$$

The other distance metrics is L_∞ *norm* that gives the maximum of the distances along individual coordinate axes, given by

$$L_\infty(\mathbf{x}, \mathbf{y}) = \max_{i=1}^{d} |x_i - y_i| \tag{5.76}$$

Suppose we compute the distance between the projections of **x** and **y** onto each of the d coordinate axis, the L_∞ is distance between **x** and **y** corresponds to the maximum of these projected distances as described in Fig. 5.11.

The other distance measures are also available when clustering will be performed such as *cosine* distance and *correlation* distance given by

$$d_{\cos}(\mathbf{x}, \mathbf{y}) = \left(1 - \frac{\mathbf{x}\mathbf{y}'}{(\mathbf{x}'\mathbf{x})^{1/2}(\mathbf{y}'\mathbf{y})^{1/2}}\right) \tag{5.77}$$

$$d_{\mathrm{corr}}(\mathbf{x}, \mathbf{y}) = \left(1 - \frac{(\mathbf{x} - \bar{\mathbf{x}})(\mathbf{y} - \bar{\mathbf{y}})'}{[(\mathbf{x} - \bar{\mathbf{x}})(\mathbf{x} - \bar{\mathbf{x}})']^{1/2}[(\mathbf{y} - \bar{\mathbf{y}})(\mathbf{y} - \bar{\mathbf{y}})']^{1/2}}\right) \tag{5.78}$$

Example 5.9 Calculate the distance metrics of **X** using *Mikowski*, *Euclidean*, *City block,* and *Cosine* measures.

$$\mathbf{X} = \begin{bmatrix} 1 & 2 \\ 1 & 3 \\ 2 & 2 \\ 3 & 1 \end{bmatrix}$$

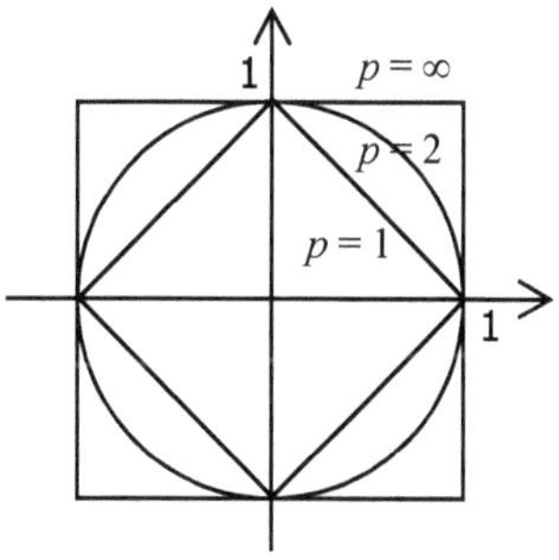

Fig. 5.11 Each shape consists of points at a distance 1.0 from the origin, measured using different values of p in the *Mikowski* L_p metric

Solution: Function pdist.m in MATLAB computes the distance between objects in the data matrix, **X**, using the method specified by "metric," where "metric" can be any of the following character strings that identify ways to compute the distance.

```
>> X = [1 2; 1 3; 2 2; 3 1]
X =
     1     2
     1     3
     2     2
     3     1
>> % Mikowski, p =3
>>Y1 = pdist(X,'minkowski',3)
Y1 =
    1.0000    1.0000    2.0801    1.2599    2.5198    1.2599
>> Y2 = pdist(X,'euclidean')
Y2 =
    1.0000    1.0000    2.2361    1.4142    2.8284    1.4142
>> Y3 = pdist(X,'cityblock')
Y3 =
     1     1     3     2     4     2
>> Y4 = pdist(X,'cosine')
Y4 =
    0.0101    0.0513    0.2929    0.1056    0.4000    0.1056
```

5.11.1 k-Centers Clustering

k-centers method is categorized in boundary method which covers the data set with *k* small balls with equal radii (Ypma and Duin 1998). This description has a resemblance to the covering numbers of Kolmogorov (Kolmogorov and Tikhomirov 1961) and it is used to characterize the complexity of an (unlabeled) data set. Unlike the covering numbers, here the ball centers' μ_k is placed on training objects such that the maximum distance of all minimum distances between training objects and the centers is minimized. In the fitting of the method to the training data, the following error is minimized

$$\varepsilon_{k\text{-ctr}} = \max_i(\min_k |x_i - \mu_k|^2) \tag{5.79}$$

The k-centers method uses a forward search strategy starting from a random initialization. The radius is determined by the maximum distance to the objects that the corresponding ball should capture. By this construction, the method is sensitive to the outliers in the training set, but it will work well when clear clusters are present in the data.

When the centers have been trained, the distance from a test object z to the target set can be calculated. This distance is now defined as

$$d_{k\text{-ctr}} = \min_k |z - \mu_k|^2 \tag{5.80}$$

To avoid a suboptimal solution during training, several random initializations are tried and the best solution (in terms of the smallest $\varepsilon_{k\text{-ctr}}$) is used. The number of parameters which is optimized in the minimization of error in Eq. (5.80) may seem to be kd at first sight, for the k-centers μ_k in d dimensions. But the centers of the balls are constrained to training objects, and therefore, only k indices out of N indices have to be selected. The number of free parameters is therefore

$$n_{\text{free}} = k$$

The user has to supply both the number of balls k and the maximum number of retries.

Example 5.10 Find the clustering of iris data using k-centers with respect to Euclidean distance and city block distance. The iris data contain 4 classes are available. We can perform clustering with corresponding to the Euclidean distance measures as shown in Fig. 5.12.

Fig. 5.12 k-centers of riply data using Euclidean distance, $k = 3$

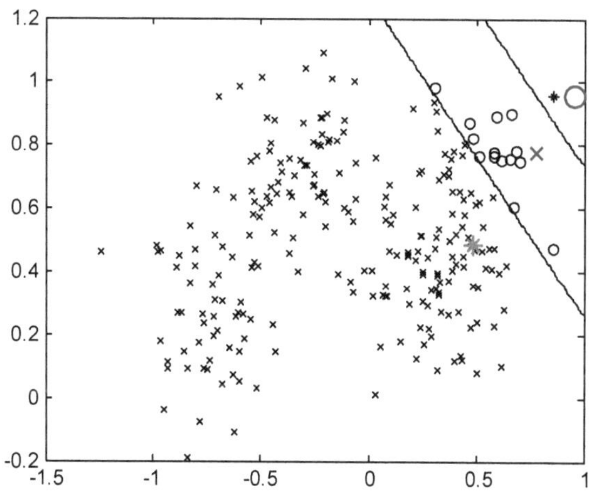

5.11.2 k-Means Clustering

k-means clustering is categorized into simplest reconstruction method that is proposed by Bishop (1995). In this method, it is assumed that the data are clustered and be characterized by a few prototype objects or vector μ_k. Most of target objects are represented by nearest prototype vector measured by the Euclidean distance. In the k-means clustering, the placing of the prototypes is optimized by minimizing the following error.

$$\varepsilon_{k-m} = \sum_i \left(\min_k |x_i - \mu_k|^2\right) \tag{5.81}$$

The k-means clustering method resembles the k-centers method, but the important difference is the error which is minimized. The k-centers method focuses on the worst case objects and tries to optimize the centers and the radii of the balls to accept all data. In the k-means method, the distance of the prototype of all objects is averaged, and therefore, the method is more robust against remote outliers. Furthermore, in the k-centers method the centers are placed, per definition, on some of the training objects, while in the k-means method, all center positions are free.

In Fig. 5.13, the examples of the boundaries of the k-means method and the k-centers method are shown. The placing of the centers of the hyperspheres by both methods is very similar. The exact positions are determined by the extreme objects for the k-centers and by means of the different clusters for the k-means clustering. The most appealing difference is that the k-centers method places a hypersphere on the objects in the lower right of the data set, while the k-means method treats is as an outlier.

The distance d of an objects z to the target set is then defined as the squared distance of that object to the nearest prototype.

$$d_{k-m}(z) = \min_k |z - \mu_k|^2 \tag{5.82}$$

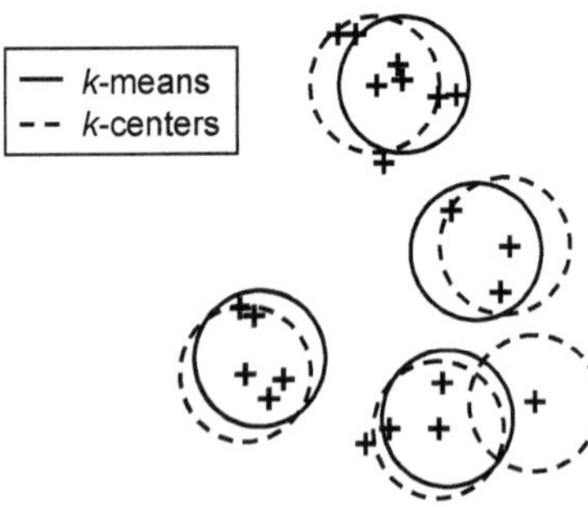

Fig. 5.13 Comparison between the boundary obtained by k-means and k-centers methods

Example 5.11 Find the clustering of iris data using *k*-means with respect to Euclidean distance and city block distance.

Solution: Using kmeans.m function we can perform clustering with corresponding to the distance measures. "sqEuclidean" and "cityblock" are Euclidean and city block measures, respectively.

```
≫ data = load('riply_trn');
≫ X=data.X';
≫[cidx, ctrs]=kmeans(X, 4, 'dist','sqEuclidean','rep', 5,'disp','final'); % Using k = 4
7 iterations, total sum of distances = 12.3798
11 iterations, total sum of distances = 12.3798
8 iterations, total sum of distances = 12.3798
10 iterations, total sum of distances = 12.3798
11 iterations, total sum of distances = 12.3798
≫ plot(X(cidx==1, 1), X(cidx==1,2),'r',...
   X(cidx==2, 1), X(cidx==2,2),'bo',...
   X(cidx==3, 1), X(cidx==3,2),'c*',...
   X(cidx==4, 1), X(cidx==4,2),'ms');
≫ hold on; plot(ctrs(:,1), ctrs(:,2),'ko','MarkerSize', 12);
```

Figure 5.14 shows the *k*-means clustering using Euclidean distance measure.

By replacing "sqEuclidean" with "cityblock," or "cosine" in kmeans.m function, we can perform *k*-means clustering with respect to related distance measure. Results of *city block* and *cosine* clustering are shown in Figs. 5.15 and 5.16. These figures

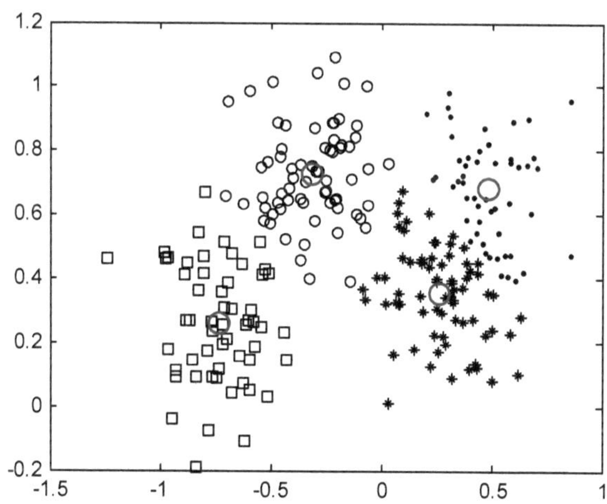

Fig. 5.14 *k*-means of *riply data* with Euclidean distance, $k = 4$

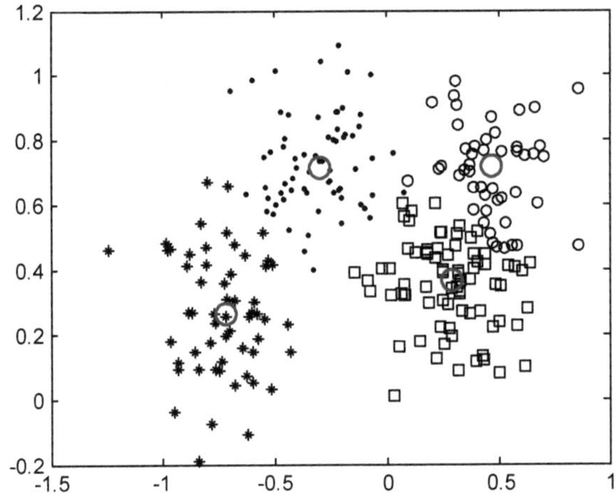

Fig. 5.15 *k*-means of *riply data* using city block distance, $k = 4$

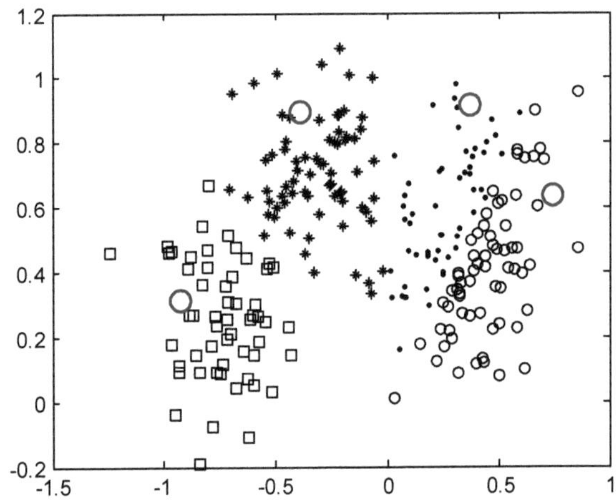

Fig. 5.16 *k*-means of *riply data* using cosine distance, $k = 4$

show that the centers are relatively similar when *k*-means use *Euclidean* and *city block* distances; however, it shows difference center if it uses *cosine* distance measure of clustering (Figs. 5.15 and 5.16).

5.11.3 Hierarchical Clustering

Let us consider a sequence of partitions of the n samples into c clusters. The first of these is a partition into n clusters, each cluster containing exactly one sample. The next is a partition into $n - 1$ clusters, the next a partition into $n - 2$, and so on until the nth, in which all the samples form one cluster. We shall say that we are at level

k in the sequence when $c = n - k - 1$. Thus, level one corresponds to n clusters and level n corresponds to one cluster. Given any two samples $\mathbf{x}$ and $\mathbf{x}'$, and at some level that will be grouped together in same cluster. If the sequence has the property that whenever two samples are in the same cluster at level k they remain together at all higher levels, and then the sequence is said to be *hierarchical clustering.*

The most natural representation of *hierarchical clustering* is a corresponding tree, called a dendrogram, which shows how the samples are grouped. Figure 5.17 shows a dendrogram for a simple problem involving eight samples. Level $k = 1$ shows the eight samples as singleton clusters. At level 2, samples x_6 and x_7 have been grouped to form a cluster, and they stay together at all subsequent levels. If it is possible to measure the similarity between clusters, then the dendrogram is usually drawn to scale to show the similarity between the clusters that are grouped. In Fig. 5.17, for example, the similarity between two groups of samples that are merged at level 5 has a value of roughly 60.

Another representation for *hierarchical clustering* is based on sets, in which each level of cluster may contain sets that are subclusters, as shown in Fig. 5.18. Yet another, textual, representation uses brackets, such as $\{\{x_1, \{x_2, x_3\}\}, \{\{\{x_4, x_5\}, \{x_6, x_7\}, x_8\}\}$. While such representations may reveal the *hierarchical* structure of the data, they do not naturally represent the similarities quantitatively. For this reason, dendrograms are generally preferred.

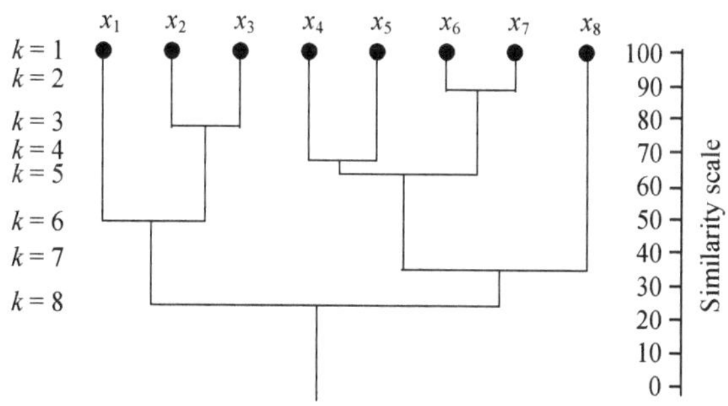

Fig. 5.17 Dendrogram can represent hierarchical clustering algorithm

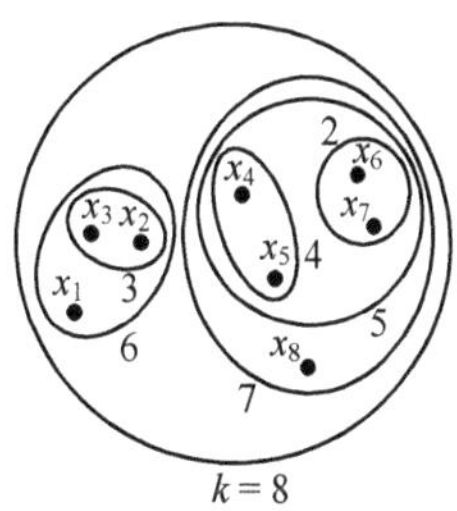

Fig. 5.18 A set of Venn diagram representation of two-dimensional data which used in the dendrogram of Fig. 5.17

Usually, dendrogram using linkage function specifies the algorithm used to generate the hierarchical cluster tree information. These linkage algorithms are based on different ways of measuring proximity between two groups of objects. For example, If n_r is the number of objects in cluster r and n_s is the number of objects in cluster s, and x_{ri} is the ith object in cluster r_i, then the definitions of these various measurements are as follows:

- Single linkage, also called *nearest neighbor*, uses the smallest distance between objects in the two groups.

$$d(r,s) = \min(\mathrm{dist}(x_{ri}.x_{sj})), \quad i \in (i,\ldots,n_r), j \in (1,\ldots,n_s) \tag{5.83}$$

- Complete linkage, also called furthest neighbor, uses the largest distance between objects in the two groups.

$$d(r,s) = \max(\mathrm{dist}(x_{ri}.x_{sj})), \quad i \in (i,\ldots,n_r), j \in (1,\ldots,n_s) \tag{5.84}$$

- Average linkage uses the average distance between all pairs of objects in cluster r and cluster s.

$$d(r,s) = \frac{1}{n_r n_s} \sum_{i=1}^{n_r} \sum_{j=1}^{n_s} \mathrm{dist}(x_{ri}, x_{sj}) \tag{5.85}$$

- Centroid linkage uses the distance between the centroids of the two groups.

$$d(r,s) = \mathrm{dist}(\bar{x}_r, \bar{x}_s) \tag{5.86}$$

  where $\bar{x}_r = \frac{1}{n_r} \sum_{i=1}^{n_r} x_{ri}$ and $\bar{x}_s$ is defined similarly.

  The centroid method can produce a cluster tree that is not monotonic. This occurs when the distance from the union of two clusters $r \cup s$ to a third cluster is less than the distance from either r or s to that third cluster. In this case, sections of the dendrogram change direction. This is an indication that you should use another method.
- Ward linkage uses the incremental sum of squares, that is, the increase in the total within-group sum of squares as a result of joining groups r and s. It is given by

$$d(r,s) = n_r n_s d_{rs}^2 (n_r + n_s)$$

  where d_{rs} is distance between cluster r and cluster s in the centroid linkage. The within-group sum of squares of a cluster is defined as the sum of the squares of the distance between all objects in the cluster and the centroid of the cluster.

Interested readers are suggested to refer to Duda et al. (2001) for detailed explanation.

Example 5.12 Calculate the hierarchical cluster tree, using the *single linkage* algorithm of matrix **X**. Use Euclidean distance for distance measure.

$$\mathbf{X} = \begin{bmatrix} 3 & 1.7 \\ 1 & 1 \\ 2 & 3 \\ 2 & 2.5 \\ 1.2 & 1 \\ 1.1 & 1.5 \\ 3 & 1 \end{bmatrix}$$

Solution: In statistical toolbox that is available in MATLAB package, we can find function linkage.m that is useful for this calculation.

```
≫ X = [31.7; 11; 23; 22.5; 1.21; 1.1 1.5; 31];
Y = pdist(X); % Default Euclidean distance
Z = linkage(Y,'single')
Z =
2.0000   5.0000   0.2000
3.0000   4.0000   0.5000
6.0000   8.0000   0.5099
1.0000   7.0000   0.7000
9.0000   11.0000  1.2806
10.0000  12.0000  1.3454
```

Example 5.13 Find the dendrogram of random matrix **X** = rand(100, 2) using *city block* distance measure.

Solution: Using function dendrogram.m of statistical toolbox that is available in MATLAB package, we can calculate the dendrogram for clustering (Fig. 5.19).

```
≫ X = rand(100,2);
Y=pdist(X,'cityblock');
Z=linkage(Y,'average')
[H,T] = dendrogram(Z,'orientation','bottom');
```

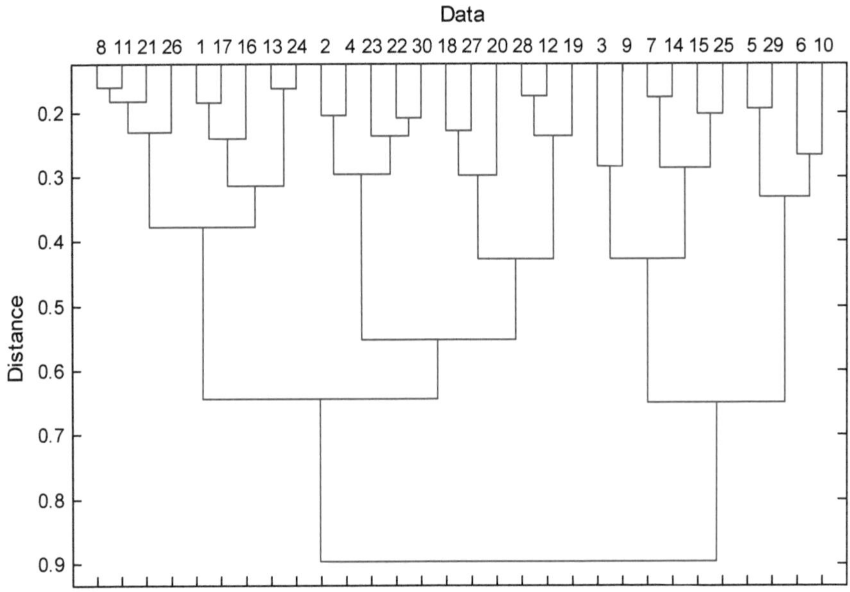

Fig. 5.19 A dendrogram for hierarchical clustering

5.12 Other Techniques

Neural networks can also be used directly for feature extraction. The self-organizing map (SOM), or Kohonen map, can also be used for nonlinear feature extraction. In SOM, neurons are arranged in an m-dimensional grid, where m is usually 1, 2, or 3. Each neuron is connected to all the d input units. The weights on the connections for each neuron form a d-dimensional weight vector. During training, features are presented to the network in a random order. At each presentation, the winner whose weight vector is the closest to the input vector is first identified. All the neurons in the neighborhood of the winner are updated such that their weight vectors move toward the input vector. Consequently, after training is done, the weight vectors of neighboring neurons in the grid are likely to represent input patterns which are close in the original feature space. Thus, a topology-preserving map is formed. SOM offers an m-dimensional map with a spatial connectivity that can be interpreted as feature extraction. The properties of these feature extraction methods are illustrated in Table 5.2.

Table 5.2 Feature extraction methods

Method	Property	Comments
PCA	Linear mapping, fast, eigenvalue-based and unsupervised, uncorrelated	Known as Karhunen–Loeve expansion, good for Gaussian data
ICA	Linear mapping, fast, eigenvalue-based and unsupervised, independent	Originally known as blind signal separation, non-Gaussian data
KPCA	Nonlinear mapping, eigenvalue-based and unsupervised	PCA-based method, using kernel functions to transform feature space
KICA	Nonlinear mapping, fast, eigenvalue-based and unsupervised, independent	ICA-based method, using kernel functions to transform feature space
LDA	Linear mapping, fast, eigenvalue-based, supervised	Better than PCA for classification, limited to $c-1$ components.
GDA	Nonlinear mapping, fast, eigenvalue-based, supervised	Better performance in separation compared with PCA, KPCA, and LDA
SOM	Nonlinear map and iterative	Based on a grid of neurons in the feature space, suitable for extracting spaces of low dimensionality

References

Antonini G, Popovici V, Thiran JP (2006) Independent component analysis and support vector machine for face feature extraction. Signal Processing Institute, Swiss Federal Institute of Technology Lausanne. http://ltsww.epfl.ch

Bach FR, Jordan MI (2002) Kernel independent component analysis. J Mach Learn Res 3:1–48

Back AD, Weigend AS (1998) A first application of independent components analysis to extracting structure from stock returns. Int J Neural Syst 8(4):473–484

Baudat G, Anouar F (2000) Generalized discriminant analysis using a kernel approach. Neural Comput 12:2385–2404

Bishop CM (1995) Neural network for pattern recognition. Clarenderon Press, Oxford

Biswall BB, Ulmer JL (1999) Blind source separation of multiple signal sources of MRI data sets using independent components analysis. J Comput Assist Tomogr 23(2):265–271

Cardoso JF (1998) Blind signal separation: statistical principles. Proc IEEE 86(10):2009–2020

Cortes C, Vapnik V (1995) Support-vector networks. Mach Learn 20(3):273–297

Duda RO, Hart PE, Stork DG (2001) Pattern classification. Wiley-Interscience, New York

Han T, Son JD, Yang BS (2005) Fault diagnosis system of induction motors using feature extraction, feature selection and classification algorithm. In: Proceedings of VS Tech2005

Harmeling S, Ziehe A, Kawanabe M, Blankertz B, Muller K (2001) Nonlinear blind source separation using kernel feature spaces. In: Proceedings of international workshop on independent component analysis and blind signal separation, pp 102–107

Hyvärinen A (1999) Fast and robust fixed-point algorithms for independent component analysis. IEEE Trans Neural Networks 10:626–634

Hyvärinen A, Oja E (2000) Independent component analysis: algorithms and applications. Neural Networks 13(4–5):411–430

Jolliffe IJ (1986) Principal component analysis. Springer, New York

Kohonen T (1990) The self-organizing map. Proc IEEE 1464–1480

Kolmogorov A, Tikhomirov V (1961) ε-entropy and ε-capacity of sets in function spaces. Trans Am Math Soc 17:277–364

Liu H, Motoda H (eds) (1998) Feature extraction, construction and selection: a data mining perspective. Kluwer Academic Publishers, London

Mao J, Jain A (1995) Artificial neural networks for feature extraction and multivariate data projection. IEEE Trans Neural Networks 6(2):296–317

Perez-Jimenez AJ, Perez-Cortes JC (2006) Genetic algorithm for linear feature extraction. Pattern Recogn Lett 27:1508–1514

Raudys SJ, Jain AK (1991) Small sample size effects in statistical pattern recognition: recommendations for practitioners. IEEE Trans Pattern Anal Mach Intell 13(3):252–264

Rosipal R, Trejo LJ (2000) Kernel partial least squares regression in reproducing kernel Hilbert space. J Mach Learn Res 2:97–123

Sammon J (1969) A non-linear mapping for data structure analysis. IEEE Trans Comput 18 (5):401–409

Schölkopf B, Smola A, Müller KR (1998) Nonlinear component analysis as a kernel eigenvalue problem. Neural Comput 10:1299–1319

Sohn H, Czarnecki JA, Farrar CR (2000) Structural health monitoring using statistical process control. J Struct Eng 126(1):1356–1363

Sung CK, Tai HM, Chen CW (2000) Locating defects of a gear system by the technique of wavelet transform. Mech Mach Theory 35:1169–1182

Trunk GV (1979) A problem of dimensionality: A simple example. IEEE Trans Pattern Anal Mach Intell 1(3):306–307

Vigario R (1997) Extraction of ocular artifacts from EEG using independent components analysis. Electroencephalogr Clin Neurophysiol 103(3):395–404

Yang BS, Han T, An JL (2004) ART-KOHONEN neural network for fault diagnosis of rotating machinery. Mech Syst Signal Process 18:645–657

Yang BS, Han T, Hwang WW (2005) Fault diagnosis of rotating machinery based on multi-class support vector machines. J Mech Sci Technol 19(3):846–859

Ypma A, Duin R (1998) Support objects for domain approximation, ICANN'98. Skovde, Sweden

Ypma A, Pajunen AP (1999) Rotating machine vibration analysis with second order independent components analysis. In: Proceeding of the workshop on ICA and signal separation, pp 37–42

Worden K, Manson G (1999) Visualization and dimension reduction of high-dimensional data for damage detection. IMAC 17:1576–1585

Zang C, Friswell MI, Imregun M (2004) Structural damage detection using independent components analysis. Struct Health Monit 3(1):69–83

Chapter 6
Feature Selection Optimization

6.1 Introduction

Reduction of feature dimensionality is a considerable importance data-driven PHM/CBM. The reason for being so is twofold: to reduce the computational complexity and to improve the algorithm's generalization ability. The first motivation is quite evident, since fewer features require less run time to train and to apply the classifier. The second motivation is low-dimensional representation reducing the risk of *overfitting*. As a rule of thumb, a minimum of 10 $d \times c$ training samples is required for a d-dimensional classification problem of c classes. When it is impractical and even impossible to obtain the required number of training samples, the reduction of feature dimensionality helps decrease the size of the training samples and consequently improves the generalization performance of the classification algorithm.

Feature extraction and *feature selection* are two different approaches for the reduction of dimensionality. Feature extraction involves linear or nonlinear transformation from the original feature space to a new one of lower dimensionality. Although it does reduce the dimensionality of the vectors fed to the algorithm, the number of features that must be measured remains the same. Feature selection, on the other hand, directly reduces the number of original features by selecting a subset of them that still retains sufficient information for classification and prediction.

Feature selection is the important issue in pattern classification for equipment fault diagnosis, even prognosis. It has three goals:

- Reducing the cost of extracting features,
- Improving the classification accuracy, and
- Improving the reliability of the estimate of performance.

G. Niu, *Data-Driven Technology for Engineering Systems Health Management*,
DOI 10.1007/978-981-10-2032-2_6

Usually, a large number of features often include many *garbage* features. Such features are not only useless in classification, but also sometimes degrade the performance of a classifier designed on the basis of a finite number of training samples. In such a case, removing the *garbage* features can improve the classification accuracy.

The choice of an algorithm for selecting features, for example, from an initial set X depends on the number of features in X. We can say that the feature selection problem is small scale, medium scale, or large scale if the number of features belongs to (0–19), (20–49), or (50–∞), respectively.

The problem of feature selection is defined as follows: Given a set of d features, select a subset of size m that leads to the smallest classification error. There has been a resurgence of interest in applying feature selection methods due to the large number of features encountered the following situation:

- Multi-sensor fusion: Features, computed from different sensor modalities, are concatenated to form a feature vector with large number of components.
- Integration of multiple data models: Sensor data can be modeled using different approaches, where the model parameters serve as features, and the parameters from different models can be pooled to yield a high-dimensional feature vector.

Moreover, the feature selection approach can solve the problem of irrelevant information in feature space. The benefits of feature selection include a reduction in the amount of data needed to achieve learning, improving classification accuracy, more compact and easily understanding knowledge base, and reducing execution time (Kumar et al. 2005). These are some feature selection methods such as conditional entropy (Lehrman et al. 1997), genetic algorithm (GA) (Jack and Nandi 2002), and distance evaluation technique (Yang et al. 2004, 2005), which have been implemented in real application.

Usually, feature selection methods can be divided into three categories: exponential algorithms, sequential algorithms, and randomized algorithms (see Table 6.1). In addition to these selection strategies, another way is that the users define an evaluation criterion for features and select them based on it. Feature distribution ranking algorithm belongs to this category. GA is one of the randomized algorithms. Exponential algorithms include *exhaustive search*, *brand and bound*, and *beam search*. These algorithms can find the best feature subsets due to exponential search. However, the calculation time is very expensive for the same reason. Therefore, this strategy is suitable for the off-line feature selection and small-size feature selection. For example, exhaustive evaluation of 10 out of 29 features involves 184,756 feature subsets, whereas exhaustive evaluation of 10 out of 100 involves more than 10^{13} feature subsets.

Sequential search takes feature dependencies into account, which has the characteristic of fast speed. But the best optimal solution usually cannot be obtained. The representative algorithms are sequential forward selection (SFS) and sequential backward selection (SBS). In order to make up the disadvantages of both algorithms, more sophisticated algorithms are developed, such as plus-l minus-

Table 6.1 Categories of feature selection methods (Mineichi and Jack 2000)

	Accuracy	Complexity	Advantages	Disadvantages
Exponential	Always finds the optimal solution	Exponential	High accuracy	High complexity
Sequential	Good if no backtracking needed	Quadratic	Simple and fast	Cannot backtrack
Randomized	Good with proper parameter selection	Generally low	Designed to escape local minima	Difficult parameters selection

R selection (LRS), bidirectional search (BDS), and sequential floating selection (SFFS or SFBS). Some of the feature selection techniques will be discussed in the following section.

In the machine learning literature, the feature selection algorithms are often categorized as "filter" or "wrapper" approaches. Filter approaches evaluate the "goodness" of each feature based on an evaluation criterion independent from the classification design. Chi-square test, *t* test, and signal-to-noise ratio present some examples of popular univariate filter approach. In contrast, in the wrapper approach, the "goodness" of a subset of features is directly evaluated by their corresponding classification performance. Wrapper approaches consider the combinatorial effects of a subset of features and are multivariate (Blum and Langely 1997). Exhaustive search and GA are examples of the wrapper approach.

For the purpose of predictive analysis, a feature selection algorithm can be assessed based on the time complexity, the resulting classification accuracy and generalizability of the identified feature subset. Although an exhaustive wrapper approach feature selection algorithm results in the highest classification accuracies, with small sample numbers, the identified feature subset may not "generalize" well to unseen samples. Additionally, this approach is computationally expensive and may be intractable.

6.2 Individual Feature Evaluation (IFE)

This algorithm is suitable for feature selection of supervised classification, which consists of two parts. First part is to classify subclasses for each known class based on the histogram statistics and then create new labels for the subclasses. Second part is to evaluate individual feature according to the space distribution. The detail procedure can be summarized as follows:

Step 1: extracting same class data according to the given labels and, then for each class, calculating the frequencies within the limited region using histogram:

$$h_s = \sum_{j=0}^{n} \frac{1}{n} r_s(x_{i,j}), \quad s = 1, 2, \ldots, b. \tag{6.1}$$

$$r_s(x) = \begin{cases} 1, & \text{if } \frac{s(\max(x_{i,j}) - \min(x_{i,j}))}{b} \leq x < \frac{(s+1)(\max(x_{i,j}) - \min(x_{i,j}))}{b} \\ 0, & \text{otherwise} \end{cases} \tag{6.2}$$

where $x_{i,j}$ is the feature value, i and j are the ith feature and sample numbers, respectively, and h_s is the frequencies within the bins we wish to divide the ranges.

Step 2: estimating the number of subclass for each class based on the peaks and bottoms of the histogram using differential calculation:

$$\begin{aligned} & h_f = \operatorname{diff}(h_s) \\ & \text{ind\_p} = \operatorname{find}(([h_f\, 0] < 0)\ \&\ ([0\, h_f] \geq 0)); \\ & \text{ind\_b} = \operatorname{find}(([h_f\, 0] \geq 0)\ \&\ ([0\, h_f] < 0)); \end{aligned} \tag{6.3}$$

Step 3: redefining the labels for subclasses.

Step 4: calculating the mean value of each subclass $m_{i,n}$ and then getting the total mean value m_i.

$$m_{i,n} = \frac{1}{M} \sum_{j=1}^{M} x_{i,j}, \quad \text{then } m_i = \frac{1}{c} \sum_{n=1}^{c} d_{i,n} \tag{6.4}$$

where M is the number of subclass sample data and c is the number of subclasses for ith feature.

Step 5: calculating the average distance d_f between different classes ($m_{i,n}$).

$$d_f = \frac{1}{c \times (c-1)} \sum_{p,q=1}^{c} \left| m_{i,p} - m_{i,q} \right|; \quad (p, q = 1, 2, \ldots, c; p \neq q) \tag{6.5}$$

Step 6: calculating the ratio of the average distance and total mean value for each feature and then selecting the good classification features.

$$\alpha_i = d_f / m_i \tag{6.6}$$

Example 6.1 Select the optimal features using individual feature evaluation (IFV) method of vibration features collected from the induction motor. This feature data set consists of matrix [180 × 63] obtained from 20 vibration measurements of 9 conditions of induction motors using 3 directions of accelerometers. In feature representation, 21 features are calculated from each accelerometer, so total 63 features were collected. Plot of two selected features (features 15 and 18) using IFB method is shown in Fig. 6.1. From this figure, each class can be clustered even though still contains overlapping between classes.

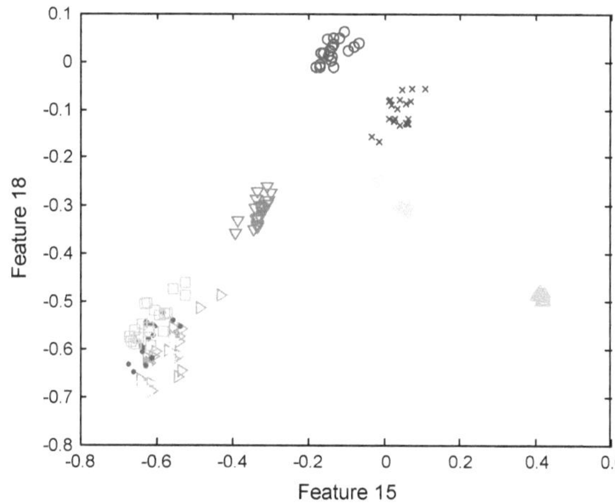

Fig. 6.1 Feature selection using individual feature evaluation method

6.3 Conditional Entropy

A good feature should map two different classes to two separate locations in feature space. Such class separability can be analyzed by the *mutual information* $I(X; C)$ between the set of feature values X and the set of classes C. The mutual information $I(X; C)$ measures the interdependence between two random variables X and C. It can be computed as follows:

$$I(X; C) = E_s(C) - E_s(C \mid X) \tag{6.7}$$

Entropy can measure the degree of uncertainty in the system. For class C the entropy is given by

$$E_s(C) = -\sum p(c) \ln p(c) \tag{6.8}$$

where $p(c)$ is the probability density function (pdf) of c.

The *conditional entropy* $E_s(C|X)$ measures the degree of uncertainty entailed by the set of classes C given the set of feature value X, and can be computed as

$$E_s(C \mid X) = \sum_{c \in C} \int_{x \in X} p(x, c) \ln \frac{p(x \mid c) p(c)}{p(x)} \tag{6.9}$$

The integration in the expression above signifies that the feature space is continuous. However, in order to implement the computation, the feature space should be discretized into intervals of width Δx, so the integration is replaced with a summation. Thus, the discrete form of Eq. (6.9) is as follows:

$$E_s(C \mid X) = \sum_{c \in C} \sum_{x \in X} p(x, c) \ln \frac{p(x \mid c)p(c)}{p(x)} \Delta x, \tag{6.10}$$

but $p(x, c) = p(x \mid c)p(c)$, so

$$E_s(C \mid X) = \sum_{c \in C} p(c) \sum_{x \in X} p(x \mid c) \Delta x \ln \frac{p(x \mid c)p(c)}{p(x)} \tag{6.11}$$

The first term in expression Eq. (6.7) is the entropy $E_s(C)$. It depends only on the classes and does not depend on the design or selection of features. It provides an upper bound of the mutual information $I\,(X;\,C)$ since $0 \leq E_s(C \mid X) \leq E_s(C)$. The second term in expression Eq. (6.7), $-E_s(C \mid X)$, can be interpreted as the decrease in the uncertainty. That is, with a higher interdependence between the feature values X and the classes C, one has a higher certainty (i.e., lower uncertainty) in classifying is given by its feature value. The mutual information $I\,(X;\,C)$ is at maximum when X and C are totally dependent on each other. Conversely, it is at minimum when there is no relationship between X and C. Thus, in order to maximize the class separability, one objective is to maximize the mutual information $I\,(X;\,C)$ between the feature values and the classes.

Example 6.2 Select the optimal features using entropy information method of vibration features collected from the induction motor in Example 6.1. Plot of two selected features (features 5 and 27) using conditional entropy is shown in Fig. 6.2. From this figure, the clustering process is shown even though it still contains overlapping between classes.

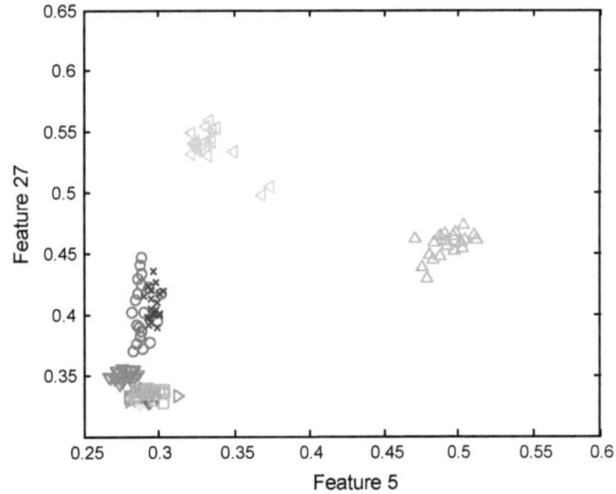

Fig. 6.2 Feature selection using conditional entropy

6.4 Backward Feature Selection

For explanation, we assume that the features are functions of all the original variables, thus preserving as much classification information as possible contained in the original variables. This assumption is reasonable in many applications of pattern recognition, particularly for the classification of random processes. A feature selector for random processes may be interpreted as a filter, with the output time-sampled values to be functions of all input time-sampled values. However, in some other applications such as machine diagnosis, we need to evaluate the effectiveness of individual tests (variables) or their combinations for classification, and select only the effective ones. This problem is called *feature subset selection.* The best subset of m variables out of n may be found by evaluating a criterion of class separability for all possible combinations of m variables. However, the number of all possible combinations $\binom{n}{m}$ becomes prohibitive even for modest values of m and n. For example, with $n = 24$ and $m = 12$, the number becomes 2,704,156. Therefore, we need some procedures to avoid the *exhaustive search.*

Let us study a simple example in which two features are chosen out of four as shown in Fig. 6.3. The subset of features a, b, and c is denoted by (a, b, c). A criterion of class separability is selected, and its value for (a, b, c) is expressed by $J_3(a, b, c)$ where the subscript indicates the number of features in the subset.

The *backward selection* procedure starts from the full set (1, 2, 3, 4). Then, eliminating one feature from four, all possible subsets of three features are obtained and their criterion values are evaluated. If $J_3(1, 3, 4)$ is the largest among the four J_3 subset, (1, 3, 4) is selected as the best subset of three variables. Then, eliminating one more feature only from (1, 3, 4), we can obtain the subsets of two features, among which the subset with the largest J_2 is chosen as the best solution of the problem.

Example 6.3 Given a random data set, demonstrate the feature selection process using *backward* selection. The original features and selected features are shown in Fig. 6.4a, b. From the figures, although the features were selected, the clustering was not performed well using this technique. The performance of feature selection process is evaluated based on 1-nearest neighbor and resulted $J = 0.7$.

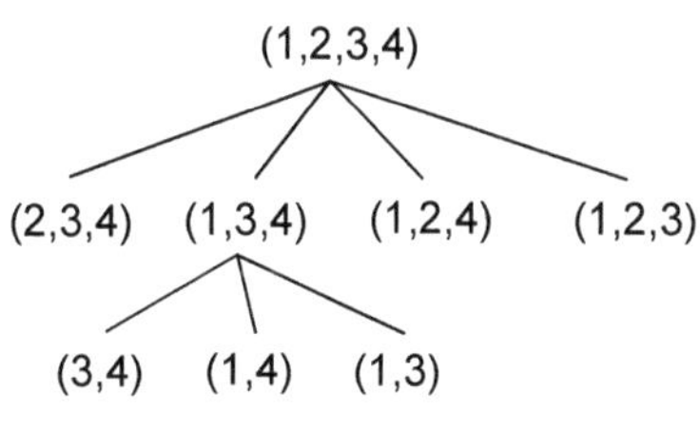

Fig. 6.3 Backward selection

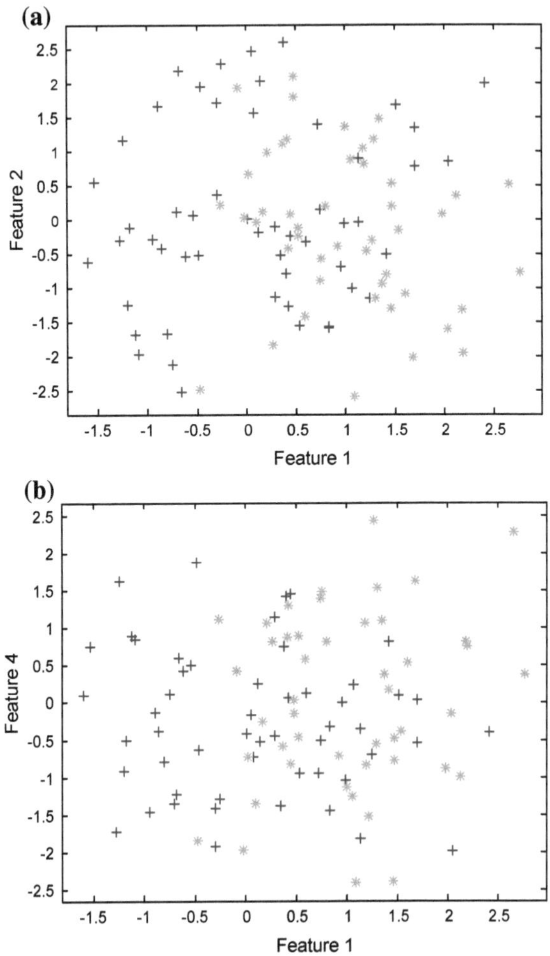

Fig. 6.4 Backward selection. **a** Before selection and **b** after selection

6.5 Forward Feature Selection

The forward selection procedure starts from the evaluation of individual features as shown in Fig. 6.5. Let the problem be to select three features out of four. If $J_1(2)$ is the largest among all $J_1(\cdot)$ subset, one feature is added to feature 2 to form the subset of two features. If $J_2(2, 3)$ is the largest among all $J_2(2, \cdot)$ subset, one more feature is added to (2, 3) to form the subset of three features. Among all possible (2, 3, ·) subset, the best subset is the one which gives the largest $J_3(2, 3, \cdot)$.

Example 6.4 Demonstrate the feature selection process using *forward* selection of a given data in Example 6.3. In this example, features 5 and 7 are the best of 5 selected features. Figure 6.6a, b is the plot of original and best selected features,

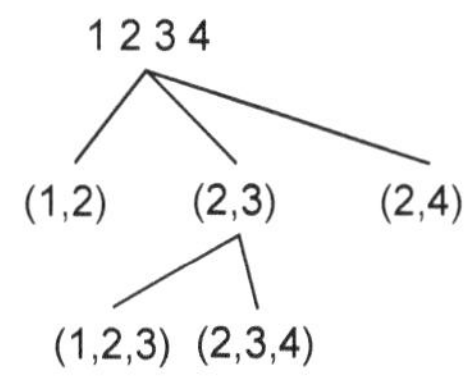

Fig. 6.5 Forward selection

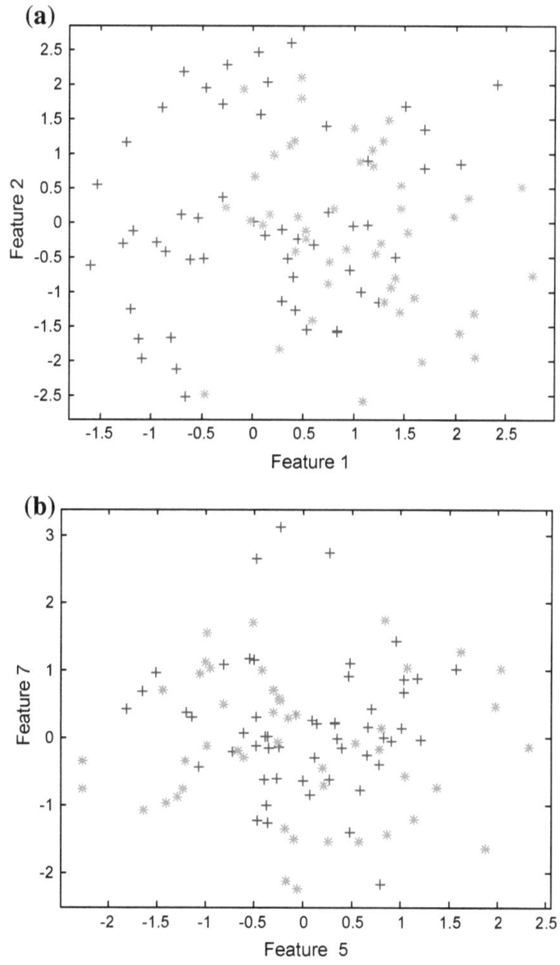

Fig. 6.6 Forward selection. **a** Before selection and **b** after selection

respectively. From the figures, although the features were selected, the clustering was not performed well using this technique. The performance of feature selection process is evaluated based on 1-nearest neighbor and resulted $J = 0.67$.

The reason why this method cannot necessarily select the optimum subset may be understood by observing the forward selection procedure. Suppose that $\mathbf{x}_1$ and $\mathbf{x}_2$

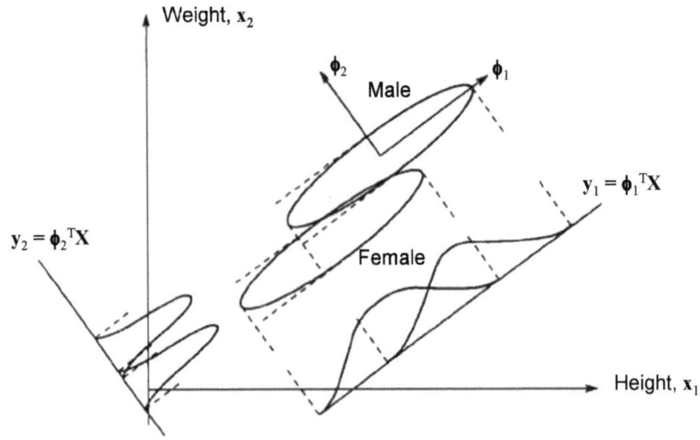

Fig. 6.7 An example of feature extraction for classification

of Fig. 6.6 are two features among n, from which m features are to be selected (Fukunaga 1990). Since the ω_1 and ω_2 marginal density function of $\mathbf{x}_1$ are heavily overlapped, $\mathbf{x}_1$ drops out when individual features are evaluated. The same is true for $\mathbf{x}_2$. Thus, one of the other features, say $\mathbf{x}_5$, and others are examined, and so on. As a result, the combination of features including both $\mathbf{x}_1$ and $\mathbf{x}_2$ might not come up for evaluation at the later stages. As shown in Fig. 6.6, although $\mathbf{x}_1$ and $\mathbf{x}_2$ are poor features individually, their combination gives a good feature space in which the ω_1 and ω_2 distributions are well separated. The forward selection procedure could fail to pick that information. This phenomenon is observed frequently when two features are highly correlated. In general, it is true for signal representation that we can eliminate one feature when two features are highly correlated. This is due to the fact the second feature gives little extra information for representing signals. For example, in Fig. 6.7, knowing one's height ($\mathbf{x}_1$) we can well guess the weight ($\mathbf{x}_2$). On the other hand, highly correlated features could enhance the class separability significantly, as shown in Fig. 6.7. Eliminating one, we might lose vital information for classification.

Thus, both backward and forward selection procedures give simple search techniques which avoid exhaustive enumeration. However, the selection of the optimal subset is not guaranteed.

6.6 Branch and Bound Feature Selection

Branch and bound methods have been developed to obtain optimal solutions to combinatorial problems without involving exhaustive enumeration (Golomb and Baumert 1965; Nilsson 1971; Narendra and Fukunaga 1977). In this section, we formulate the feature subset selection as a combinatorial optimization problem and develop an efficient algorithm for selecting feature subsets. The subset selected by this procedure is guaranteed to be the best among all the possible combinations of features.

Rather than enumerating the subsets of m features, we will find it more convenient to enumerate the subsets of $\bar{m} = n - m$ features *discarded* from the n feature set. Let $(z_1, \ldots, z_{\bar{m}})$ denote the set of those $\bar{m}$ discarded features. Each variable z_i can take on integer values in $\{1, \ldots, n\}$. However, since the order of the z_i is immaterial, every permutation of the sequence $(z_1, \ldots, z_{\bar{m}})$ will yield the same value of the criterion. Moreover, all the z_i should be different. Hence, it is sufficient to consider the sequences which satisfy

$$z_1 < z_2 < \cdots < z_{\bar{m}} \tag{6.12}$$

The feature selection criterion is a function of the m features obtained by deleting the $\bar{m}$ features from the n feature set. However, for notational convenience, we write the criterion as $J_{\bar{m}}(z_1, \ldots, z_{\bar{m}})$. Then, the subset selection problem is to find the optimum sequence $(z_1^*, \ldots, z_{\bar{m}}^*)$ such that

$$J_{\bar{m}}(z_1^*, \ldots, z_{\bar{m}}^*) = \max_{z_1, \ldots, z_{\bar{m}}} J_{\bar{m}}(z_1, \ldots, z_{\bar{m}}) \tag{6.13}$$

If the criterion were to be minimized instead, all the inequalities in the following discussion would be reversed.

Enumeration of the sequences $(z_1, \ldots, z_{\bar{m}})$ satisfying Eq. (6.12) can be illustrated by a solution tree. Figure 6.8 is an illustration for solution tree corresponding to $n = 6$ and $m = 2$ ($\bar{m} = 4$). A node at level i is labeled by the value of z_i. Also, each node can be identified by the sequence of discarded features, for example, (1, 4) for node A. At level 1, z_i can only assume values 1, 2, or 3, because with z_i greater than 3, it would not be possible to have sequences $(z_1, \ldots, z_4)$ satisfying Eq. (6.12). Similar considerations govern the enumeration at other levels of the tree, and the largest value for z_i must be $(m + i)$ in general.

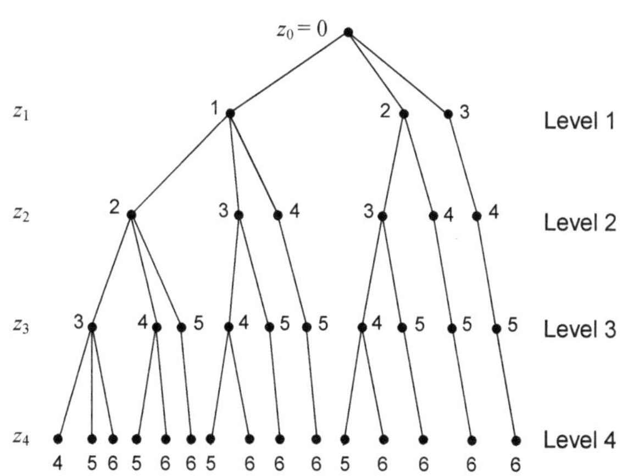

Fig. 6.8 The solution tree for the basic algorithm ($n = 6$, $m = 2$, and $\bar{m} = 4$)

Let us assume that the criterion J satisfies *monotonicity*, which is defined as

$$J_1(z_1) \geq J_2(z_1, z_2) \geq \cdots \geq J_{\bar{m}}(z_1, \ldots, z_{\bar{m}}) \tag{6.14}$$

The monotonocity is not particularly restrictive, as it merely implies that a subject of features should be no better than any larger set that contains the subset. Indeed, a large variety of feature selection criteria do satisfy the monotonicity relation. They are the Bayes error, asymptotic k-NN error, and the function of the scatter matrices.

Let α be the best (maximum) value of $J_{\bar{m}}(z_1, \ldots, z_{\bar{m}})$ found so far in the search. If

$$J_k(z_1, \ldots, z_k) \leq \alpha \quad \text{for } k < \bar{m} \tag{6.15}$$

then by Eq. (6.14)

$$J_{\bar{m}}(z_1, \ldots, z_k, z_{k+1}, \ldots, z_{\bar{m}}) \leq J_k(z_1, \ldots, z_k) \leq \alpha \tag{6.16}$$

for all possible $\{z_{k+1}, \ldots, z_{\bar{m}}\}$.

This means that whenever the criterion evaluated for any node is less than α, all nodes that are successors of that node also have criterion values less than α and therefore cannot be optimal. This forms the basis for the branch and bound algorithm.

The branch and bound algorithm successively generates portions of the solution tree and computes the criterion for the nodes explored. Whenever a suboptimal partial sequence of nodes is found to satisfy Eq. (6.16), the subtree under the node is implicitly rejected, and enumeration begins on partial sequences which have not yet been explored.

We shall give a simple procedure for enumerating the partial sequences with $z_1 < \cdots < z_2$ as follows:

Basic algorithm

Step 1: Initialization

$$\text{Set } \alpha = -\infty, \text{the level } i = 1, \text{ and } z_0 = 0. \tag{6.17}$$

Step 2: Generate successors

Initialize *LIST*(i) which is the list of the feature values that can assume given the values of (z_1, …, z_{i-1}). That is,

$$LIST(i) = \{z_{i-1} + 1, z_{i-1} + 2, \ldots, m + i\} \quad (i = 1, \ldots, \bar{m}) \tag{6.18}$$

Step 3: Select new node

If *LIST*(i) is empty, go to step 5.
Otherwise, set $z_i = k$, where

$$J_i(z_1, \ldots, z_{i-1}, k) = \max_{j \in LIST(i)} J_i(z_1, \ldots, z_{i-1}, j) \tag{6.19}$$

Delete k from *LIST*(i).

Step 4: Check bound

$$\text{if } J_i(z_1, \ldots, z_i) < a, \text{ go to step 5.} \tag{6.20}$$

If the level $i = \bar{m}$, go to step 6

$$\text{Otherwise, set } i = i + 1 \text{ \{advance to a new level\}} \tag{6.21}$$

Go to step 2.

Step 5: Backtrack to lower level

$$\text{Set } i = i - 1, \text{ if } i = 0, \text{ \{terminate the algorithm\}} \tag{6.22}$$

Otherwise go to step 3.

Step 6: Last level

$$\text{Set } a = J_{\bar{m}}(z_1, \ldots, z_{\bar{m}}) \text{ and set } (z_1^*, \ldots, z_{\bar{m}}^*) = (z_1, \ldots, z_{\bar{m}}). \tag{6.23}$$

Go to step 5.

The functioning of the algorithm is as follows. Starting from the root of the tree, the successors of the current node are enumerated in *LIST*(i). The successor, for which the partial criterion $J_i(z_1, \ldots, z_i)$ is maximum (the most promising node), is picked as the new current node, and the algorithm moves on to the next higher level. The lists *LIST*(i) at each level i keep track of the nodes that have not been explored. Whenever the partial criterion is found to be less than α, the algorithm backtracks to the previous level and selects a hitherto unexplored node for expansion. Whenever the algorithm reaches the last level $\bar{m}$, α is updated to be the current value of $J_{\bar{m}}(z_1, \ldots, z_{\bar{m}})$ and the current sequence $(z_1, \ldots, z_{\bar{m}})$ is saved as $(z_1^*, \ldots, z_{\bar{m}}^*)$. When all the nodes in *LIST*(i) for a given i are exhausted, the algorithm backtracks to the previous level. When the algorithm backtracks to level 0, it terminates. Upon termination, the current value of $(z_1^*, \ldots, z_{\bar{m}}^*)$ gives the complement of the optimum set of m features and the current value of α gives the optimum value of the criterion. The procedure guarantees that all sequences are either explicitly evaluated or implicitly rejected, and thus, the sequence $(z_1^*, \ldots, z_{\bar{m}}^*)$ gives the best subset of features among all possible subsets.

Example 6.5 Demonstrate the feature selection process using branch and bound selection of a given random data in Example 6.1.

Figure 6.9a, b is the plot of original and best selected features, respectively. From the figures, although the features were selected, the clustering was not performed well using this technique. The performance of feature selection process is evaluated based on the 1-nearest neighbor and resulted $J = 0.56$.

Fig. 6.9 Branch and bound selection. **a** Before selection and **b** after selection

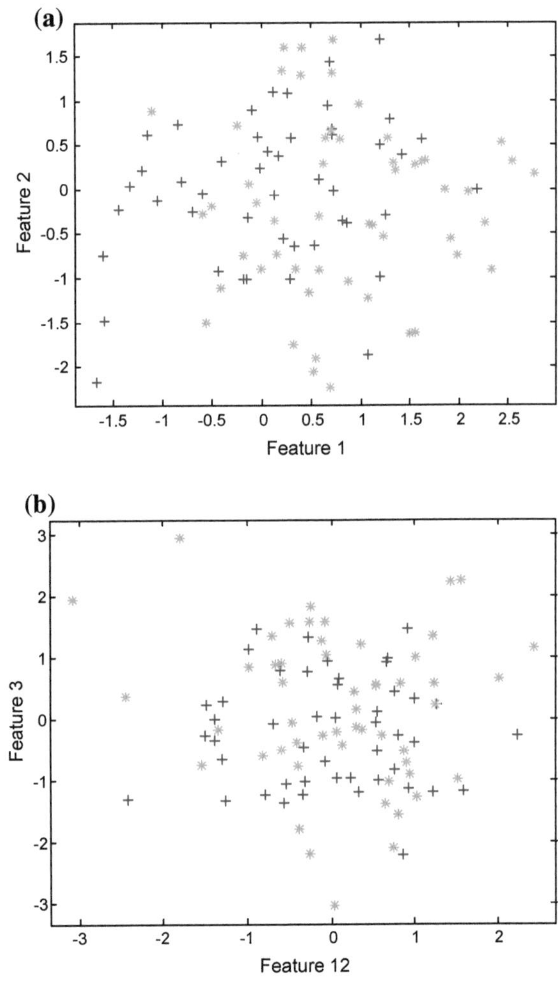

6.7 Plus *l*-Take Away *r* Feature Selection

The well-known SFS and SBS are step optimal only since the best (the worst) feature is always added (discarded) in SFS and SBS, respectively. These results are nested feature subsets without any chance to correct the decision in later steps, causing the performance to be often far from optimal.

A definitive improvement can be obtained by combining SFS and SBS to avoid the nesting effect. The *plus l-take away r method* (Stearns 1976) consists of applying SFS l times followed by r steps of SBS with this fixed cycle of forward and backward selection repeated until the required number of features is reached. Therefore, it results in a forward method if $l > r$ or a backward one if $l < r$. In this case, features that have been previously added can be removed in posterior steps, thus avoiding the nesting effect. In this case, the method allows a fixed backtracking defined by the values of l or r depending whether the search is top down or bottom up.

The plus l-take away r algorithm and, consequently, the SFS [(1, 0)-search] and SBS [(0, 1)-search] algorithms can be described in an algorithmic way as follows:

Step 1: Input

$$Y = \{y_j \,|\, j = 1, \ldots, D\} \quad \text{//available measurement//} \tag{6.24}$$

Step 2: Output

$$X_k = \{x_j \,|\, j = 1, \ldots k, x_j \in Y\}, \quad k = 0, 1, \ldots, D \tag{6.25}$$

Step 3: Initialization

$$\text{if}\, l > r \quad \text{then}\, k := 0;\, X_0 := 0; \tag{6.26}$$

Go to step 5.

$$\text{else}\, k := D; \quad X_D := Y; \tag{6.27}$$

Go to step 6.

Step 4: Termination, stop when k equals the number of features required
Step 5: Inclusion, repeat l times

$$x^+ := \arg \max_{x \in Y - X_k} J(X_k + x). \tag{6.28}$$

{The most significant features with respect to X_k}

$$X_{k+1} := X_k + x^+; \quad k := k + 1 \tag{6.29}$$

G o to step 6.

Step 6: Exclusion, repeat r times

$$x^{-} := \arg\max_{x \in X_k} J(X_k - x), \tag{6.30}$$

{The least significant features in X_k}

$$X_{k-1} := X_k - x^{-}; \quad k := k - 1 \tag{6.31}$$

Go to step 5.

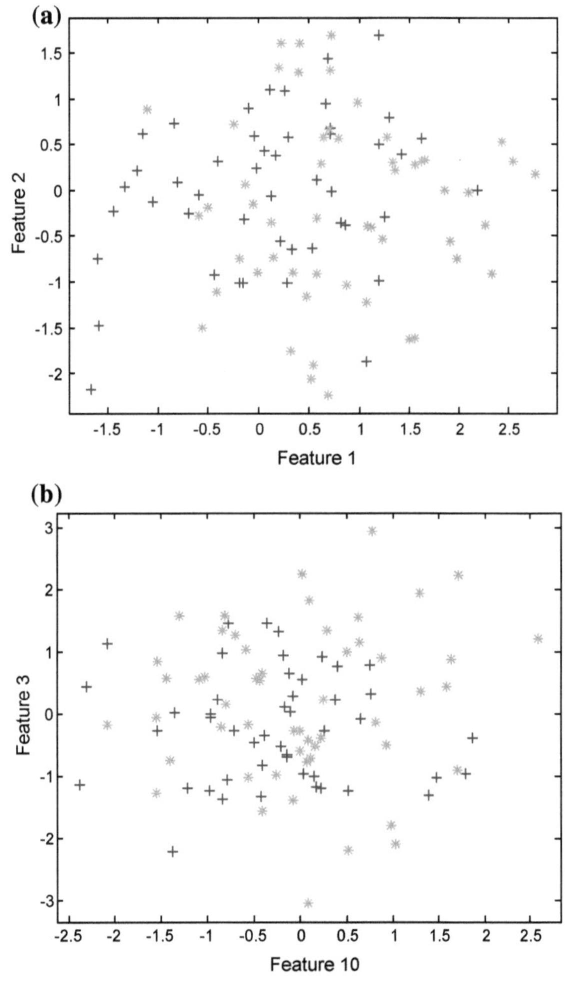

Fig. 6.10 Plus l-take away r selection. **a** Before selection and **b** after selection

This procedure can be generalized to search a larger region of the space by substituting l steps forwarded by the search of the best superset that can be formed by adding l features (the same for backward). Even though the problem of nested features can be partially overcome with this procedure, another problem arises; there is no way of predicting the best values of l and r to obtain good enough solutions with a moderate amount of computation.

Example 6.6 Demonstrate the feature selection process using plus l-take away r selection of a given random data in Example 6.1.

In this case, features 10 and 3 are the best of 5 selected features. The original features and selected features are plotted in Fig. 6.10a, b. In those figures presenting the selected features, the clustering was not performed well using this technique. The performance of feature selection process is evaluated based on 1-nearest neighbor and resulted $J = 0.61$.

6.8 Floating Forward Feature Selection

Floating search method is very efficient and effective even on the problems of high dimensionality involving non-monotonic feature selection criterion functions (Pudil et al. 1994). The floating search method is related to the *plus l-take away r* algorithm, but in contrast to the latter, the number of forward and backtracking steps is dynamically controlled instead of being fixed beforehand. Procedures of this method are constructed in parallel with feature sets of all dimensionalities up to a specified threshold. By means of sequential forward and backward selection, these sets are updated whenever the modification results in a better performance. In consequence, the resulting feature sets, as in the case of the (l, r) sequential algorithm, are not necessarily nested. By the same token, the selection process can correct for any effects caused by non-monotonicity of the feature selection criterion.

The sequential floating forward selection procedure consists of applying after each forward step a number of backward steps as long as the resulting subsets are better than the previously evaluated ones at that level. Consequently, there are no backward steps at all if the performance cannot be improved. This method can be described algorithmically in a similar way to the previous method as follows:

Step 1: Input

$$Y = \{y_j \,|\, j = 1, \ldots, D\} \quad //\text{available measurement}// \tag{6.32}$$

Step 2: Output

$$X_k = \{x_j \,|\, j = 1,\ldots,k, x_j \in Y\}, \quad k = 0,1,\ldots,D \tag{6.33}$$

Step 3: Termination, stop when k equals the number of features required
Step 4: Inclusion

$$x^+ := \arg \max_{x \in Y - X_k} J(X_k + x), \tag{6.34}$$

{The most *significant features* with respect to X_k}

$$\left\{ \textit{significant feature}: x^+ \text{ satisfies } J(X_k + x^+) = \max_{x \in Y - X_k} J(X_k + x) \right\} \tag{6.35}$$

$$X_{k+1} := X_k + x^+; \quad k := k + 1 \tag{6.36}$$

Step 5: Conditional inclusion

$$x^- := \arg \max_{x \in X_k} J(X_k - x), \tag{6.37}$$

{The least significant features in X_k}

$$\text{if } J(X_k - \{x^-\}) > J(X_k - 1) \text{ then} \tag{6.38}$$

$$X_{k-1} := X_k - x^-; \quad k := k - 1 \tag{6.39}$$

Go to step 5
Else
Go to step 6.

The backward counterpart of this algorithm can be obtained in a straightforward manner by substituting inclusion by exclusion and initializing in the previous description. Both algorithms allow a self-controlled backtracking, so they can eventually find solutions by adjusting the trade-off between forward and backward steps dynamically. It is possible to say that in a certain way, they compute only what they need without any parameter setting.

Example 6.7 Demonstrate the feature selection process using floating forward selection of a given random data.

In this case, features 10 and 3 are the best of 6 selected features. The original features and selected features are plotted in Fig. 6.11a, b. Examining the figures, although the features were selected, the clustering was not performed well using this technique. The performance of feature selection process is evaluated based on the 1-nearest neighbor and resulted $J = 0.67$.

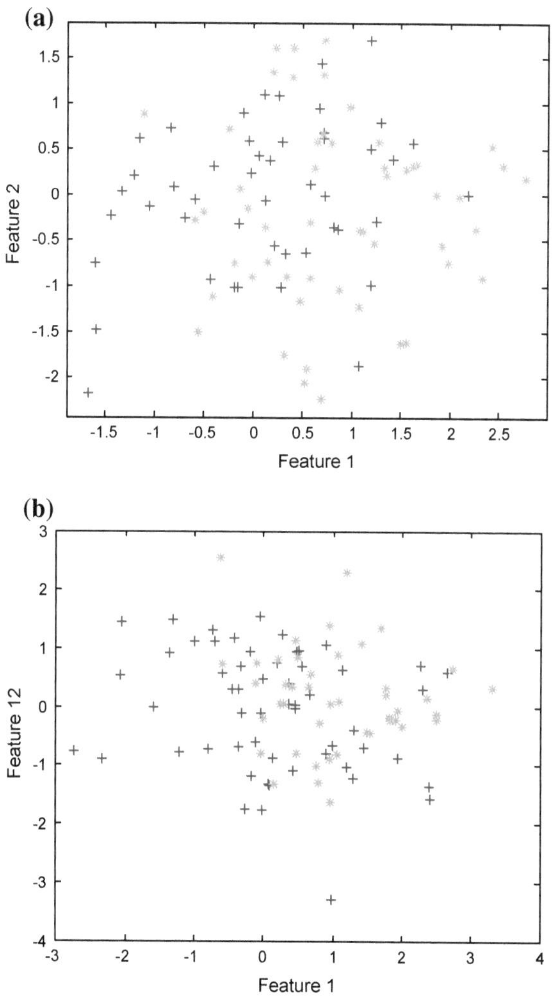

Fig. 6.11 Floating forward feature selection. **a** Before selection and **b** after selection

6.9 Distance-Based Evaluation Technique

Step 1: Calculate the relative average value of the sampling data for the same class $d_{i,j}$ and then obtain the average distance of class d_{ai}. The equation can be defined as follows:

$$d_{i,j} = \frac{1}{N \times (N-1)} \sum_{m,n=1}^{N} \left| p_{i,j}(m) - p_{i,j}(n) \right| \quad (m, n = 1, 2, \ldots, N, m \neq n) \quad (6.40)$$

where N is the sampling number of each class and $p_{i,j}$ is the value of ith feature under jth class.

$$d_{ai} = \frac{1}{M} \sum_{j=1}^{M} d_{i,j} \quad (6.41)$$

where M is the number of class.

Step 2: Calculate the average distance of interclass d'_{ai}

$$d'_{ai} = \frac{1}{M \times (M-1)} \sum_{m,n=1}^{M} \left| p_{ai,m} - p_{ai,n} \right| \quad (m, n = 1, 2, \ldots, M; m \neq n) \quad (6.42)$$

where $p_{ai,m}$ and $p_{ai,n}$ are the average values of the sampling data under different class.

$$p_{ai,j} = \frac{1}{N} \sum_{n=1}^{N} p_{i,j}(n) \quad (n = 1, 2, \ldots, N) \quad (6.43)$$

Step 3: Calculate the ratio d_{ai}/d'_{ai}

Step 4: Select the eight largest feature parameters α_i. Bigger α_i represent a well-selected feature. This requires a small d_{ai} and a large d'_{ai}.

$$\alpha_i = d'_{ai}/d_{ai} \quad (6.44)$$

where α_i (i = 1, …, k) is the effectiveness factor of the features and k is the number of selected features.

Given α_i, one can now establish a raking methodology among the individual feature components. The useful features are expected to show high values of α_i, indicating a good interclass spread in the classifier.

6.10 Taguchi Method-Based Feature Selection

The Taguchi method, based in part on the Fisher's experimental methods, was applied for the robust design of products by Taguchi in the early 1950s. One basic ingredient of the Taguchi method is the orthogonal array (OA) which is used to find the important control variables that influence the performance of a product among many candidate variables based on the experiments and to assign them appropriate values. First, some control variables that are suspected of influencing the performance of a product are selected, and then, experiments are performed by changing the values of the control variables systematically. Then, the best combinations of the values of the control variables are found. Here, Taguchi method-based selection method is reviewed briefly, which was proposed by Kwak and Choi (2002).

The following is a short explanation of how to apply the Taguchi method to the input feature selection problem (Tables 6.2 and 6.3). First, we make an OA and let each column correspond to each input feature. Each row corresponds to one training of the classifier with input features set to level 1 or level 2 in that row. After all of the rows are trained, we compare the average performance with a specific feature in the input vector and that without the feature. Then, we choose features which gives better average improvement in performance. In this algorithm when there are N inputs, we need $2^{(\log_2 N+1)}$ trainings. As we need $O(N)$ experiments, we conclude that if we choose inputs in this way, we can get good input features with a relatively small number of trainings. The computational effort increases linearly with the number of features. Therefore, the method can be applied with effect to large problems without excessive computational effort. A demonstration as to how to use the OA for the input feature selection problem will be given with the example of Table 6.2. There are three input features (F_1–F_3), and we used the OA of L_4 as shown in Table 6.2.

In Table 6.3, the first row of OA represents that all the three features are selected as the input features. The NN with these features are trained to give 90 % of the classification rate (in this case, the performance measure is the classification rate). The second row of the table shows that the network trained solely with F_1 gives 30 % of the classification rate. In this way, after all the four trainings listed in the OA have been finished, the average performance using each feature and without it is calculated. For example, for F_1 the average performance using this feature is 60 % (= (90 + 30)/2) and that without it is 50 % (= (40 + 60)/2). With this average performance, the improvement of the average performance for each feature is evaluated. It is 10 % for F_1 and 20 % for F_2, while F_3 shows a 40 % improvement.

Table 6.2 Example of orthogonal array (L_4) for input feature selection

Run	F_1	F_2	F_3
1	1	1	1
2	1	2	2
3	2	1	2
4	2	2	1

Table 6.3 Example of input feature selection (O: select and –: do not select)

Classifier training	F_1	F_2	F_3	Performance (%)
1	O	O	O	90
2	O	–	–	30
3	–	O	–	40
4	–	–	O	60
Average performance (select) (%)	60	65	75	
Average performance (not select) (%)	50	45	35	
Average improvement of performance (%)	10	20	40	
Order of selection	3	2	1	

This means that F_3 has the most influence on the performance of the network and can be regarded as the most salient feature. So we selected features F_3, F_2, and F_1 in that order.

For most OAs, the first row has a series of 1s and the others have 1s less than half the number of columns. If we use level 1 for the inclusion and level 2 for the exclusion of features, the training corresponding to the first row will always include all the features. Such an unbalance in the number of included features between the runs can degrade the performance of the selection procedure. To avoid this problem, we train the classifier all over again with input vectors replacing level 1 (level 2) by level 2 (level 1) in Table 6.3. With this additional training, we can select better features.

A relevant question to ask is what if there are hundreds of features that can make the size of OA extremely large. In such cases, we can use the Taguchi method after reducing the features by other selection algorithms such as the greedy selection algorithms described in the previous section, and in doing so, better performance can be expected.

The feature selection algorithm using the Taguchi method is summarized as follows:

Step 1 (filtering): If there are too many features, reduce the number of features to twice the number we want to select by using some algorithms.
Step 2 (obtaining the OA): Obtain the OA corresponding to the number of features which are filtered in step 1.
Step 3: Repeat the following steps with $i = 1, 2$.
Step 3-1 (form an input feature vector for classifier): For each row of the OA, form an input feature vector with features whose values are in the OA.
Step 3-2 (training the classifier): For each row, train the classifier with the training data and store the performance for the test data.
Step 3-3 (analysis of the mean): Calculate the average performance of each feature for its inclusion and exclusion in the input feature vector. Then, for each feature, evaluate the performance increment for its inclusion case over the exclusion case.

Step 4 (selecting the input features): For each feature, average out the two increments in performance for $i = 1$ and $i = 2$ cases in step 3 and select input features by the order of these averaged terms.

6.11 Genetic Algorithm

6.11.1 General Concept

Genetic algorithm simulates a probabilistic search technique that is analogous to the biological evolutionary process. The GA consists of three main strategies: reproduction, crossover, and mutation. The performance of the GA depends on the operating parameters, namely crossover, mutation, and reproduction.

The GA consists of three main strategies (reproduction, crossover, and mutation). Using reproduction in the GA, individuals are selected from the population and recombined, producing offspring, which will comprise the next generation. Two individuals are selected, and their chromosomes are recombined. Crossover is the operation when two individuals are taken and their chromosomes are cut at some randomly chosen position, to produce two head and tail segments. These segments are swapped to reproduce two new full-length chromosomes. The offspring inherits some genes from each parent. Mutation is the technique used to randomly alter the genes with a small probability and is typically applied after crossover. Crossover is more important for rapidly exploring a search space. Usually, mutation provides a small amount of random search.

A GA for a particular problem must have the following five components:

- A genetic representation for potential solutions to the problem,
- A way to create an initial population of potential solutions,
- An evaluation function that plays the role of the environment, rating solution in terms of their *fitness*,
- Genetic operator that alters the composition of children from parents, and
- Values for various parameters such as population size and probabilities of applying genetic operators.

The parameters to be optimized are usually represented in a string (or chromosome) form since genetic operators are suitable for this type of representation. The method of representation has a major impact on the performance of the GAs. Different representation schemes might cause different performances in terms of accuracy and calculating time. There are two common representation methods for numerical optimization problems.

It seems that there are two important issues in the evolution process of the GA: One is population diversity and the other is selective pressure. These factors are strongly related: An increase in the selective pressure decreases the diversity of the population and vice versa. In other words, strong selective pressure supports the premature convergence of the GA and a weak selective pressure can make the search ineffective. Thus, it is important to keep a balance between these two factors.

There are three common genetic operators: selection, crossover, and mutation. An additional reproduction operator, inversion, is sometime also applied.

(1) *Selection*

The aim of the selection procedure is to reproduce more copies of individuals whose fitness values are higher than those whose fitness values are low. The selection procedure has a significant influence on driving the search toward a promising area and finding food solutions in a short time. However, the diversity of the population must be maintained to avoid the premature convergence and to reach the global optimal solution. Selection determines the reproduction probability of each individual in selection pool. This probability depends on the own objective value and the objective value of all other individuals. Three kinds of selection method are usually used in application.

Roulette wheel selection (Holland 1975): The mechanism of this selection is reminiscent of the operation of a roulette wheel. Fitness values of individuals represent the widths of slots on the wheel. After a random spinning of the wheel to select an individual for the next generation, individuals in slots with large widths representing high fitness values will have a higher chance to be selected.

Rank-based selection (Baker 1985): According to this procedure, each individual generates an expected number of offspring, which is based on the rank of its fitness value and not on the actual evaluation values. This strategy is similar to roulette wheel selection, excluding the application of uniform region and controlling better the selective pressure than that of the roulette wheel strategy.

Tournament selection (Goldberg 1990): This method selects randomly a group, *k*, of individuals from a beginning population, and from this group, the most fitness individual is chosen to move on to the next population. This process is repeated population-size number of times. It is clear that large value of *k* increases selective pressure of this procedure.

(2) *Crossover*

This operator is considered the one that makes the GA different from other algorithms, such as dynamic programming. It is used to create two new individuals (children or offspring) from two existing individuals (parent) picked from the current population by the selection operation. There are several ways of doing this.

Simple crossover: Simple crossover is two kinds of crossover, single or one point crossover and multi-point crossover. First, two individuals are randomly selected as parents from the pool of individuals formed by the selection procedures. Second, they are cut at a randomly chosen point. Finally, the tails, which are the parts after the cutting point, are swapped, and then, two new individuals (offspring) are produced.

Uniform crossover: Uniform crossover is proposed to overcome the problem that the process of simple crossover may be lost a sequence of string, which is schema.

This is important in that an individual solution is coded as a string. In the concept, uniform crossover is similar to multi-point crossover. The difference between simple crossover and uniform crossover is in the way that a swapping point is selected.

(3) *Mutation*

The part of mutation is that the initial individuals are widely distributed in the search space and prevented the initial local convergence. In this procedure, all individuals in the population are checked bit by bit and the bit values are randomly reversed according to a specified rate. Unlike crossover, this is a monadic operation. That is, a child string is produced from a single parent string. The mutation operator forced the algorithm to search new areas. Eventually, it helps the GA avoid premature convergence and find the global optimal solution. In the binary coding, this simply means changing 1 to 0 and vice versa and is the occasional random alteration of the value of a string position.

Classical mutation: Goldberg (1989) proposed this strategy in a basic GA, which was modeled by Holland (1975). A genotype of selected parent is exchanged by mutation rate, which is similarly small (or smaller) in natural population.

Uniform mutation: It is similar to the definition of the classical version, which searches a new point with a uniform probability distribution. This operator requires a single parent $\mathbf{x}$ and produces a single offspring $\mathbf{x}'$. The operator selects a random component $k \in (1, \ldots, q)$ of the vector $\mathbf{x} = (x_1, \ldots, x_k, \ldots, x_q)$ and

$$\text{Produces } \mathbf{x}' = (x_1, \ldots, x_k', \ldots, x_q)$$

$$x_k' = x_k^L + n(x_k^U - x_k^L), \quad n \in R[0\,1] \tag{6.45}$$

where x_k^L and x_k^U are the lower and upper boundaries of the parameter x_k, respectively, and n is the real value selected randomly from 0 to 1.

Dynamical mutation: If a high mutation rate is applied to the all stages, we may be lost the searched good candidates for optimum solutions at the previous generation. In order to avoid this problem, the elite preservation strategy and the dynamical mutation are applied. The one conserves the individuals that have higher fitness with a certain proportion rate and the other guarantees that the search point (initial candidates) is wildly distributed in the search space.

Equation (6.46) shows the dynamical mutation, which is considered in the global search steps.

$$M_i = \exp\left(-\frac{D}{5(d_i + 1)} - \frac{4g_i}{G}\right) \tag{6.46}$$

where g_i is the ith generation number, G is the total generation number, d_i is the reproduced offspring number at the ith generation, and D is the population number.

The feature of the dynamical mutation decreases exponentially at once with the generation increasing and is fluctuated by the reproduction rate, which is a total population number to a generation number.

6.11.2 Differences from Other Traditional Methods

Goldberg had summarized the characteristic of GA in comparison with conventional optimizations as follows (Goldberg 1989):

- GA is a multi-point search algorithm using a population, which is a set of random solutions, not using a potential solution;
- GA works with a coding of candidate set, not solutions themselves;
- GA uses only fitness function, not derivative or other auxiliary knowledge;
- GA is a stochastic search algorithm based on the mechanism of natural world and natural genetic. GA starts with an initial set of random solutions called population; and
- GA uses probabilistic transition rules, not deterministic rules.

GA does not have much mathematical requirements about the optimization problems. Due to their evolutionary nature, GAs will search for solutions without regard to the specific inner workings of the problem. GAs can handle any kind of objective functions and any kind of constraint (i.e., linear or nonlinear) defined on discrete, continuous, of mixed search spaces. GA does not associate with an initial point problem. To be precise, because GAs compose randomly a group of potential solutions, GAs do not have the notion of an initial point problem. That provides us with the great belief that GAs can find out global optimum solutions and a flexibility to hybridize with domain-dependent heuristics to make an efficient implementation for a specific problem.

However, GAs have also the following drawbacks or limitations:

- A binary code is not free to make a genotype of individuals.
- The fittest individual may be lost during the selection process due to its stochastic nature.
- Fit individuals may be copied several times, and a fit individual may quickly dominate the population at an early stage, especially, if the population size is small.
- The selection operation alone explores no new points in a search space. In other words, if cannot create new schemata.
- Different genetic parameters such as population size, crossover probability, and mutation probability greatly affect the accuracy and calculation time of optimum solution.

6.11.3 Simple Genetic Algorithm (SGA)

Figure 6.12 shows the flowchart of simple GA. A simple GA randomly generates an initial population. The GA proceeds for a fixed number of generations or until it satisfies some stopping criterion. During each generation, the GA performs fitness proportionate selection, followed by single-point crossover and mutation.

Figure 6.13 illustrates closely a process of evolution at k generation. First, fitness proportionate selection assigns each individual structure in the population $\bar{P}$, according to the ratio of fitness and the probability of selection. Second, using the single-point crossover $\tilde{P}$ is composed. After the crossover stage has finished, the mutation stage begins. For every string that advances to the mutation stage, each of its bits is flipped with probability (mutation rate). The population resulting from the mutation stage then overwrites the old population (the one prior to selection), completing one generation ($k + 1$). Subsequent generations follow the same cycle of selection, crossover, and mutation.

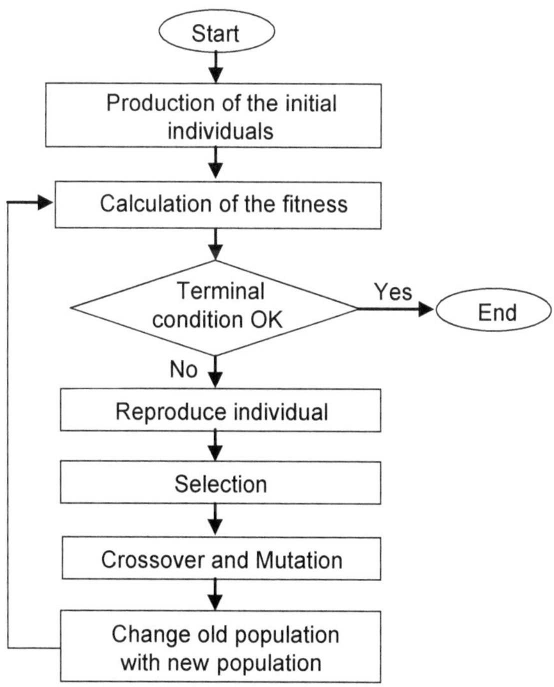

Fig. 6.12 Flowchart of the simple genetic algorithm

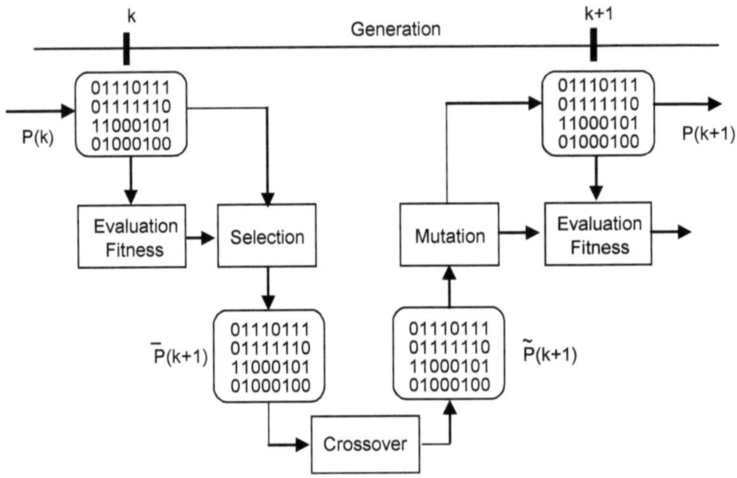

Fig. 6.13 Structure of the simple genetic algorithm

6.11.4 *Feature Selection Using GA*

A standard GA is used as a feature selector in this chapter. GA has been known to have superior performance to other search algorithms for data sets with high dimensionality. A GA is a parallel search procedure that simulates the evolutionary process by applying genetic operators. Here, a brief introduction of a standard GA is given for completeness. More extensive discussion on GA can be found in Goldberg.

The string length of the simple binary-based chromosome is determined by two parts: One is the number of features indicating whether the corresponding feature is selected or not (1 if a feature is selected and 0 otherwise) and the other is a structure parameter of ANN which varies with selected feature subset. Once fitness values are assigned to each chromosome in the current population, the GA proceeds to the next generation through three genetic operators: selection (or reproduction), crossover, and mutation.

Selection: This operator is to pick up excellent parents to reproduce promising individuals whose characteristics may be better than the last generation. This procedure has a significant influence on driving the search toward a promising area and finding good solutions in a short time. The roulette wheel selection is used for individual selection. The selection probability $P_s(s_i)$ of the *i*th individual is expressed as follows:

$$P_s(s_i) = \frac{f(s_i)}{\sum_{j=1}^{N} f(s_j)}, \quad (i = 1, 2, \ldots, N) \tag{6.47}$$

where s is an individual, $f(s_i)$ is the fitness value of the ith individual s_i, and N is the number of individual. According to the values of $P_s(s)$, each individual is defined for the widths of slots on the wheel.

Crossover: The crossover operator is used to create two new individuals (children or offspring) from two existing individuals (parents) picked from the current population by the selection operation. This operator combines a part of one string with a part of another string and is controlled by a crossover probability P_c. Many different kinds of crossover operators have been proposed including single-point, two-point, and uniform crossover. Single-point crossover is used for this process. After that, all individuals in the population are checked bit by bit and the bit values are randomly reversed according to a specified rate.

Mutation: This operator assigns a new value to a randomly chosen gene and is controlled by a mutation probability P_m. It prevents to converge to a local optimum and can be find the global optimal solution. In the binary coding, this simply means changing 1–0 and vice versa. In the standard GA, the mutation probability is set equal to a constant. However, it is clear in examining the convergence characteristics of GA that what is actually desired is a probability of mutation which varies during generational processing. In early generations, the population is diverse and mutation may actually destroy some of the benefits gained by crossover. Thus, in early generations, it would be desired to have a low probability of mutation. In later generations, the population is losing diversity as all members move "close" to the optimal solution, and thus, a higher probability of mutation is needed to maintain the search over the entire design space. Thus, the selection of the probability of mutation must carefully balance these two conflicting requirements. The mutation probability $P_m(s_i)$ is then tied to the diversity measure through an exponential function:

$$P_m(s_i) = 1 - 0.99\exp(-4 \times N_i/N_t) \tag{6.48}$$

where N_i and N_t are the number of the ith generation and total generation, respectively.

Evaluation metrics: The goal of feature subset selection is to use fewer features to achieve the same or better performance. Therefore, the fitness evaluation usually contains two terms: (i) accuracies which are the higher the better from the training and validation data and (ii) fewer number of features used. Since GA is also used to optimize the ANN, the structure parameter ρ is added to consider in the fitness evaluation and the value is the smaller the better which is explained in the next section. The objective is aimed at finding solutions by minimizing the fitness value as follows:

$$\text{fitness} = \frac{1}{10^4 \times (\eta_t + \eta_v)/2 + 10^2/\rho - N_0} \tag{6.49}$$

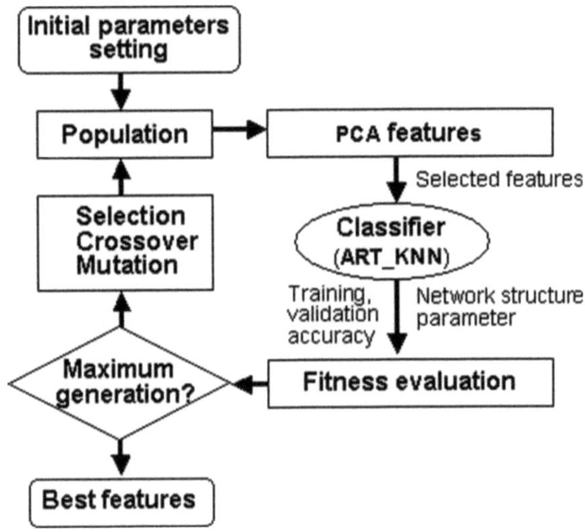

Fig. 6.14 Flowchart of feature selection by GA

where η_t and η_v represent the classification success rates of training data and validation data, respectively. ρ is a structure parameter of ANN explained in next section. N_0 is the number of selected features in the chromosome. The coefficients of variances are set by the importance degree.

The flowchart of the GA procedure for feature selection is shown in Fig. 6.14. After setting the initial parameters of GA, first-generation population is generated. Corresponding to the population strings, "1" features are selected from PCA-transformed feature base. The selected features are evaluated by the classifier, and the results are sent to the fitness function. The fitness value is calculated. If the maximum generation does not reach, produce new population through selection, crossover, and mutation and repeat the above process. At last, the smallest fitness value is selected; the corresponding feature subset is considered as the best one.

Example 6.8 Select the optimal features using GA method of vibration and current features collected from induction motor.

Twenty generation is performed during feature selection process by GA. This parameter is adjusted from setting function. From feature selection, 11 features are selected as best features.

The performance of feature selection is shown in Figs. 6.15 and 6.16. Two from the best features (features 8 and 14) are plotted in Fig. 6.15. From this figure, each class can be clustered even though it still contains of overlapping between classes. The user should try adjusting parameters of GA to reach the better performance in feature selection.

Figure 6.16 shows the convergence of fitness function according to the generation. In this example, fitness function reaches convergence at 15 generation.

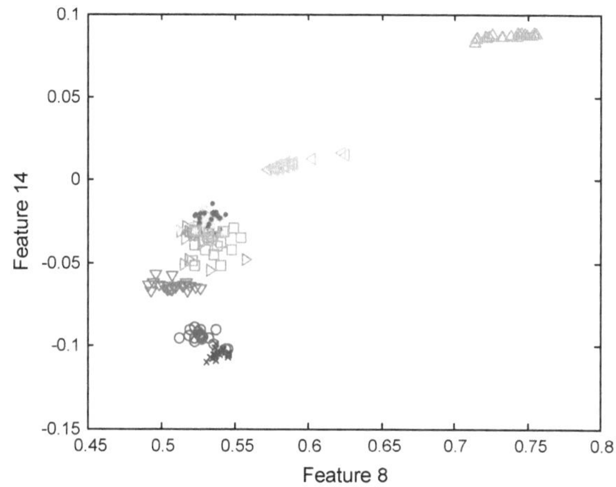

Fig. 6.15 Feature selection using GA: best features 8 and 14

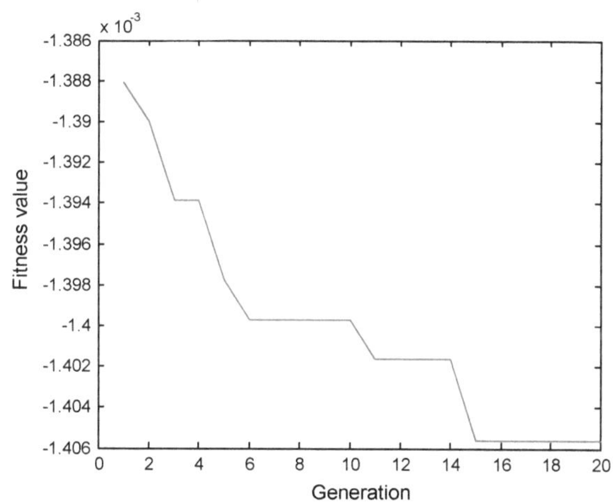

Fig. 6.16 Convergence trending of fitness versus generation

6.12 Summary

A fundamental problem of data-driven PHM/CBM is to find a reasonable number of useful features, which will then be used for diagnosis and prognosis. Although there does not exists a general theory for the design of such features, if one has got a fixed set of feature $X = \{x_i \mid i = 1, 2, \ldots, n\}$, there are methods to estimate their quality. Unfortunately, it is not sufficient to just evaluate the discriminatory power of the features individually, but all 2^n subsets must be considered to find the best combination. Because the number of subsets grows exponentially with n, various suboptimal algorithms have been suggested to find an at least "good" solution. All of the algorithms consist of two main components: the feature selection algorithm itself and the separability criterion which measures the discriminatory power of the

Table 6.4 Summary of feature selection methods (Jain et al. 2000)

Method	Property	Remark
Exhaustive search	Evaluate all possible subset	Guaranteed to find the optimal subset; not feasible for even moderately large values of m and d
Sequential forward selection (SFS)	Select the best single feature and then add the one feature at a time which in combination with the selected features maximizes the criterion function	Once feature is retained, it cannot be discarded; computationally attractive since to select a subset of size 2, it examines only $(d - 1)$ possible subset
Sequential backward selection (SBS)	Start with all the d features and successively delete one feature at a time	Once feature is deleted, it cannot be brought back into the optimal subset and requires more computation than SFS
Branch and bound search	Uses well-known branch and bound search method; only a fraction of all possible feature subsets needs to be enumerated to find the optimal subset	Guaranteed to find the optimal subset provided the criterion function satisfies the monotonicity property; the worst-case complexity of this algorithm in exponential
Plus-l take away r selection	First, enlarge the feature set by l features using forward selection and then delete r features using backward selection	Avoid the problem of feature set nesting encountered in SFS and SBS methods, need to select values of l and r $(l > r)$
Floating forward selection	A generalization of plus-l take away r method; the values of l and r are determined automatically and updated dynamically	Provide close to optimal solution at an affordable computational cost

subset under consideration. In this chapter, several methods of feature selection such as SFS, SBS, plus l-take away r, sequential forward floating selection, and sequential backward floating selection have been presented. All these algorithms use stepwise inclusion and exclusions of features into/from the subset of consideration, but they differ in their strategy of applying them. Although the floating methods are considered to be more intelligent, they are still suboptimal and even more there is no warranty that they yield better results. Also, the examples are included to illustrate the process of feature selection.

Table 6.4 summarizes the feature selection methods and their characteristics. However, how reliable are the feature selection results when ratio of the available number of training samples and number of features is depending on the criterion used. In this chapter, several examples used 1-NN as criterion for performance measure of feature selection method, but the effect of feature selection can be seen clearly when classification process is performed. We can plot the classification error and the number of features for different of number of training patterns, and then, the effect of feature selection can be highlighted.

References

Baker JE (1985) Adaptive selection method for genetic algorithm. In: Proceedings of international conference on genetic algorithm and their application, pp 101–111
Blum AL, Langely P (1997) Selection of relevant features and example in machine learning. Artif Intell 1–2:245–271
Fukunaga K (1990) Introduction to pattern recognition. Morgan Kaufmann, San Diego
Goldberg DE (1989) Genetic algorithm in search. Optimization and Machine Learning, Addison-Wesley, New York
Goldberg DE (1990) A note on Bopltzman tournament selection for genetic algorithm and population-oriented simulated annealing. Complex Syst 4:445–460
Golomb SW, Baumert LD (1965) Backtrack programming. J ACM 12:516–526
Holland JH (1975) Adaptation in natural and artificial system. The University of Michigan Press, Michigan
Jack LB, Nandi AK (2002) Fault detection using support vector machines and artificial neural network, augmented by genetic algorithms. Mech Syst Signal Process 16:373–390
Jain AK, Duin RPW, Mao J (2000) Statistical pattern recognition: a review. IEEE Trans Pattern Anal Mach Intell 22(1):4–36
Kumar R, Jayaraman VK, Kulkarni BD (2005) An SVM classifier incorporating simultaneous noise reduction and feature selection: illustrative case examples. Pattern Recogn 38:41–49
Kwak N, Choi CH (2002) Input feature selection for classification problems. IEEE Trans Neural Netw 13(1):143–159
Lehrman M, Rechester AB, White RB (1997) Symbolic analysis of chaotic signals and turbulent fluctuation. Phys Rev Lett 78(1):645–657
Mineichi K, Jack S (2000) Comparison of algorithm that select features for pattern classifiers. Pattern Recogn 33:25–41
Narendra PM, Fukunaga K (1977) A branch and bound algorithm for feature subset selection. IEEE Trans Comput 26:917–922
Nilsson NJ (1971) Problem solving methods in artificial intelligence. McGraw-Hill, New York
Pudil P, Ferri FJ, Kittler J (1994) Floating search for feature selection with nonmonotonic criterion functions. In: Proceedings of the 12th IAPR international conference on pattern recognition, vol 2, pp 279–283
Stearns SD (1976) On selecting features for pattern classifiers. In: Proceedings of international conference on pattern recognition, pp 71–75
Yang BS, Han T, An JL (2004) ART-Kohonen neural network for faults diagnosis of rotating machinery. Mech Syst Signal Process 18:645–657
Yang BS, Han T, Hwang WW (2005) Fault diagnosis of rotating machinery based on multi-class support vector machines. J Mech Sci Technol 19:845–858

Chapter 7
Intelligent Fault Diagnosis Methodology

7.1 Introduction

There are many approaches that can be used to design a data-driven diagnosis algorithm. Usually, we divide them into three strategies. The simplest and the most intuitive approach to classifier design is based on the concept of similarity: Patterns that are similar should be assigned to the same class. So, once a good metric has been established to define similarity, patterns can be classified by template matching or the minimum-distance classifier using a few prototypes per class. The choice of the metric and the prototypes is crucial to the success of this approach. One of the representative algorithms is k-nearest neighbor (k-NN).

The second main strategy is based on the probabilistic approach. The optimal Bayes decision rule assigns a pattern to the class with the maximum posterior probability. This rule can be modified to take into account costs associated with different types of misclassifications. For known class-conditional densities, the Bayes decision rule gives the optimum classifier, in the sense that for given prior probabilities, loss function, and class-conditional densities, no other decision rule will have a lower risk. The third category of classifiers is to construct decision boundaries directly by optimizing certain error criterion. While this approach depends on the chosen metric, sometimes classifiers of this type may approximate the Bayesian classifier. The driving force of the training procedure is the minimization of a criterion such as the apparent classification error or the mean-squared error (MSE) between the classifier output and preset target value.

A special type of classifier is the decision tree that is trained by an iterative selection of individual features, which are most salient at each node of the tree. The criteria for feature selection and tree generation include the information content and the node purity. During classification, just those features are under consideration, so feature selection is implicitly built-in. The most commonly used decision tree classifiers are binary in nature and use a single feature at each node, resulting in decision boundaries that are parallel to the feature axes.

G. Niu, *Data-Driven Technology for Engineering Systems Health Management*,
DOI 10.1007/978-981-10-2032-2_7

In this chapter, several classifier algorithms will be reviewed based on similarity of patterns and probabilistic approaches. Moreover, after presenting the basic theory, some examples are also presented to give better understanding of classifier algorithms. Finally, the end of this chapter will discuss several case studies of fault diagnosis of induction motors based on reviewed classifier algorithms.

7.2 Linear Classifier

The goal of classification is to group items that have similar feature values, into groups. A linear classifier achieves this by making a classification decision based on the value of the linear combination of the features.

The linear classification rule of n-dimensional input space can be given as follows:

$$q : X \subseteq R^n \to Y = \{1, 2, \ldots, c\} \tag{7.1}$$

that is composed of a set of discriminant functions

$$f_y(\mathbf{x}) = \langle \mathbf{w}_y \cdot \mathbf{x} \rangle + b_y, \quad \forall y \in Y \tag{7.2}$$

which are linear with respect to both the input vector $\mathbf{x} \in R^n$ and their parameter vector $\mathbf{w} \in R^n$. The scalar b_y, $\forall y \in R^n$ introduces bias to the discriminant function. The input vector $\mathbf{x} \in R^n$ is assigned to the class $y \in Y$ and its corresponding discriminant function f_y attains the maximal rule

$$y = \arg \max_{y \in Y} f_y(x) = \arg \max_{y \in Y} (\langle \mathbf{w}_y \cdot \mathbf{x} \rangle + b_y) \tag{7.3}$$

In the particular binary case, i.e., $Y = \{1, 2\}$, the linear classifier is represented by a single discriminant function

$$f(\mathbf{x}) = \langle \mathbf{w} \cdot \mathbf{x} \rangle + b \tag{7.4}$$

given by parameter vector $\mathbf{w} \in R^n$ and bias $b \in R$.

The input vector $\mathbf{x} \in R^n$ is assigned to class $y \in \{1, 2\}$ as follows:

$$q(\mathbf{x}) = \begin{cases} 1 & \text{if } f(\mathbf{x}) = \langle \mathbf{w} \cdot \mathbf{x} \rangle + b \geq 0 \\ 2 & \text{if } f(\mathbf{x}) = \langle \mathbf{w} \cdot \mathbf{x} \rangle + b < 0 \end{cases} \tag{7.5}$$

The data type used to describe the linear classifier and the implementation of the classifier is described by a finite set $T = \{(x_1, y_1), \ldots, (x_l, y_l)\}$ containing pairs of observations $\mathbf{x} \in R^n$ and the corresponding class label $y_i \in Y$.

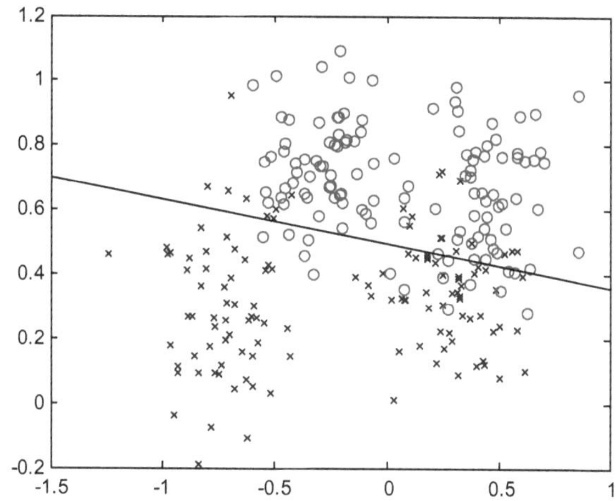

Fig. 7.1 Linear classification of *riply data set*

Example 7.1 Demonstrate the linear classifier using given train and test data sets based on Fisher linear discriminant theory.

According to the test result, the accuracy of the classification is low. This phenomenon also is shown in Fig. 7.1.

7.2.1 *Linear Separation of Finite Set of Vectors*

The input training data $T = \{(x_1, y_1), \ldots, (x_l, y_l)\}$ consist of pairs of observations $\mathbf{x} \in R^n$ and the corresponding class labels $y_i \in Y = \{1, \ldots, c\}$. The implemented methods solve the task of

(1) Training separating hyperplane for binary case $c = 2$,
(2) Training optimal hyperplane, and
(3) Training multi-class linear classifier. The definitions of this task are given below.

The problem of training the binary linear classifier in Eq. (7.5) with zero training error is equivalent to finding the hyperplane

$$H = \{\mathbf{x} : \langle \mathbf{w} \cdot \mathbf{x} \rangle + b = 0\} \tag{7.6}$$

which separates the training vectors of the first $y = 1$ and the second $y = 2$ class. The problem is formally defined as solving the set of linear inequalities

$$\begin{aligned} \langle \mathbf{w} \cdot \mathbf{x}_i \rangle + b \geq 0, \quad & y_i = 1 \\ \langle \mathbf{w} \cdot \mathbf{x}_i \rangle + b < 0, \quad & y_i = 2 \end{aligned} \tag{7.7}$$

with respect to the vector $\mathbf{w} \in R^n$ and the scalar $b \in R$. If the inequalities in Eq. (7.7) have a solution, then the training data T are linearly separable, and the optimal separating hyperplane can be defined as the solution of the following task:

$$\begin{aligned}(\mathbf{w}^*, b^*) &= \arg\max_{\mathbf{w},b} m(\mathbf{w}, b) \\ &= \arg\max_{\mathbf{w},b} \min\left(\min_{i \in y_1} \frac{\langle \mathbf{w} \cdot \mathbf{x}_i \rangle + b}{||\mathbf{w}||}, \min_{i \in y_2} - \frac{\langle \mathbf{w} \cdot \mathbf{x}_i \rangle + b}{||\mathbf{w}||}\right)\end{aligned} \tag{7.8}$$

The optimal separating hyperplane $H^* = \{\mathbf{x} : \langle \mathbf{w}^* \cdot \mathbf{x} \rangle + b^* = 0\}$ separates the training data T with maximal margin $m(\mathbf{w}*, b*)$. The optimal separating hyperplane cannot be found exactly except the special case. Therefore, the numerical algorithms seeking the approximate solution are applied instead. The ε-optimal solution is defined as the vector $\mathbf{w}$ and the scalar b such that the inequality

$$m(\mathbf{w}^*, b^*) - m(\mathbf{w}, b) \leq \varepsilon \tag{7.9}$$

holds. The parameter $\varepsilon \leq 0$ defines the closeness to the optimal solution in terms of the margin. The problem of training the multi-class linear classifier $c > 2$ with zero training error is formally stated as the problem of solving the set of linear inequalities

$$\langle \mathbf{w}_{yi} \cdot \mathbf{x}_i \rangle + b_{yi} > \langle \mathbf{w}_y \cdot \mathbf{x}_i \rangle + b_y, \quad i = 1, \ldots, l, \quad y_i \neq y \tag{7.10}$$

with respect to the vectors $\mathbf{w}_y \in R^n$, $y \in Y$ and scalars $b_y \in R^n$, $y \in Y$.

The task of Eq. (7.10) can be solved by the perceptron and Kozinec's algorithm described in the following section.

7.2.2 Perceptron Algorithm

The input is the data set $T = \{(x_1, y_1), \ldots, (x_l, y_l)\}$ of binary labeled $y_i = \{1, 2\}$ training vectors $\mathbf{x}_i \in R^n$. The problem of training the separating hyperplane in Eq. (7.5) can be formally rewritten to a simpler form

$$\langle \mathbf{v} \cdot \mathbf{z}_i \rangle > 0, \quad i = 1, \ldots, l \tag{7.11}$$

where the vector $\mathbf{v} \in R^{n+1}$ is constructed as follows:

$$\mathbf{v} = [\mathbf{w}; b] \tag{7.12}$$

and transformed training data $\mathbf{z} = \{z_1, \ldots, z_l\}$ are defined as follows:

$$\mathbf{z}_i = \begin{cases} [\mathbf{x}_i; 1] & \text{if } y_i = 1 \\ -[\mathbf{x}_i; 1] & \text{if } y_i = 2 \end{cases} \tag{7.13}$$

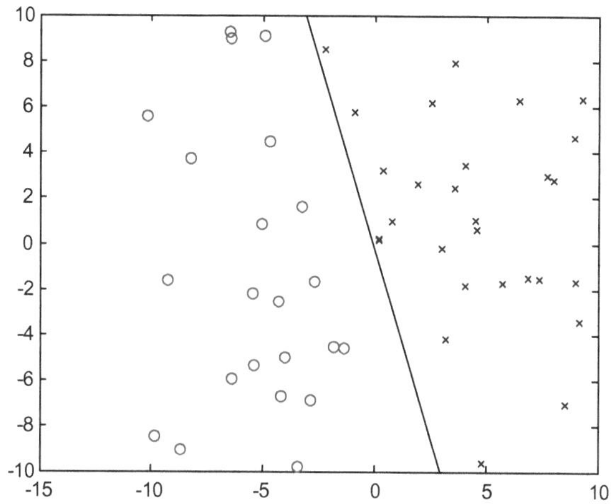

Fig. 7.2 Perceptron algorithm for linear classification

The problem of solving Eq. (7.11) with respect to the unknown vector $\mathbf{v} \in R^{n+1}$ is equivalent to the original task Eq. (7.5). The parameters ($\mathbf{w}$, b) of the linear classifier are obtained from the found vector $\mathbf{v}$ by inverting the transformation Eq. (7.12).

The perceptron algorithm is an iterative procedure which build a series of vectors $\mathbf{v}^{(0)}, \mathbf{v}^{(1)}, \ldots, \mathbf{v}^{(t)}$ until the set of inequalities Eq. (7.11) is satisfied. The initial vector $\mathbf{v}^{(0)}$ can be set arbitrarily (usually $\mathbf{v} = 0$). The Novikoff's theorem (Novikoff 1962) ensures that the perceptron algorithm stops after finite number of iterations t if the training data are linearly separable.

Example 7.2 Apply the perceptron algorithm to find the binary linear classifier for synthetically generated 2D data using random data simulation as shown in Fig. 7.2.

7.2.3 Kozinec's Algorithm

The input is data set $T = \{(x_1, y_1), \ldots, (x_l, y_l)\}$ of binary labeled $y_i = \{1, 2\}$ training vectors $\mathbf{x}_i \in R^n$. The Kozinec's algorithm builds a series of vectors $\mathbf{w}_1^{(0)}, \mathbf{w}_1^{(1)}, \ldots, \mathbf{w}_1^{(t)}$ and $\mathbf{w}_2^{(0)}, \mathbf{w}_2^{(1)}, \ldots, \mathbf{w}_2^{(t)}$ which converge to the vector $\mathbf{w}_1^*$ and $\mathbf{w}_2^*$, respectively. The vectors $\mathbf{w}_1^*$ and $\mathbf{w}_2^*$ are the solutions of the following task:

$$\mathbf{w}_1^*, \mathbf{w}_2^* = \underset{\mathbf{w}_1 \in X_1, \mathbf{w}_2 \in X_2}{\arg\min} \|\mathbf{w}_1 - \mathbf{w}_2\| \tag{7.14}$$

where X_1 stands for the convex hull of the training vectors of the first class $X_1 = \{\mathbf{x}_i: y_i = 1\}$ and X_2 for the convex hull of the second class likewise. The vector $\mathbf{w}^* = \mathbf{w}_1^* - \mathbf{w}_2^*$ and the bias $b^* = \frac{1}{2}(\|\mathbf{w}_2^*\|^2 - \|\mathbf{w}_1^*\|^2)$ determine the optimal hyperplane Eq. (7.8) separating the training data from the maximal margin.

The Kozinec's algorithm is proven to converge to the vectors $\mathbf{w}_1^*$ and $\mathbf{w}_2^*$ in infinite number of iterations $t = \infty$. If the ε-optimal optimality stopping condition is

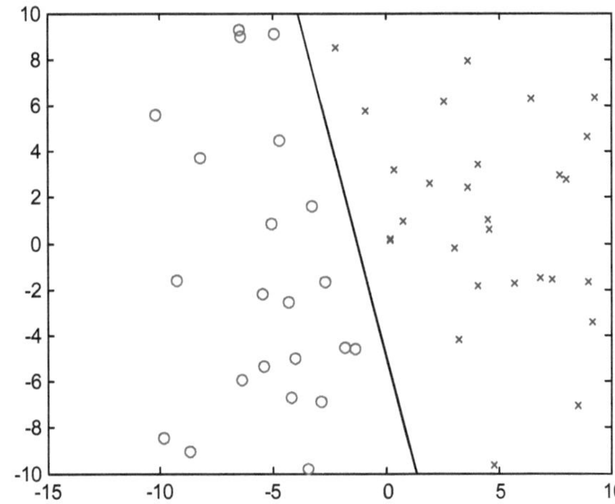

Fig. 7.3 Kozinec's algorithm for linear classification

used, then the Kozinec's algorithm converges in the finite number of iterations. The Kozinec's algorithm can also be used to solve a simpler problem of finding the separating hyperplane Eq. (7.5). Therefore, the following two stopping conditions are implemented:

- The separating hyperplane Eq. (7.5) is sought for $\varepsilon < 0$. The Kozinec's algorithm is proven to converge in a finite number of iterations of the separating hyperplane that exists.
- The ε-optimal hyperplane Eq. (7.9) is sought for $\varepsilon \geq 0$. Note that setting $\varepsilon = 0$ forces the algorithm to seek the optimal hyperplane which is generally ensured to be found in an infinite number of iterations $t = \infty$.

Example 7.3 Same as Example 7.2, apply the Kozinec's algorithm for linear classification as shown in Fig. 7.3.

7.2.4 Multi-class Linear Classifier

The input training data $T = \{(x_1, y_1), \ldots, (x_l, y_l)\}$ consist of pairs of observations $\mathbf{x}_i \in R^n$ and the corresponding class labels $y_i \in Y = \{1, \ldots, c\}$. The task is to design the linear classifier Eq. (7.3) which classifies the training data T without error. The problem of training in Eq. (7.3) is equivalent to solving the set of linear inequalities in Eq. (7.10).

The inequalities in Eq. (7.10) can be formally transformed to a simpler set

$$\langle \mathbf{v} \cdot \mathbf{z}_i^y \rangle > 0, \quad i = 1, \ldots, l, \quad y = Y \backslash \{y_i\} \tag{7.15}$$

where the vector $\mathbf{v} \in R^{(n+1)c}$ contains the originally optimized parameters

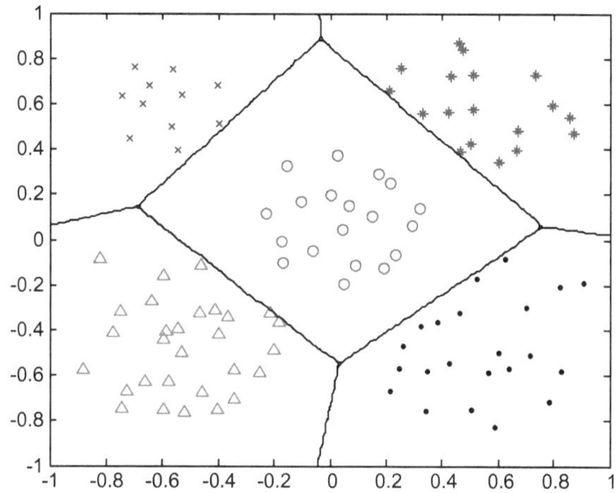

Fig. 7.4 Perceptron algorithm for multi-class linear classification

$$\mathbf{v} = [\mathbf{w}_1; b_1; \mathbf{w}_2; b_2; \ldots; \mathbf{w}_c, b_c] \tag{7.16}$$

The test of vectors $\mathbf{z}_i^y \in R^{(n+1)c}, \quad i = 1, \ldots, l, \quad y \in Y \backslash \{y_i\}$ is created from the training set T such that

$$\mathbf{z}_i^y(j) = \begin{cases} [\mathbf{x}_i; 1], & \text{for } j = y_i \\ -[\mathbf{x}_i; 1], & \text{for } j = y \\ 0, & \text{otherwise} \end{cases} \tag{7.17}$$

where $\mathbf{z}_i^y(j)$ stands for the jth slot between coordinates $(n + 1)(j - 1) + 1$ and $(n + 1)j$. The described transformation is known as the Kesler's construction.

The transformed task in Eq. (7.15) can be solved by perceptron algorithm. The simple form of the perceptron updating rule allows us to implement the transformation implicitly without mapping the training data T into $(n + 1)c$-dimensional space.

Example 7.4 Apply the perceptron algorithm to find the multi-class linear classifier for synthetically generated 2-D simulated data as shown in Fig. 7.4.

7.3 Quadratic Classifier

The quadratic classification rule of n-dimensional input space can be given as follows:

$$q : X \subseteq R^n \rightarrow Y = \{1, 2, \ldots, c\} \tag{7.18}$$

which is composed of a set of discriminant functions

$$f_y(\mathbf{x}) = \langle \mathbf{x} \cdot \mathbf{A}_y\mathbf{x} \rangle + \langle \mathbf{b}_y \cdot \mathbf{x} \rangle + c_y, \quad \forall y \in Y \tag{7.19}$$

which are quadratic with respect to the input vector $\mathbf{x}_i \in R^n$. The quadratic discriminant function f_y is determined by a matrix $\mathbf{A}_y$ $[n{\times}n]$, a vector $\mathbf{b}_y$ $[n{\times}1]$, and a scalar c_y $[1{\times}1]$. The input vector $\mathbf{x}_i \in R^n$ is assigned to the class $y \in Y$ and its corresponding discriminant function f_y attains the maximal value

$$y = \arg\max_{y \in Y} f_y(\mathbf{x}) = \arg\max_{y \in Y}(\langle \mathbf{x} \cdot \mathbf{A}_y\mathbf{x} \rangle + \langle \mathbf{b}_y \cdot \mathbf{x} \rangle + c_y) \tag{7.20}$$

In the particular binary case $Y = \{1, 2\}$, the quadratic classifier is represented by a single discriminant function

$$f(\mathbf{x}) = \langle \mathbf{x} \cdot \mathbf{A}\mathbf{x} \rangle + \langle \mathbf{b} \cdot \mathbf{x} \rangle + c \tag{7.21}$$

It is given by a matrix $\mathbf{A}$ $[n{\times}n]$, a vector $\mathbf{b}$ $[n{\times}1]$, and a scalar c $[1{\times}1]$. The input vector $\mathbf{x}_i \in R^n$ is assigned to the class $y \in \{1, 2\}$ as follows:

$$q(\mathbf{x}) = \begin{cases} 1 & \text{if} \quad f(\mathbf{x}) = \langle \mathbf{x} \cdot \mathbf{A}\mathbf{x} \rangle + \langle \mathbf{b} \cdot \mathbf{x} \rangle + c \geq 0 \\ 2 & \text{if} \quad f(\mathbf{x}) = \langle \mathbf{x} \cdot \mathbf{A}\mathbf{x} \rangle + \langle \mathbf{b} \cdot \mathbf{x} \rangle + c < 0 \end{cases} \tag{7.22}$$

Example 7.5 Perform classification using quadratic classifier to a given training data, and then evaluate the classification process as shown in Fig. 7.5.

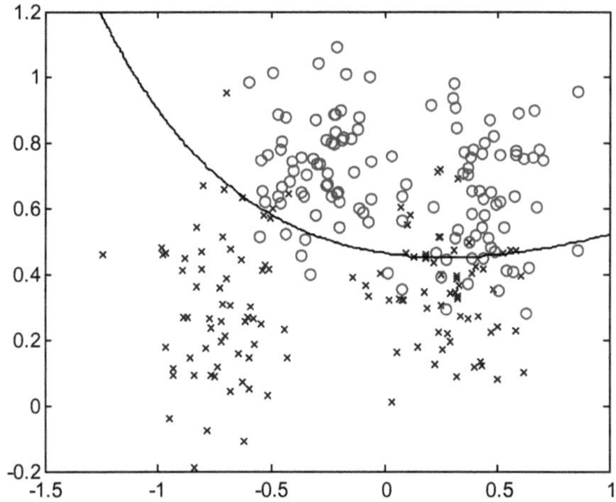

Fig. 7.5 Quadratic classifier algorithm for binary classification

7.4 Bayesian Classifier

The object under study is assumed to be described by a vector of observations $\mathbf{x} \in X$ and a hidden state $y \in Y$. The $\mathbf{x}$ and y are realizations of random variables with a joint probability distribution $P_{XY}(\mathbf{x}, y)$. A decision rule q: $X \rightarrow D$ takes a decision $d \in D$ based on the observation $\mathbf{x} \in X$. Let W: $D \times Y \rightarrow R$ be a loss function with penalizes the decision $q(\mathbf{x}) \in D$ when the true hidden state is $y \in Y$, and let $X \subseteq R^n$ and the sets Y and D are finite. The Bayesian risk $R(q)$ is an expectation of the value of the loss function W when the decision rule q is applied, i.e.,

$$R(q) = \int_X \sum_{y \in Y} P_{XY}(\mathbf{x}, y) W(q(\mathbf{x}), y)\, \mathbf{dx} \tag{7.23}$$

The optimal rule q^* which minimizes the Bayesian risk of Eq. (7.23) is referred to as the Bayesian rule

$$q^*(\mathbf{x}) = \arg\min_{y} \sum_{y \in Y} P_{XY}(\mathbf{x}, y) W(q(\mathbf{x}), y), \forall \mathbf{x} \in X \tag{7.24}$$

For minimization of classification task, the set of decision D coincides with the set of hidden states $Y = \{1, \ldots, c\}$. The loss function given by:

$$W(q(\mathbf{x}), y) = \begin{cases} 0 & \text{for} \quad q(\mathbf{x}) = y \\ 1 & \text{for} \quad q(\mathbf{x}) \neq y \end{cases} \tag{7.25}$$

is used. The Bayesian risk in Eq. (7.23) with the loss function corresponds to the expectation of misclassification. The rule q: $X \rightarrow Y$ which minimizes the expectation of misclassification is defined by

$$q(\mathbf{x}) = \arg\max_{y \in Y} P_{Y|X}(y|\mathbf{x}) = \arg\max_{y \in Y} P_{X|Y}(\mathbf{x}|y) P_Y(y) \tag{7.26}$$

For classification task with reject point, the set of dimension D is assumed to be $D = Y \cup \{\text{don't\_know}\}$. The loss function is defined as follows:

$$W\varepsilon(q(x), y) = \begin{cases} 0 & \text{for} \quad q(\mathbf{x}) = y \\ 1 & \text{for} \quad q(\mathbf{x}) \neq y \\ \varepsilon & \text{for} \quad = \text{dont\_know} \end{cases} \tag{7.27}$$

where ε is penalty for the decision don't_know.

The rule q: $X \rightarrow Y$ which minimizes the Bayesian risk with the loss function Eq. (7.27) is defined as follows:

$$q(\mathbf{x}) = \begin{cases} \operatorname{argmax}_{y \in Y} P_{X|Y}(\mathbf{x}|y) P_Y(y) & \text{if} \quad 1 - \max_{y \in Y} P_{Y|X}(y|\mathbf{x}) < \varepsilon \\ \text{dont\_know} & \text{if} \quad 1 - \max_{y \in Y} P_{Y|X}(y|\mathbf{x}) \geq \varepsilon \end{cases} \tag{7.28}$$

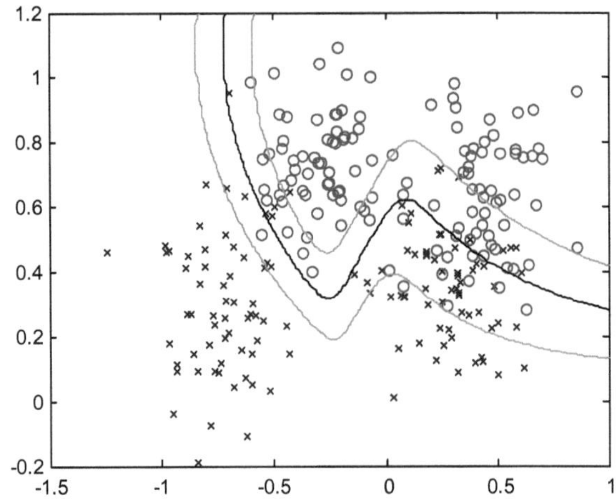

Fig. 7.6 Bayesian classifier algorithm for binary classification

To apply the optimal classification rules, one has to know the class-conditional distributions $P_{X|Y}$ and the prior distribution P_Y (or their estimates).

Example 7.6 Apply Bayesian classifier to the given data. The Bayesian classifier needs class-conditional distribution modeled by the Gaussian mixture model (GMM).

Figure 7.6 shows how to visualize the decision boundary of the found Bayesian classifier minimizing the misclassification (solid black line). The Bayesian classifier splits the feature space into two regions corresponding to the first class and the second class. The decision boundary of the reject option rule (dashed line) is displayed for comparison. The reject option rule splits the feature space into three regions corresponding to the first class, the second class, and the region in between corresponding to the don't know decision.

7.5 *k*-Nearest Neighbors (*k*-NN)

Nearest-neighbor methods can be used as an important pattern recognition tool (Duda et al. 2001). In such methods, the aim is to find the nearest neighbors of an undefined test pattern within a hypersphere of predefined radius in order to determine its true class. The traditional NN rule has been described as follows:

- Out of N training vectors, identify the k-NN, irrespective of the class label.
- Out of these k samples, identify the number of vectors k_i that belong to class c_i, $i = 1, 2, \ldots, l$. Obviously, $\Sigma\ k_i = k$.
- Assign x to the class c_i with the maximum number k_i of samples.

Nearest-neighbor methods can detect a single or multiple numbers of nearest neighbors. A single nearest-neighbor method is primarily suited to recognize data

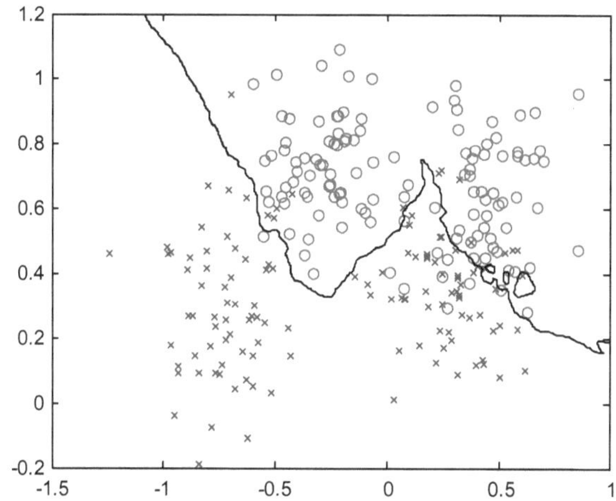

Fig. 7.7 *k*-nearest-neighbor algorithm (k = 10) for binary classification

where we have sufficient confidence in the fact that class distributions are non-overlapping and the features used are discriminatory. In most practical applications, however, the data distributions for various classes are overlapping and more than one nearest neighbors are used for majority voting.

For illustration, let us define the set of training data $T = \{(x_1, y_1), \ldots, (x_l, y_l)\}$ to be a set of prototype vector $\mathbf{x}_i \in X \in R^n$ and hidden states $y_i \in Y = \{1, \ldots, c\}$. Let $R^n(\mathbf{x}) = \{\mathbf{x}' | \|\mathbf{x} - \mathbf{x}'\| \leq r^2\}$ be a ball centered in the vector $\mathbf{x}$ in which lie k prototype vectors $\mathbf{x}_i$, $i \in \{1, \ldots, l\}$, i.e.,$|\{\mathbf{x}_i : \mathbf{x}_i \in R^n\ (\mathbf{x})\}| = k$. The k-nearest-neighbor classification rule q: $X \rightarrow Y$ is defined as follows:

$$q(\mathbf{x}) = \arg\max_{y \in Y} v(\mathbf{x}, y) \tag{7.29}$$

where $v(\mathbf{x}, y)$ is the number of prototype vectors $\mathbf{x}_i$ with hidden state $y_i = y$ which lies in the ball $\mathbf{x}_i \in R^n(\mathbf{x})$. The classification rule Eq. (7.29) is computationally demanding if the set of prototype vectors is large.

Example 7.7 Perform classification using k-nearest neighbor (k = 10) to labeled training data, and then evaluate the classification process using test data as shown in Fig. 7.7.

7.6 Self-Organizing Feature Map (SOFM) Neural Network

The SOFM is a neural network model that implements a characteristic nonlinear projection from a high-dimensional space of sensory or other input signals onto a low-dimensional array of neurons. The SOFM is an unsupervised learning algorithm which produces a map pattern feature on its output layer (Kohonen 1995).

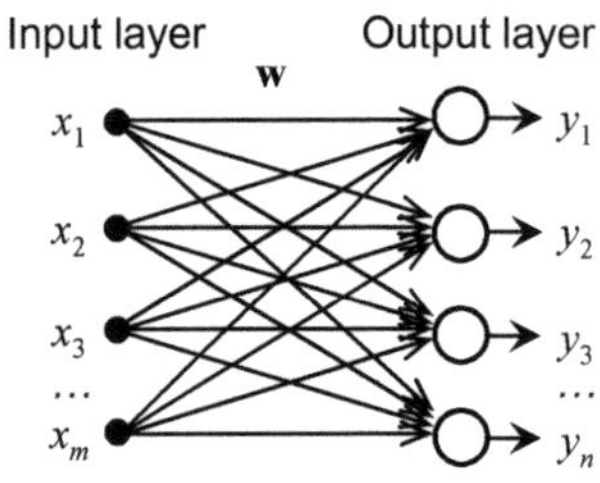

Fig. 7.8 Structure of the SOFM network

Input patterns with similar features are mapped onto neighboring output nodes. The network consists of an input layer with m neurons and an output layer with n neurons as shown in Fig. 7.8.

The image of the signal space tends to manifest the clusters of input information and their relationship on the feature map. With every node j, a parametric reference vector $\mathbf{w}_j \in R^n$ is associated. In an abstract scheme, it may be imagined that at time k, the input $\mathbf{x}(k)$, by means of some parallel computing mechanisms, is compared with all $\mathbf{w}_j(k)$, and the location of the best match in some metric is defined as the location of the response. However, in many practical applications, the Euclidean distances $\|\mathbf{x}(k) - \mathbf{w}_j(k)\|$ can be made to define the best-matching node (neuron) i which is signified by the index $i(\mathbf{x})$ (Fig. 7.9):

$$i(\mathbf{x}) = \arg\min\left|\left|\mathbf{x}(k) - \mathbf{w}_j(k)\right|\right| = \arg\min\{\left|\left|\mathbf{e}_j(k)\right|\right|\}, \tag{7.30}$$

where $j = 1, 2, \ldots, N$; $k = 1, 2, \ldots, M$; $\mathbf{x}(k) = \{x_1, x_2, \ldots, x_M\}^{\mathrm{T}}$ represents the input vectors, and $\mathbf{w}_j\ (k) = [w_{j1}, w_{j2}, \ldots, w_{jM}]^{\mathrm{T}}$ $(j = 1, 2, \ldots, N)$ represents the weight vectors of the jth output vectors. $\mathbf{e}_j(k)$ is the error which is given by $\|\mathbf{x}(k) - \mathbf{w}_{,i(\mathbf{x})}(k)\| = \min\{\|\mathbf{e}_j(k)\|\}$.

Initially, the reference vectors $\mathbf{w}_j(0)$ are random. Self-organization is achieved by using the following learning process:

$$\mathbf{w}_j(k+1) = \mathbf{w}_j(k) + h(k)h_{j,i(\mathbf{x})}(k)\mathbf{e}_j(k) \tag{7.31}$$

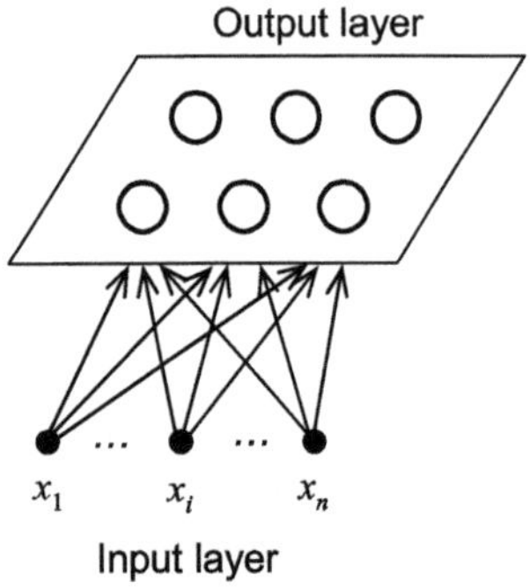

Fig. 7.9 Two-dimensional (2D) mapping of SOFM network

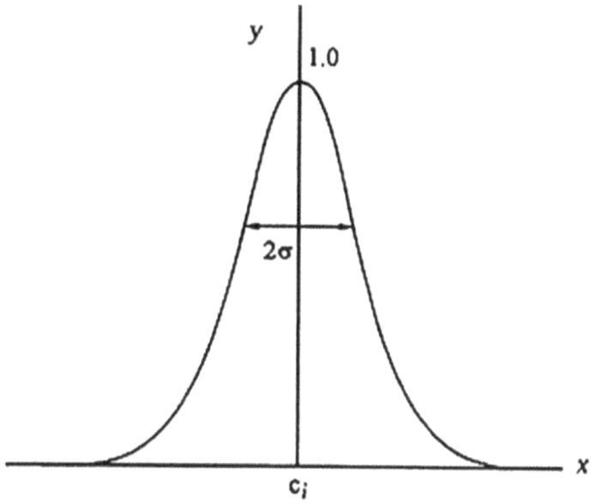

Fig. 7.10 Gaussian neighborhood function

where $k = 0, 1, 2, \ldots$, and M is an integer and discrete-time coordinate. $\eta(k)$ is the learning rate factor which decreases with increasing iterations ($0 < \eta(k) < 1$). $h_{j,i(\mathbf{x})}$ is the neighborhood function and plays an important role in the learning process; $h_{j,i(\mathbf{x})}(k) \to 0$ when $k \to \infty$ (Fig. 7.10).

The Gaussian distribution function is usually selected for the neighborhood function and is shown in Eq. (7.32).

$$h_{j,i(\mathbf{x})}(k) = \exp\{-d_j^2/2\sigma^2(k)\} \tag{7.32}$$

where d_j is the grid distance between the best-matching neuron i and the j neighborhood neurons. $\sigma(k)$ is the effective width of the neighborhood and is defined as a monotonically decreasing function of the time with σ_t being a constant value,

$$\sigma(k) = \sigma_0 \exp(-k/\sigma_t) \tag{7.33}$$

The algorithm is repeated using the successive data samples, and as the algorithm proceeds, different neurons start representing specific clusters of the input space. It can be shown that the algorithm can be interpreted as a gradient approach to minimize the energy function $E_j(k)$ in each neuron j (Figs. 7.11 and 7.12):

$$E_j(k) = \sum h_{j,i(\mathbf{x})}(\kappa)(\mathbf{e}_j(\kappa)^{\mathrm{T}}\mathbf{e}_j(\kappa)), \quad \kappa = 0, 1, \ldots, k \tag{7.34}$$

The process of SOFM can be summarized as follows (Fig. 7.13).

Example 7.8 Classify the iris data according to their classes using SOFM method (Fig. 7.14).

From the U-matrix, it is easy to see that the top three rows of the SOFM are from a very clear cluster. By looking at the labels, it is immediately seen that this corresponds to the Setosa subspecies. The two other subspecies Versicolor and Virginia form the other cluster. The U-matrix shows no clear separation between them, but from the labels, it seems they correspond to two different parts of the cluster. From the component planes, it can be seen that the petal length and the petal

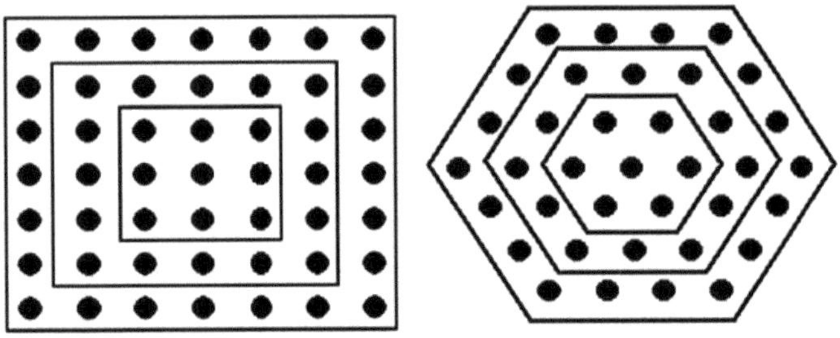

Fig. 7.11 Two different output neuron type of two-dimensional map

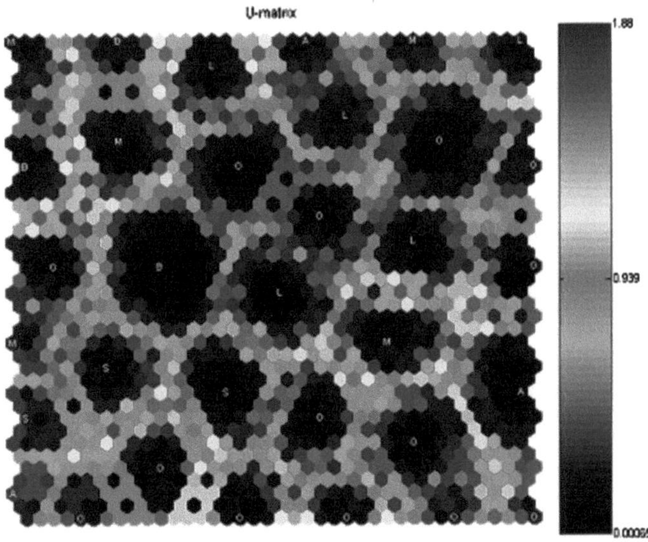

Fig. 7.12 Unified map of all components after training

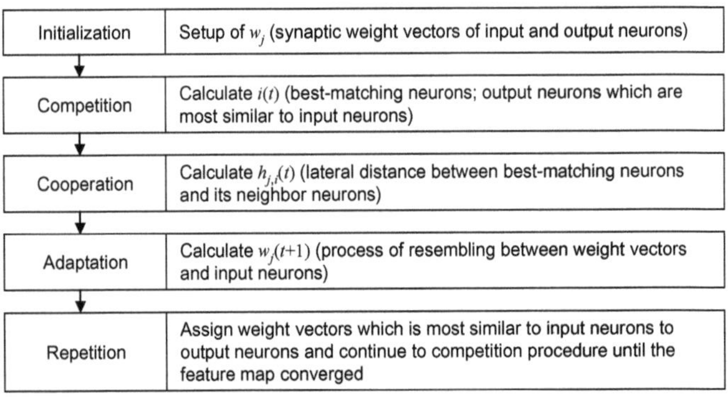

Fig. 7.13 The flowchart of SOFM

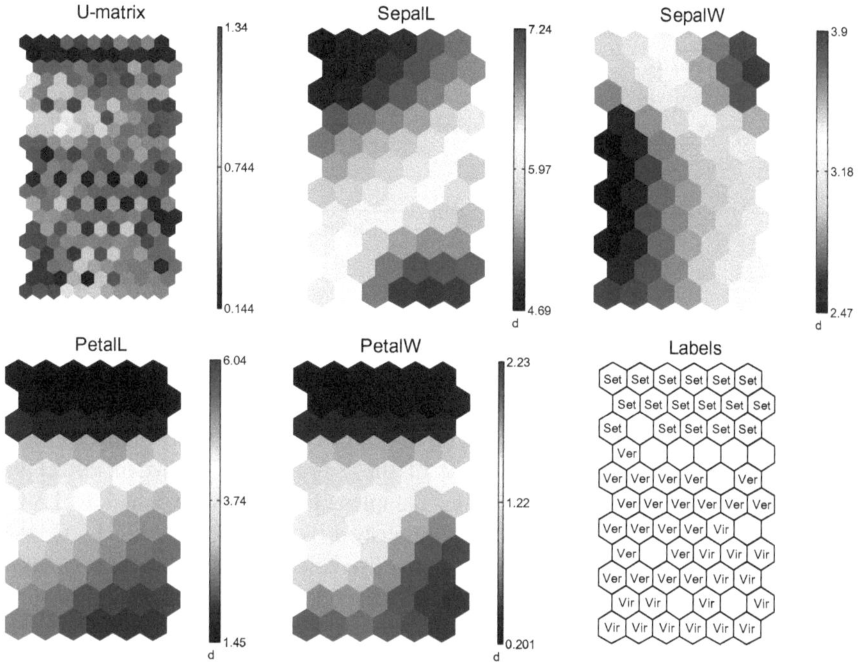

Fig. 7.14 Visualization classification of the SOFM of iris data: U-matrix (*top left*), then component planes, and map unit labels on *bottom right*

width are very closely related to each other. Also, some correlations exist between them and sepal length. The Setosa subspecies exhibits small petals and short but wide sepals. The separating factor between Versicolor and Virginia is that the latter has bigger leaves.

7.7 Learning Vector Quantization (LVQ) Neural Network

LVQ as proposed by Kohonen (1992) is a simple and intuitive, though very successful prototype-based clustering algorithm. It combines the simplicity of self-organizing learning with the accuracy of supervised training algorithms. Successful applications can be found in widespread areas such as data mining, robotics, or linguistics.

LVQ consists of two layers: competitive layer and linear layer, shown in Fig. 7.15. The competitive layer selects the winner based on the distance calculation between the input vector and *codebook vectors*. The linear layer transforms the competitive layer's classes into target classifications defined by the user. We refer to the classes learned by the competitive layer as subclasses and the classes of the linear layer as target classes (Fig. 7.15).

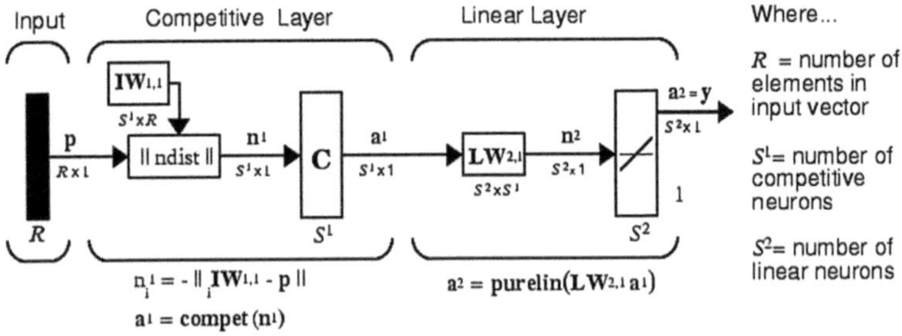

Fig. 7.15 LVQ network architecture

Learning vector quantization (LVQ1, LVQ2, LVQ2.1, and LVQ3) is a nearest-neighbor classification method in which a fixed number of prototype vectors are progressively modified to cover the input space. The LVQ family of algorithms is concerned with the optimal placement of these prototypes, so as to reflect the probability distribution of the training samples. From now, we review LVQ3 algorithm briefly which was previously published by Kohonen (1992). LVQ does phase classification which enhances the characteristics of classification boundaries via supervised learning using input vectors (or codebook vectors) $\mathbf{x}(k)$. The input layer of an LVQ network is connected directly to the output layer. Each node in the output layer has a weight vector (or prototype) attached to it. Assume that a number of codebook vectors $\mathbf{m}_i(k)$ (free parameter vectors) are placed into the input space to approximate various domains of the input vector by their quantized values. Some parts of the entire codebook vectors can be used for reducing calculation speed and memory allocation. It is assumed that the output vector of LVQ is $\mathbf{m}_i(k)$.

If $\mathbf{m}_i(k)$ is the most similar vector to $\mathbf{x}(k)$, $\mathbf{m}_c$ is determined by the k-nearest-neighbor rule,

$$\mathbf{m}_c(k) = \arg\min||\mathbf{x}(k) - \mathbf{m}_i(k)||, \quad (i = 1, 2, \ldots, N) \tag{7.35}$$

Equation (7.35) defines the nearest $\mathbf{m}_i(k)$ to $\mathbf{x}(k)$ and is denoted by $\mathbf{m}_c(k)$. The values for $\mathbf{m}_i(k)$ that approximately minimize the misclassification errors in the above nearest-neighbor classification can be found as asymptotic values in the following learning process. Let $\mathbf{x}(k)$ be a sample of input, and let the $\mathbf{m}_i(k)$ represent the sequences of the $\mathbf{m}_i(k)$ in the discrete-time domain.

Starting with properly defined initial values, the following equations define the basic LVQ process. If $\mathbf{x}(k)$ and $\mathbf{m}_c(k)$ belong to the same class,

$$\mathbf{m}_c(k+1) = \mathbf{m}_c(k) + a(k)(\mathbf{x}(k) - \mathbf{m}_c(k)) \tag{7.36}$$

If $\mathbf{x}(k)$ and $\mathbf{m}_c(k)$ belong to the different class,

$$\mathbf{m}_c(k+1) = \mathbf{m}_c(k) - a(k)(\mathbf{x}(k) - \mathbf{m}_c(k)) \tag{7.37}$$

Otherwise, for $i \neq c$,

$$\mathbf{m}_c(k+1) = \mathbf{m}_i(k) \tag{7.38}$$

Here, $0 < \alpha(k) < 1$, and learning rate $\alpha(k)$ may be constant or monotonically decreases with time. It is recommended that the initial value of $\alpha(k)$ is not to be taken more than 0.1 in accordance with the basic LVQ algorithm.

The data can be learned in a self-organized fashion by only choosing the learning rate and the number of output neurons. Also, the feature map capability has an added advantage, and it can visualize the classification result.

Example 7.9 Classify the iris data according to their classes using LVQ method as shown in Fig. 7.16.

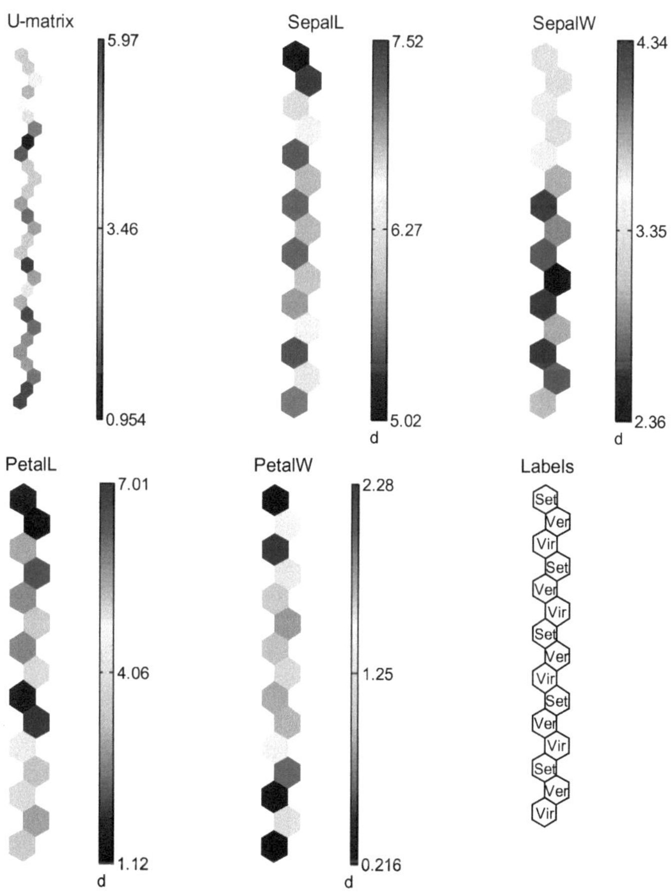

Fig. 7.16 Visualization classification of the LVQ3 of *iris data*: U-matrix (*top left*), then component planes, and map unit labels on *bottom right*

7.8 Radial Basis Function (RBF) Neural Network

Currently, the RBFN is attracting a lot of interest due to its rapid learning, generality, and simplicity (Yang et al. 2002). A RBFN can be trained fast while exhibiting none of the conventional back-propagation algorithm's training problems such as saturation and local minima. This network utilizes the overlapping localized regions formed by simple kernel functions (usually a Gaussian function) to handle complex decision regions. There are two main subdivisions of the regression problems in statistics: parametric and nonparametric. In the parametric regression, the form of the functional relationship between the dependent and independent variables is known. On the other hand, in the nonparametric regression, there is none or very little a priori knowledge about the form of the true function that is being estimated. In the RBFN, if the outputs of the estimated function are interpreted as being proportional to the probability that the input belongs to the corresponding class, classification problems can be made to look like nonparametric regression. The training output values are vectors of length equal to the number of classes and containing a single one and otherwise zeros. After training, the network responds to a new pattern with continuous values in each component of the output vector that can be interpreted as being proportional to the class probability.

The most basic form of the RBFN involves three layers with entirely different roles (Fig. 7.17). The input layer is made up of source nodes (sensory units) that connect the network to its environment. The middle layer, the only hidden layer in the network, applies a nonlinear transformation from the input space to the hidden space; in most applications, the hidden space has high dimensionality. The output layer is linear, supplying the response of the network to the information applied to the input layer. Each middle node evaluates a kernel function on the incoming input, and the classifier output is simply a weighted linear summation of the kernel function. The kernel function gives the highest output when the input is close to its center and a monotonically decreasing output as the input moves away from the center. The center that is the weights between the input and the hidden layers is usually computed by a clustering algorithm such as *k*-means and the SOFM method. The weights between the hidden and output layers are computed by the gradient descent method.

For the centers in the hidden layer, there are two methods of making centers: the direct use of training data and the clustering method. It determines the function in the hidden layer with a spread constant σ which determines the property of the RBFN. Usually, it is calculated as the standard deviation $std(\mathbf{x})$ of input vectors $\boldsymbol{x}$ in the training data.

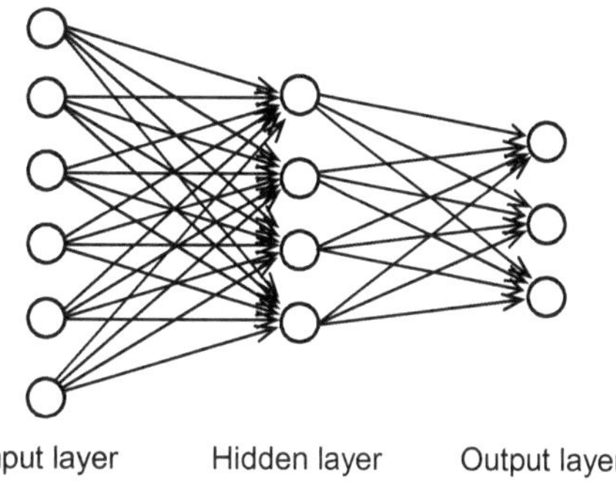

Fig. 7.17 The structure of the RBFN

$$\sigma = \gamma\, std(\mathbf{x}) \tag{7.39}$$

where γ is an arbitrary constant with an initial value of 1.

It makes an important effect on the classification capability of the RBFN.

$$h(\mathbf{x}, c_i, \sigma) = \exp(-||\mathbf{x}-c_i||^2/\sigma^2) \tag{7.40}$$

where $h()$ is a radial basis function and c_i is the center of the ith cluster.

The weights of the neuron between the hidden neuron and the output neuron are calculated by using Eq. (7.41), where, the pseudo inverse has the same role as the network with the least mean-square (LMS) method.

$$\mathbf{y} = \mathbf{W}\mathbf{A}(\mathbf{x}, c, \sigma) \rightarrow \mathbf{W} = \mathbf{y}\mathbf{A}^T(\mathbf{A}\mathbf{A}^T)^{-1} \tag{7.41}$$

where **W** is the weight matrix, **A** is the output of the hidden layer, and **y** is the output of the output layer.

The classification procedure using the trained RBFN is as follows:

Step 1: Calculation of the Euclidean distances between the centers and the inputs,
Step 2: Calculation of the output of the hidden layer by the radial basis function, Eq. (7.40),
Step 3: Linear combination of hidden layer outputs with the weighting values,
Step 4: Decision of the input data class. The class is determined by the maximum element in the output vector.

Figure 7.18 shows the flowchart of the RBFN.

Example 7.10 Perform the classification process using RBFN method as shown in Fig. 7.19.

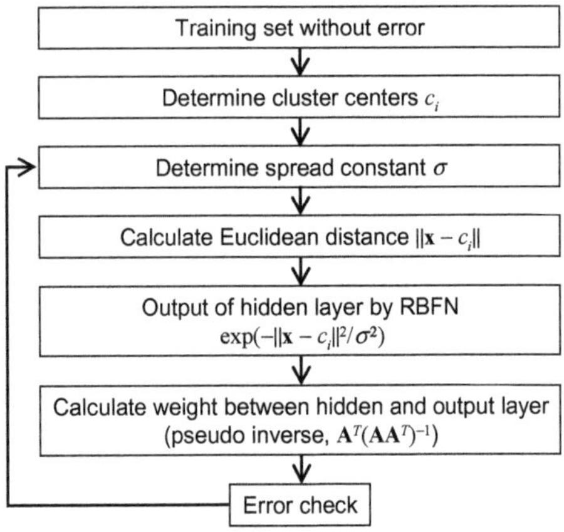

Fig. 7.18 Flowchart of the RBFN

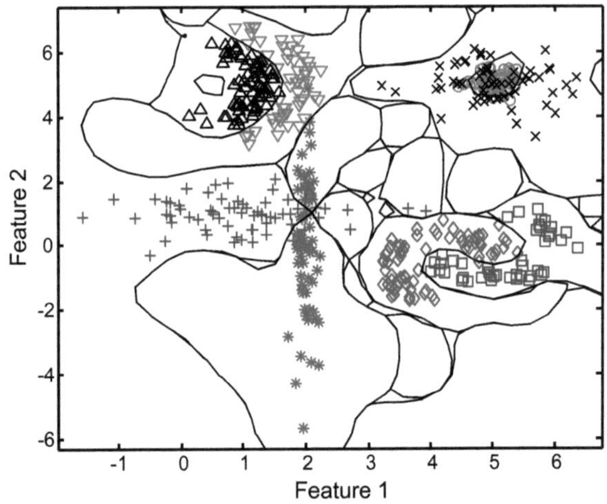

Fig. 7.19 Multi-class classification using RBF neural network

7.9 ART Kohonen Neural Network (ART-KNN)

Presently, the fault diagnosis is increasingly intelligent with wide applications of artificial neural networks (ANNs). However, "off-line" NNs are unable to well adapt to unexpected changes in the environment. Furthermore, the data of the data set used to train networks need to be added, as new fault occurs. In this case, the "off-line" network requires to be retrained using the complete data set. This can result in a time-consuming and costly process (Liobet et al. 1999). In the real world, although part of fault signals can be obtained, it is very difficult to compose the training data set representing the features of all faults. Nobody knows what will happen next time. These characteristics limit the applications of "off-line" NNs in fault diagnosis field. The NNs for fault diagnosis of machinery are required to learn gradually the knowledge in operating process, and to have the adaptive function expanding the knowledge continuously without the loss of the previous knowledge during learning new knowledge. A human brain is able to learn many new events without necessarily forgetting events that occurred in the past. So we want an intelligent system capable of adapting "online" to changes in the environment, and the system should be able to deal with the so-called the stability-plasticity dilemma (Carpenter and Grossberg 1987, 1988, 1992). That is, the system should be designed to have some degree of plasticity to learn new events in a continuous manner and should be stable enough to preserve its previous knowledge and to prevent new events destroying the memories of prior training. As a solution to this problem, the adaptive resonance theory (ART) networks were developed and have been applied with some success to real-time training and classification. The ART network is a neural network that self-organizes stable recognition codes in real time in response to arbitrary sequences of input patterns and is a vector classifier as the mathematical model for the description of fundamental behavioral functions of the biological brain such as the learning, parallel and distributed information storage, short- and long-term memory, and pattern recognition.

The Kohonen neural network (KNN) is also called self-organizing feature map (SOFM) network, and it defines a forward two-layer neural network that implements a characteristic nonlinear projection from the high-dimensional space of sensory or other input signals onto a low-dimensional array of neurons (Kohonen 1995). The KNN consists of three major steps: competition, cooperation, and adaptation. In the first step, the network compares the output values with the input vector according to a chosen discriminating function. Among the output neurons, only one particular neuron with the closest relationship to the input vector is picked up and labeled as the winning (best-matching) neuron. Once the winning neuron is picked up, the next step is to select those neurons within a predefined neighborhood. Only the weights of those neurons defined within the topological neighborhood of the winning neuron will be updated. The synaptic weights of neurons outside the neighborhood will remain unchanged. As the winning neuron best matches the input vector in the sense of the Euclidean distance, the above learning strategy is able to move the synaptic weight vectors toward the distribution of the input vectors.

In this section, it describe a fault diagnosis network, the adaptive resonance theory Kohonen neural network (ART-KNN) (Yang et al. 2004a, b), which does not destroy the initial learning and can adapt the additional training data that are suitable for fault diagnosis of rotating machinery.

The characteristics of ART networks are suitable for the condition monitoring and fault diagnosis. There are two general classes of ART networks: ART1, and

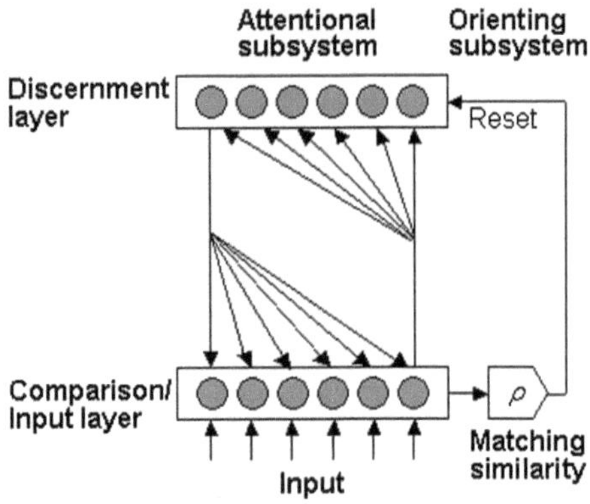

Fig. 7.20 Architecture of the ART-KNN network

ART2 and ART3. The ART1 is for classifying the binary input patterns, while the ART2 and ART3 are for the binary and decimal input patterns.

But the ART networks have some disadvantages for the fault diagnosis. The input patterns to the input layer are normalized before passing through the adaptive filter defined the pathways from input layer to discernment layer. Because the absolute values of input signals represent only the image brightness and sound level for the image and sound classifications, the relative value normalized is important to analyze the image and sound discrimination. But, the absolute value of vibration signal is important information for the fault diagnosis. When it is normalized, some important information to detect the faults may be lost. At the same time, the ART2 and ART3 adequately control the noise of the input signal; the initial signal becomes fussy after filtering. So the features of fault signals are destroyed to some degree.

In this paper, the proposed ART-KNN combines the theory of ART with Kohonen's learning strategy to realize machinery fault diagnosis. The architecture of ART-KNN is shown in Fig. 7.20. It is similar to ART1's, excluding the adaptive filter. ART-KNN is also formed by two major subsystems: the attentional subsystem and the orienting subsystem. Two interconnected layers, discernment layer and comparison layer, which are fully connected bottom-up and top-down, comprise the attentional subsystem. The application of a single input vector leads to the patterns of neural activity in both layers. The activity in discernment nodes reinforces the activity in comparison nodes due to top-down connections. The interchange of bottom-up and top-down information leads to a resonance in neural activity. As a result, critical features in comparison are reinforced and have the greatest activity. The orienting subsystem is responsible for generating a reset signal to discernment when the bottom-up input pattern and top-down template pattern mismatch at comparison, according to a similarity. In other words, once it has detected that the input pattern is novel, the orienting subsystem must prevent the previously organized category neurons in discernment from learning this pattern (via a reset signal). Otherwise, the category will become increasingly non-specific. When a mismatch is detected, the network adapts its structure by immediately storing the novelty in additional weights. The similarity criterion is set by the value of the similarity parameter. A high value of the similarity parameter means that only a slight mismatch will be tolerated before a reset signal is emitted. On the other hand, a small value means that large mismatches will be tolerated. After the resonance check, if a pattern match is detected according to the similarity parameter, the network changes the weights of the winning node.

The learning strategy is introduced by the Kohonen neural network. The Euclidean distances of all weights between input vector X and each neuron of the discernment layer are evaluated as the similarity given by Eq. (7.42), and the smallest one becomes the winning neuron.

$$||B_J - X|| < ||B_j - X||; \quad (j, J = 1, 2, \ldots, n; j \neq J) \tag{7.42}$$

where B_j is the weight of jth neuron in the discernment layer and B_J is the weight of the winning neuron.

After producing the winning neuron, input vector X returns to the comparison layer. The absolute similarity S is calculated by

$$S = \frac{||B_J|| - ||B_J - X||}{||B_J||} \tag{7.43}$$

If B_J and X in Eq. (7.43) are same, $||B_J - X||$ is equal to 0, and S is 1. The larger the Euclidean distance between B_J and X is, the smaller the S is. A parameter ρ is introduced as the evaluation criterion of similarity. If $S > \rho$, it indicates that the Jth cluster is sufficiently similar to X. So X belongs to the Jth cluster. In order to make the weight more accurate to represent the corresponding cluster, the weight of the Jth cluster is improved by the following equation:

$$B_J = (n * B_{J0} + X)/(n + 1) \tag{7.44}$$

where B_J is the enhanced weight, B_{J0} is the origin weight, and n is the changed time.

On the contrary, as $S < \rho$, it means that X is much different than the Jth cluster. Thus, there is no cluster that matches X in the original network. The network needs one more neuron to remember this new case by resetting the discernment layer. The weight of the new neuron is given by:

$$B_{n+1} = X \tag{7.45}$$

The detail procedure of ART-KNN training can be summarized as follows:

1. Start with empty network structure.
2. Input training data X ($Q \times R$).
3. Setting initial structure parameters: criterion value (p) and relative origin point (O).
4. Generating a new neuron.

If No. of N (neuron) < 1 then

N (weights) = ones ($1 \times R$), R is same size of input vector

Else

N (weights) = Input vector ($1 \times R$)

End

5. Distance calculation between input vector and neurons, and selection the winner neuron.

D = dis ($X(i)$, N);

D_winner = Min (D);

6. Similarity measurement and comparing with criterion value.

$$S = \frac{\text{dis}(N\_\text{winner},\text{O})\text{-}D\_\text{winner}}{\text{dis}(N\_\text{winner},\text{O})} < \rho$$

If $S < \rho$ then

Go to step 4

Else

Existing winner neuron weights updating

N (weights) = (N (weights) + $X(i)$)/(n + 1), n is updating time.

End

7. Training completion

7.10 Support Vector Machines (SVMs)

SVMs are a relatively new computational learning method based on the statistical learning theory presented by Vapnik (1999). In SVMs, original input space is mapped into a high-dimensional dot product space called a feature space, and in the feature space, the optimal hyperplane is determined to maximize the generalization ability of the classifier. The maximal hyperplane is found by exploiting the optimization theory and respecting insights provided by the statistical learning theory.

SVMs have the potential to handle very large feature spaces, because the training of SVMs is carried out so that the dimension of classified vectors does not have as distinct an influence on the performance of SVM as it has on the performance of conventional classifier. That is why it is noticed to be especially efficient in large classification problem. This will also benefit in fault classification, because the number of features to be the basis of fault diagnosis may not have to be limited. Also, SVM-based classifiers are claimed to have good generalization properties compared to conventional classifiers, because in training SVM classifier the so-called structural misclassification risk is to be minimized, whereas traditional classifiers are usually trained so that the empirical risk is minimized. The performance of SVMs in various classification task is reviewed, e.g., in Christianini and Shawe-Taylor (2000).

Given the data input $\mathbf{x}_i$ (i = 1, 2, …, M), M is the number of samples. The samples are assumed to have two classes, namely positive class and negative class. Each of the classes is associated with labels $y_i = 1$ for positive class and $y_i = -1$ for negative class, respectively. In the case of linear data, it is possible to determine the hyperplane $f(\mathbf{x}) = 0$ that separates the given data

$$f(\mathbf{x}) = \mathbf{w}^T\mathbf{x} + b = \sum_{j=1}^{M} w_j x_j + b = 0 \tag{7.46}$$

where $\mathbf{w}$ is the M-dimensional vector and b is a scalar. The vector $\mathbf{w}$ and scalar b are used to define the position of separating hyperplane. The decision function is made using sign $f(x)$ to create separating hyperplane that classifies the input data in either positive class or negative class.

A distinctly separating hyperplane should satisfy the constraints

$$\left.\begin{aligned} f(x_i) &= 1 \quad \text{if } y_i = 1 \\ f(x_i) &= -1 \quad \text{if } y_i = -1 \end{aligned}\right\} \tag{7.47}$$

or it can be presented in complete equation

$$y_i f(\mathbf{x}_i) = y_i(\mathbf{w}^T\mathbf{x}_i + b) \geq 1 \quad \text{for } i = 1, 2, \ldots, M \tag{7.48}$$

The separating hyperplane that creates the maximum distance between the plane and the nearest data, i.e., the maximum margin, is called the optimal separating hyperplane. An example of the optimal hyperplane of two data sets is presented in Fig. 7.21.

In Fig. 7.21, a series data points for two different classes of data are shown, black squares for negative class and white circles for positive class. The SVMs try to place a linear boundary between the two different classes and orientate it in such a way that the margin represented by the dotted line is maximized. Furthermore, SVMs attempt to orientate the boundary to ensure that the distance between the boundary and the nearest data point in each class is maximal. Then, the boundary is

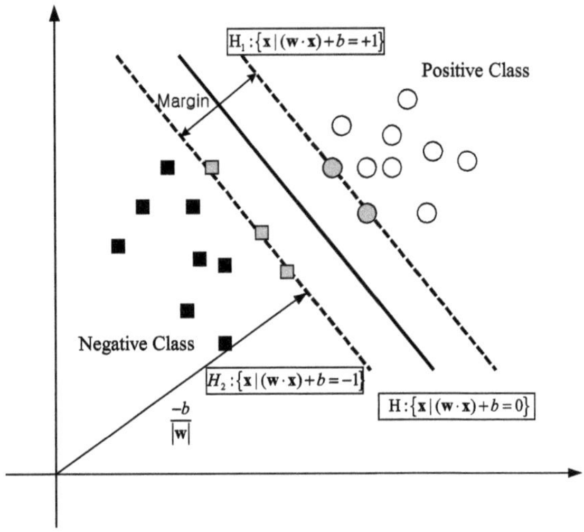

Fig. 7.21 Classification of two classes using SVM

placed in the middle of this margin between two points. The nearest data points that used to define the margin are called support vectors, represented by the gray circles and squares. When the support vectors have been selected, the rest of the feature set is not required, as the support vectors can contain all the information based need to define the classifier. From the geometry, the geometrical margin is found to be $\|\mathbf{w}\|^{-2}$.

Taking into account the noise with slack variables ξ_i and the error penalty C, the optimal hyperplane separating the data can be obtained as a solution to the following optimization problem:

$$\text{Minimize } \frac{1}{2}\|\mathbf{w}\|^2 + C\sum_{i=1}^{M}\xi_i \tag{7.49}$$

$$\text{subject to} \begin{cases} y_i(\mathbf{w}^T\mathbf{x}_i + b) \geq 1 - \xi_i, & i = 1, \ldots, M \\ \xi_i \geq 0 & i = 1, \ldots, M \end{cases} \tag{7.50}$$

where ξ_i measures the distance between the margin and the examples $\mathbf{x}_i$ that lie on the wrong side of the margin. The calculation can be simplified by converting the problem with Kuhn–Tucker condition into the equivalent Lagrangian dual problem, which will be

$$\text{Minimize } L(\mathbf{w}, b, \boldsymbol{\alpha}) = \frac{1}{2}\|\mathbf{w}\|^2 - \sum_{i=1}^{M}\alpha_i y_i(\mathbf{w}\cdot\mathbf{x}_i + b) + \sum_{i=1}^{M}\alpha_i \tag{7.51}$$

The task is to minimize Eq. (7.51) with respect to **w** and b, while requiring the derivatives of L to α to vanish. At optimal point, we have the following saddle point equations:

$$\frac{\partial L}{\partial \mathbf{w}} = 0, \quad \frac{\partial L}{\partial b} = 0 \tag{7.52}$$

which replace into the form

$$\mathbf{w} = \sum_{i=1}^{M} \alpha_i y_i \mathbf{x}_i, \quad \sum_{i=1}^{M} \alpha_i y_i = 0 \tag{7.53}$$

From Eq. (7.53), we find that **w** is contained in the subspace spanned by the $\mathbf{x}_i$. Substituting Eq. (7.53) into Eq. (7.51), we get the dual quadratic optimization problem

$$\begin{aligned} &\text{Maximize } L(\alpha) = \sum_{i=1}^{M} \alpha_i - \frac{1}{2} \sum_{i,j=0}^{M} \alpha_i \alpha_j y_i y_j \mathbf{x}_i \cdot \mathbf{x}_j \\ &\text{subject to } \alpha_i \geq 0, \quad i = 1, \ldots, M. \end{aligned} \tag{7.54}$$

$$\sum_{i=1}^{M} \alpha_i y_i = 0 \tag{7.55}$$

Thus, by solving the dual optimization problem, one obtains the coefficients α_i which are required to express the **w** to solve Eq. (7.49). This leads to the nonlinear decision function.

$$f(\mathbf{x}) = \operatorname{sign}\left(\sum_{i,j=1}^{M} \alpha_i y_i (\mathbf{x}_i \mathbf{x}_j) + b \right) \tag{7.56}$$

SVMs can also be used in nonlinear classification tasks with the application of kernel functions. The data to be classified are mapped onto a high-dimensional feature space, where the linear classification is possible. Using the nonlinear vector function $\mathbf{\Phi}(\mathbf{x}) = (\phi_1(\mathbf{x}), \ldots, \phi_l(\mathbf{x}))$ to map the n-dimensional input vector **x** onto l-dimensional feature space, the linear decision function in dual form is given by:

$$f(\mathbf{x}) = \operatorname{sign}\left(\sum_{i,j=1}^{M} \alpha_i y_i (\mathbf{\Phi}^T(\mathbf{x}_i) \cdot \mathbf{\Phi}(\mathbf{x}_j)) + b \right) \tag{7.57}$$

Working in the high-dimensional feature space enables the expression of complex functions, but it also generates the problem. Computational problem occurs due to the large vectors, and the overfitting also exists due to the high

Table 7.1 Formulation of kernel functions

Kernel	$K(\mathbf{x}, \mathbf{x}_j)$
Linear	$\mathbf{x}^{\mathrm{T}} \cdot \mathbf{x}_j$
Polynomial	$(\gamma\ \mathbf{x}^{\mathrm{T}} \cdot \mathbf{x}_j + \mathrm{r})^d,\ \gamma > 0$
Gaussian RBF	$\exp(-\ \|\mathbf{x} - \mathbf{x}_j\|^2/2\gamma\ ^2)$

dimensionality. The latter problem can be solved by using the kernel function. Kernel is a function that returns a dot product of the feature space mappings of the original data points, stated as $K(\mathbf{x}_i, \mathbf{x}_j) = (\mathbf{\Phi}^T(\mathbf{x}_i) \cdot \mathbf{\Phi}_j(\mathbf{x}_j))$. When applying a kernel function, the learning in the feature space does not require explicit evaluation of $\mathbf{\Phi}$ and the decision function will be

$$f(\mathbf{x}) = \mathrm{sign}\left(\sum_{i,j=1}^{M} \alpha_i y_i K(\mathbf{x}_i, \mathbf{x}_j) + b\right) \tag{7.58}$$

Any function that satisfies the Mercer's theorem (Vapnik 1999) can be used as a kernel function to compute a dot product in feature space. There are different kernel functions used in SVMs, such as linear, polynomial, and Gaussian RBF. The selection of the appropriate kernel function is very important, since the kernel defines the feature space in which the training set examples will be classified. The definition of legitimate kernel function is given by the Mercer's theorem. The function must be continuous and positive definite. In this work, linear, polynomial, and Gaussian RBF functions are evaluated and formulated in Table 7.1.

7.10.1 Wavelet SVM

The idea of wavelet analysis is to approach a function or signal using a family of functions which are produced by translation and dilatation of the mother wavelet function $\psi_{a,b}(x)$

$$\psi_{a,b}(x) = |a|^{-1/2}\psi\left(\frac{x-b}{a}\right) \tag{7.59}$$

where x, a, $b \in R$, a is the dilatation factor, and b is the translation factor. The wavelet transform of any function $f(x)$ can be expressed as follows:

$$W_{a,b}(f) = \langle f(x), \psi_{a,b}(x)\rangle, \quad f(x) \in L_2(R) \tag{7.60}$$

where the notation $\langle\ ,\ \rangle$ refers to the inner product in $L_2(R)$.

Equation (7.60) means that any function $f(x)$ can be decomposed on a wavelet basis $\psi_{a,b}(x)$ if it satisfies the condition (Daubechies 1990)

$$C_\psi = \int 0^\infty \frac{|H(\omega)|^2}{|\omega|} \mathrm{d}\omega < \infty \tag{7.61}$$

where $H(\omega)$ is the Fourier transform of $\psi_{a,b}(x)$.

Following (Daubechies 1990), the function $f(x)$ can be reconstructed as follows:

$$f(x) = \frac{1}{C_\psi} \int_{-\infty}^{\infty} \int_{0}^{\infty} W_{a,b}(f) \quad \psi_{a,b}(x) \frac{\mathrm{d}a}{a^2} \mathrm{d}b \tag{7.62}$$

To approximate Eq. (7.62), then the finite can be written as follows:

$$\hat{f}(x) = \sum_{i=1}^{l} W_i \psi_{a_i,b_i}(x) \tag{7.63}$$

Using Eq. (7.63), $f(x)$ can eventually be approximated by $\hat{f}(x)$.

For a common multidimensional wavelet function, the mother wavelet can be given as the product of one-dimensional (1-D) wavelet function (Zhang and Benveniste 1992)

$$\psi(x) = \prod_{i=1}^{N} \psi(x_i) \tag{7.64}$$

where $\mathbf{x} = (x_1, \ldots, x_N) \in R^N$. So, every 1-D wavelet mother $\psi(x)$ must satisfy Eq. (7.61).

Recalling the decision function for SVM in Eq. (7.58), the dot product can be replaced using kernel function as it was done by Vapnik (1995), so that $K(\mathbf{x}, \mathbf{x}') = K(\langle \mathbf{x} \cdot \mathbf{x}' \rangle)$. In SVM theory, any function which satisfies the Mercer's condition can serve as a kernel function (Christianini and Shawe-Taylor 2000).

Suppose K is a continuous symmetric function on R^N, such that integral operator T_K: $L_2(R^N) \rightarrow L_2(R^N)$,

$$(T_K)f(\cdot) = \int R^d K(\cdot, \mathbf{x}) f(\mathbf{x}) \, \mathrm{d}\mathbf{x} \tag{7.65}$$

is positive. Let $\phi_i \in L_2(R^N)$ be the eigenfunction of T_k associated with the eigenvalue $\lambda_i \geq 0$ and be normalized in such a way that $||\phi_i||_{L_2} = 1$, and then the kernel function $K(\mathbf{x}, \mathbf{x}')$ can be expanded as follows:

$$K(\mathbf{x}, \mathbf{x}') = \sum_{i=1}^{\infty} \lambda_i \phi_i(\mathbf{x}) \, \phi_i(\mathbf{x}') \tag{7.66}$$

and must satisfy the positivity condition of the following integral (Christianini and Shawe-Taylor 2000)

$$\int_{L_2 \otimes L_2}\int K(\mathbf{x},\mathbf{x}')f(\mathbf{x})f(\mathbf{x}')\mathrm{d}\mathbf{x}\,\mathrm{d}\mathbf{x}' \geq 0 \;\forall f \in L_2(R^N) \tag{7.67}$$

For building a new kernel using wavelet, it may be helpful to refer to the frame theory, introduced by Duffin and Schaeffer (1952), which is an extension of the normalized orthogonal basis. In the frame theory, one can reconstruct perfectly a function f in a Hilbert space H from its inner product $\langle , \rangle$ with family vectors $\{\psi_k\}$ if they satisfy

$$A\|f\|^2 \leq \sum_k |\langle f, \bar{\psi}_k\rangle|^2 \leq B\|f\|^2 \tag{7.68}$$

where the constants A and B satisfy the condition $0 < A \leq B < \infty$.

Any function in a Hilbert space can be decomposed as follows:

$$f = \sum_k \langle f, \bar{\psi}_k\rangle\psi_k = \sum_k \langle f, \psi_k\rangle\bar{\psi}_k \tag{7.69}$$

where $\bar{\psi}_k = (T^*T)^{-1}\psi_k$ is the dual frame of ψ_k and T is the frame operator.

In $L_2(R^N)$, if $f = \{\psi_i\}$ is a frame and $\{\lambda_i\}$ is a positive increasing sequence, a function $K(\mathbf{x},\mathbf{x}')$ can be given by:

$$K(\mathbf{x},\mathbf{x}') = \sum_{i=1}^{\infty} \lambda_i\,\psi_i(\mathbf{x})\,\psi_i(\mathbf{x}') \tag{7.70}$$

Equation (7.70) is similar to Eq. (7.66) since both of them satisfy the condition of kernel function. Moreover, a mother wavelet $\psi_{a,b}(x)$ is called a frame wavelet if $\psi \in L_2(R^N)$, $a > 1$, $b > 0$ and the family function

$$\{\psi_{mn}\} = \{D_{am}T_{nb}\psi\} \tag{7.71}$$

where D and T are unitary dilatation operator and unitary translation operator, respectively, while a is the scale parameter and b is the translation parameter.

A wavelet kernel function can be constructed by any mother wavelet which can generate frame wavelet while satisfying the Mercer's condition in Eq. (7.67). In addition to the inner product, there exists a kernel called translation-invariant kernel (Smola et al. 1998) such that

$$K(\mathbf{x},\mathbf{x}') = K(\langle\mathbf{x} - \mathbf{x}'\rangle) \tag{7.72}$$

If the translation-invariant kernel is admissible in SVM kernel function, then the necessary and sufficient condition of the Mercer's theorem must be satisfied. The other theorem stated that a translation-invariant kernel is an admissible support vector (SV) kernel if and only if the following Fourier transforms (Smola et al. 1998)

$$F[K](\omega) = (2\pi)^{-N/2} \int_{R^N} \exp(-j(\omega \cdot \mathbf{x}))K(\mathbf{x})\,d\mathbf{x} \tag{7.73}$$

are nonnegative. Based on the mother wavelet, the wavelet kernel which satisfies the translation-invariant theorem can be given as follows:

$$K(\mathbf{x}, \mathbf{x}') = K(\mathbf{x} - \mathbf{x}') = \prod_{i==1}^{N} \psi\left(\frac{x_i - x_i'}{a_i}\right) \tag{7.74}$$

The construction of wavelet kernel function using Haar, Daubechies, and Symmlet is shown in Fig. 7.22.

7.10.2 Multi-class Classification

The above discussion deals with binary classification where the class labels can take only two values: 1 and −1. In the real-world problem, however, we find more than two classes for example: In fault diagnosis of rotating machineries, there are several fault classes such as mechanical unbalance, misalignment, and bearing faults.

The earliest used implementation for SVM multi-class classification is one-against-all methods. It constructs k SVM models where k is the number of classes. The ith SVM is trained with all of the examples in the ith class with positive labels and all the other examples with negative labels. Thus, given l training data (x_1, y_1), …, (x_l, y_l), where $x_i \in R^n$, $i = 1, \ldots, l$ and $y_i \in \{1, \ldots, k\}$ is the class of x_i, the ith SVM solves the following problem:

$$\text{Minimize}: \frac{1}{2}||\mathbf{w}^i||^2 + C\sum_{i=1}^{l} \xi_j^i(\mathbf{w}^i)^T \tag{7.75}$$

$$\text{subject to}: (\mathbf{w}^i)^T\phi(\mathbf{x}_j) + b^i \geq 1 - \xi_j^i, \quad \text{if } y = i \tag{7.76}$$

$$(\mathbf{w}^i)^T\phi(\mathbf{x}_j) + b^i \leq -1 + \xi_j^i, \quad \text{if } y \neq i \tag{7.77}$$

$$\xi_J^i \geq 0, \quad j = 1, \ldots, l \tag{7.78}$$

where the training data $\mathbf{x}_i$ are mapped to a higher dimensional space by function ϕ and C is the penalty parameter.

Minimizing Eq. (7.75) means we would like to maximize $2/||\mathbf{w}_i||$, the margin between two groups of data. When data are not separable, there is a penalty term $C\sum_{i=1}^{l} \xi_{i,i}$ which can reduce the number of training errors.

Another major method is called one-against-one method. This method constructs $k(k-1)/2$ classifiers where each one is trained on data from two classes. For

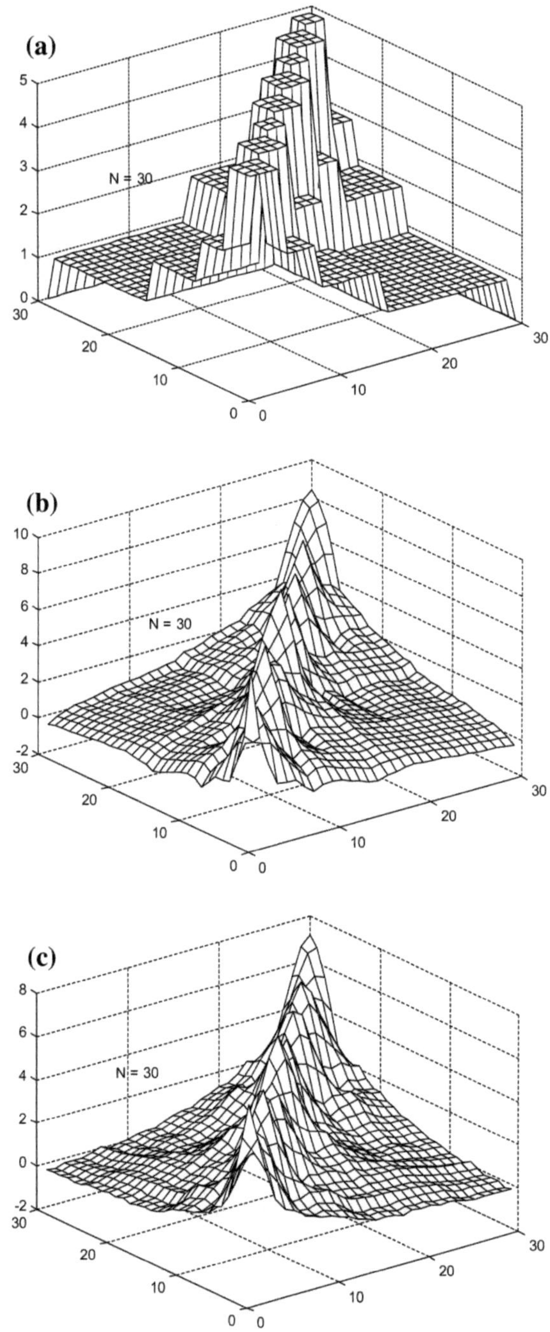

Fig. 7.22 Wavelet kernel function. **a** Haar kernel. **b** Daubechies kernel. **c** Symmlet kernel

training data from the ith and the jth classes, we solve the following binary classification problem:

$$\text{Minimize}: \frac{1}{2}||\mathbf{w}^{ij}||^2 + C\sum_t \xi_t^{ij}(\mathbf{w}^{ij})^T \quad (7.79)$$

$$\text{subject to}: (\mathbf{w}^{ij})^T\phi(\mathbf{x}_t) + b^{ij} \geq 1 - \xi_t^{ij}, \quad \text{if } y_t = i \quad (7.80)$$

$$(\mathbf{w}^{ij})^T\phi(\mathbf{x}_t) + b^{ij} \leq -1 + \xi_t^{ij}, \quad \text{if } y_t = j \quad (7.81)$$

$$\xi_t^{ij} \geq 0, \quad j = 1, \ldots, l \quad (7.82)$$

There are different methods for doing the future testing after all $k(k - 1)/2$ classifiers are constructed. After some tests, the decision is made using the following strategy: If sign $((\mathbf{w}^{ij})^T\phi(\mathbf{x}) + b^{ij})$ says $\mathbf{x}$ is in the ith class, then the vote for the ith class is added by one. Otherwise, the jth is increased by one. Then, $\mathbf{x}$ is predicted in the class using the largest vote. The voting approach described above is also called as max win strategy.

7.10.3 Sequential Minimal Optimization (SMO)

Vapnik (1982) describes a method which used the projected conjugate gradient algorithm to solve the SVM-QP problem, which has been known as *chunking*. The chunking algorithm uses the fact that the value of the quadratic form is the same if you remove the rows and columns of the matrix that corresponds to zero Lagrange multipliers. Therefore, chunking seriously reduces the size of the matrix from the number of training examples squared to approximately the number of nonzero Lagrange multipliers squared. However, chunking still cannot handle large-scale training problems, since even this reduced matrix cannot fit into memory. Osuna et al. (1997) proved a theorem which suggests a whole new set of QP algorithms for SVMs. The theorem proves that the large QP problem can be broken down into a series of smaller QP subproblems. Sequential minimal optimization (SMO) proposed by Platt (1999) is a simple algorithm that can be used to solve the SVM-QP problem without any additional matrix storage and without using the numerical QP optimization steps. This method decomposes the overall QP problem into QP subproblems using the Osuna's theorem to ensure convergence. In this paper, the SMO is used as a solver and detail descriptions can be found in Platt (1999).

In order to solve the two Lagrange multipliers α_1, α_2, SMO first computes the constraints on these multipliers and then solves for the constrained minimum. For convenience, all quantities that refer to the first multiplier will have a subscript 1, while all quantities that refer to the second multiplier will have a subscript 2. The new values of these multipliers must lie on a line in (α_1, α_2) space and in the box defined by $0 \leq \alpha_1, \alpha_2 \leq C$.

$$\alpha_1 y_1 + \alpha_2 y_2 = \alpha_1^{\text{old}} y_1 + \alpha_2^{\text{old}} y_2 = \text{constant} \tag{7.83}$$

Without loss of generality, the algorithm first computes the second Lagrange multipliers α_2^{new} and successively uses it to obtain $\alpha_1{}^{\text{new}}$. The box constraint $0 \leq \alpha_1, \alpha_2 \leq C$, together with the linear equality constraint $\Sigma \alpha_i\, y_i = 0$, provides a more restrictive constraint on the feasible values for α_2^{new}. The boundary of feasible region for α_2 can be applied as follows:

$$\text{If } y_1 \neq y_2; L = \max(0, \alpha_2^{\text{old}} - \alpha_1^{\text{old}}), H = \min(C, C + \alpha_2^{\text{old}} - \alpha_1^{\text{old}}) \tag{7.84}$$

$$\text{If } y_1 = y_2; L = \max(0, \alpha_1^{\text{old}} + \alpha_2^{\text{old}} - C), H = \min(C, C + \alpha_1^{\text{old}} + \alpha_2^{\text{old}}) \tag{7.85}$$

The second derivative of the objective function along the diagonal line can be expressed as follows:

$$\eta = K(\mathbf{x}_1, \mathbf{x}_1) + K(\mathbf{x}_2, \mathbf{x}_2) - 2K(\mathbf{x}_1, \mathbf{x}_2) \tag{7.86}$$

Under normal circumstances, the objective function will be positive definite, there will be a minimum along the direction of the linear equality constraint, and η will be greater than zero. In this case, SMO computes the minimum along the direction of the constraint:

$$\alpha_2^{\text{new}} = \alpha_2^{\text{old}} + \frac{y_2(E_1^{\text{old}} - E_2^{\text{old}})}{\eta} \tag{7.87}$$

where E_i is the prediction error on the ith training example. As a next step, the constrained minimum is found by clipping the unconstrained minimum to the ends of the line segment:

$$\alpha_2^{\text{new,clipped}} = \begin{cases} H & \text{if } \alpha_2^{\text{new}} \geq H \\ \alpha_2^{\text{new}} & \text{if } L < \alpha_2^{\text{new}} < H \\ L & \text{if } \alpha_2^{\text{new}} \leq L \end{cases} \tag{7.88}$$

Now, let $s = y_1\, y_2$. The value of α_1^{new} is computed from the new α_2^{new}:

$$\alpha_1^{\text{new}} = \alpha_1^{\text{old}} + s(\alpha_2^{\text{old}} - \alpha_2^{\text{new}}) \tag{7.89}$$

Solving Eq. (7.26) for the Lagrange multipliers does not determine the threshold b of the SVM, so b must be computed separately. The following thresholds b_1, b_2 are valid when the new α_1, α_2 are not at the each bound, because it forces the output of the SVM to be y_1, y_2 when the input is $\mathbf{x}_1$, $\mathbf{x}_2$, respectively.

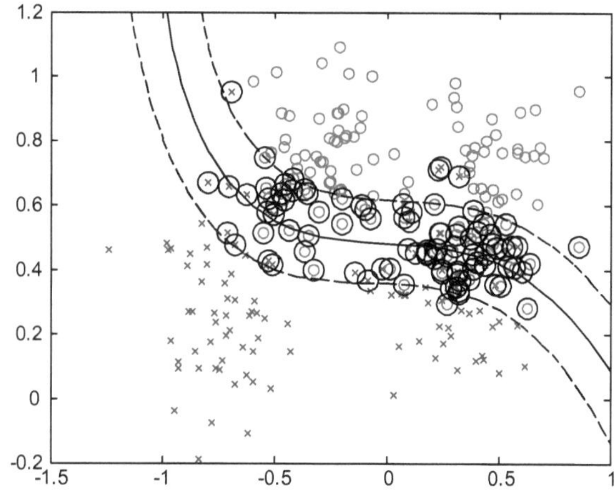

Fig. 7.23 Binary SVM classifier trained on *riply's data*

$$b_1 = E_1 + y_1(\alpha_1^{\text{new}} - \alpha_1^{\text{old}})K(\mathbf{x}_1, \mathbf{x}_1) + y_2(\alpha_2^{\text{new,clipped}} - \alpha_2^{\text{old}})K(\mathbf{x}_1, \mathbf{x}_2) + b^{\text{old}}$$
$$b_2 = E_2 + y_1(\alpha_1^{\text{new}} - \alpha_1^{\text{old}})K(\mathbf{x}_1, \mathbf{x}_2) + y_2(\alpha_2^{\text{new,clipped}} - \alpha_2^{\text{old}})K(\mathbf{x}_2, \mathbf{x}_2) + b^{\text{old}} \tag{7.90}$$

When both b_1 and b_2 are valid, they are equal. When both new Lagrange multipliers are at bound and if L is not equal to H, the interval between b_1 and b_2 is all thresholds that are consistent with the Karush–Kuhn–Tucker conditions which are necessary and sufficient for an optimal point of a positive definite QP problem. In this case, SMO chooses the threshold to be halfway between b_1 and b_2 (Platt 1999).

Example 7.11 Perform the binary classification of given data using SMO solver for classification.

The circles in Fig. 7.23 refer to the selected support vector for defining hyperplane in the classification process.

Example 7.12 Perform multi-class classification of given data using one-against-all decomposition strategy as shown in Fig. 7.24. SMO binary solver is used to train the binary SVM subtasks.

Example 7.13 Perform multi-class classification of given data using one-against-one decomposition strategy as shown in Fig. 7.25.

Example 7.14 Classify the iris data using wavelet SVM. In this example, Haar wavelet is selected as shown in Fig. 7.26. User can select the other wavelet, i.e., Daubechies and Symmlet. This procedure also needs wavelet toolbox developed by Donoho (WaveLab802) for wavelet function selection.

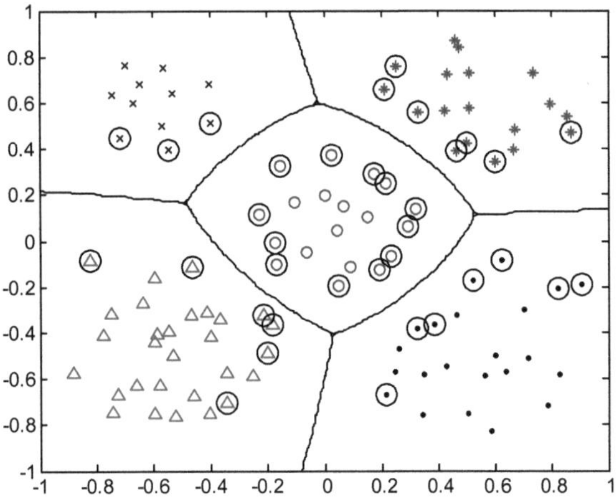

Fig. 7.24 The SVM multi-class classifier using one-against-all decomposition

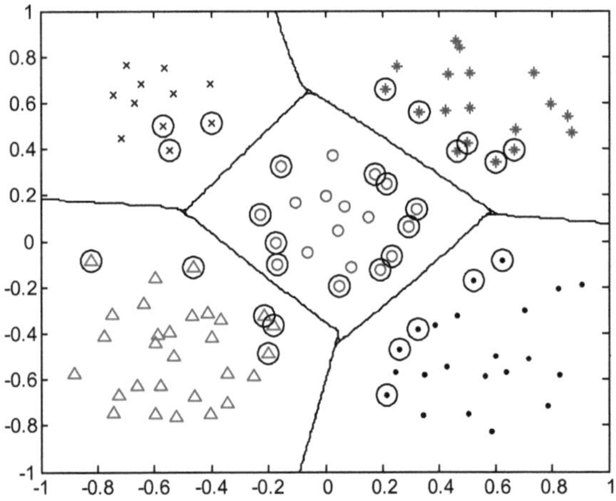

Fig. 7.25 The SVM multi-class classifier using one-against-one decomposition

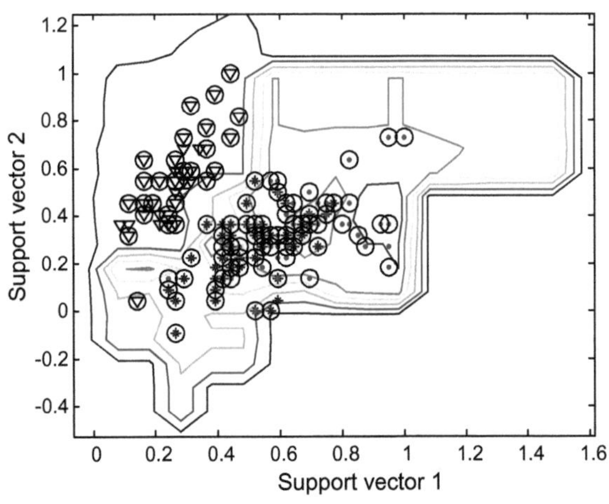

Fig. 7.26 Classification iris data using W-SVM (Haar wavelet)

7.11 Decision Tree

Decision trees are a rule-based classification model and classify the data according to the previous rule when new data enter. Decision trees are recursive algorithms that partition data according to a splitting criterion taken from predictor, or independent, variables of the data. The data are recursively partitioned so as to decrease an impurity function with respect to the dependent variable for the subsets. Consequently, as you descend the tree, the subsets that it contains are more homogeneous with respect to the impurity function chosen. Decision trees are useful tools for data exploration tasks of the following:

- Description, because they allow for data reduction and a compact description of the data,
- Classification, because they allow for the discovery of hypothesis based on separable classes that can be explained by their predictor variables, and
- Generalization, because they uncover mapping from independent variables to dependent variables.

The most common types of decision trees are binary. They have zero or two directed outgoing edges from any node of the tree. A node with zero outgoing edges is called a *leaf node*; all other nodes are called *internal nodes*. Leaf nodes are labeled with class labels: a category of the dependent variable. An internal node, n, is labeled with a predictor attribute, and this is known as the *splitting attribute*. The predictor attribute of the node has a splitting predicate associated with it. An internal node produces a splitting of the database into two disjoint subsets. The part of the original data that fits the criterion will be considered by the left subtree of the decision tree, and the part of the data that do not satisfy the splitting predicate will be considered by the right subtree. Now, a path from the root to a leaf represents a classification rule. For each left subtree taken, the predicate of the node is conjoined to the previous predicates, and for each right subtree taken, the negation of the predicate is conjoined.

Two key concerns in constructing a decision tree are the selection of split points used in the internal nodes and the growth of a right-sized tree. Decision tree construction is accomplished in two phases. The first phase is known as the growth phase of the decision tree. In this phase, an overly large decision tree is grown. While the resubstitution error for a tree grown using a consistent training set can be driven to zero, the danger is that the tree will be overfitted to the training set and will not generalize to test data. The second phase prunes the tree. Here, the aim is to keep the misclassification rate low while reducing the size of the tree in the hope of producing a tree that generalizes well.

The advantages of DT are described as follows:

- DT is easy to understand that informs the baseline for prediction and classification.
- DT is easy to find the important attributes that affect classification.
- DT demands less time than other classification models.

- DT is easy to select the proper data when there are too many attributes which are not good to construct the classification model.
- DT reduces the time and effort spent during the data conversion process among knowledge discovery processes because it can deal with continuous and nominal attributes without changing its type.

7.11.1 Building Decision Tree

Decision tree is described well in Fig. 7.27. Table 7.2 explains each term which is used in building decision tree. X_1 and X_2 are also called parent node, and *A*, *B*, and *C* are called child nodes. According to attribute values, some cases are determined what classes they belong to.

When constructing the decision tree, we should estimate all instances to divide and conquer. Criterion has an important role at this step. The construction of the tree is determined how to take some attributes into node to split the instances. Consequently, criterion searches the best attribute to split the instances and stop the growth of the tree. There are many decision tree inducers such as ID3, C4.5, and CART. ID3 which is proposed by Quinlan uses information gain as split criterion. The growing stops when all instances belong to a single value of target feature or when the best information gain is not greater than zero. ID3 does not consider numeric attributes and missing values.

C4.5 is a developed inducer of ID3, presented by the same author. It uses gain ratio as split criterion. The split stops when the number of instances to be split is

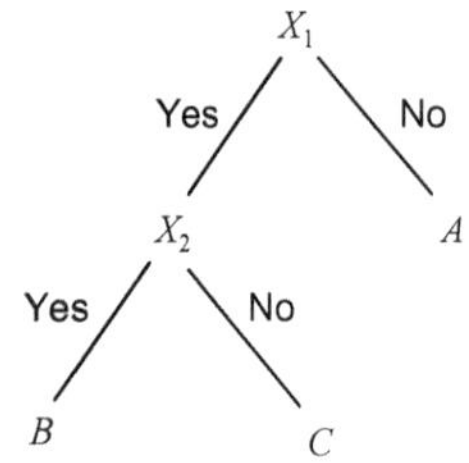

Fig. 7.27 Description of decision tree

Table 7.2 Definition of decision tree

Symbols	Name	Example
X_1, X_2	Attribute name	What is the predominant frequency?
Yes, No	Attribute value	1X, 2X, unknown
A, *B*, *C*	Class	Unbalance, misalignment

below a certain threshold. It is also capable of dealing with numeric attributes. It can induce from a training set that incorporates missing values by correcting gain ratio criterion. CART is developed by Breiman and is characterized by the fact that it constructs binary trees, namely each internal node has exactly two outgoing edges.

The tree growth phase involves choosing a splitting selection that minimizes the overall misclassification error. The most common form of the split selection methods is the impurity-based selection methods. These methods find the splitting criterion by minimizing a concave impurity function, $i(t)$, at a node t. Examples of impurity functions are the well-known entropy measure, Gini index, and twoing rule:

$$\text{Entropy measure: } i(t) = -\sum_{j=1}^{c} P_j \log P_j \tag{7.91}$$

$$\text{Gini index: } i(t) = 1 - \sum_{j=1}^{c} P_j^2 \tag{7.92}$$

where P_j is the probability for class ω_j and c is the total number of classes.

Gini index metric is based on the Gini criterion by Breiman, but modified as in OC1 by Murthy. The Gini index measures the probability of misclassifying a set of instances. Twoing rule is also proposed by Breiman and used in Murthy's OC1 and compares the number of examples in each category on each side of the proposed split (Fig. 7.28).

Gini index looks for the largest class in database (e.g., class A) and strives to isolate it from all other classes. For example, with four classes, A, B, C, and D, representing 40, 30, 20, and 10 % of the data, respectively, the Gini rule would immediately attempt to pull out the class A records into one node. Of course, such a separation may not be possible using the available data, but if it is, the Gini opts for that split. The diagram below shows the best possible Gini split for these data.

Once the first split is made, Gini index continues attempting to split the data that require further segmentation, i.e., the right child node that contains classes B, C, and D. Using the same strategy, Gini index attempts to pull out all the class B records, separating them from the other classes in the node. Gini index then tackles

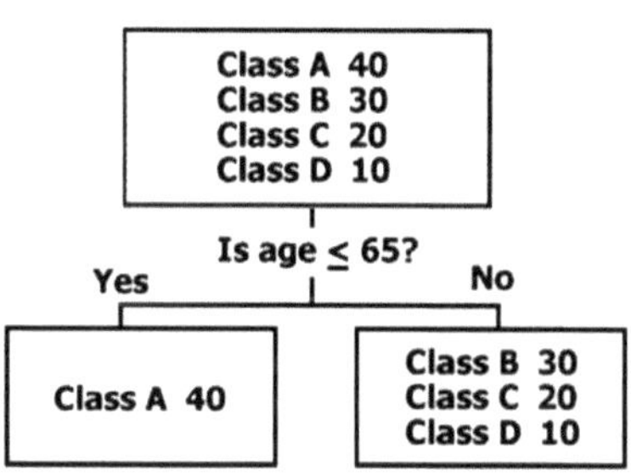

Fig. 7.28 The best possible Gini split

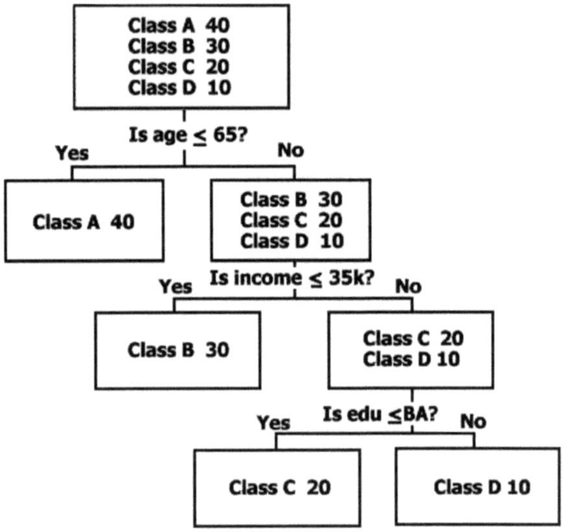

Fig. 7.29 Final tree using Gini splitting rule

the last heterogeneous node, striving to separate class C from class D. If the Gini rule is successful, the final tree would contain four "pure" child nodes.

A pure decision tree such as the above is attainable only in very rare circumstances; in most real-world applications, database fields that clearly partition class from class are not available. If they were, no one would ever receive an unwelcome direct mail piece and bank losses on bad debts would be zero. Therefore, we cannot expect to grow trees like the ones above routinely; however, Gini index will try to come as close as possible to this ideal. A hypothetical example of a more realistic decision tree grown by Gini index is displayed in Fig. 7.29. While imperfect, it is still an unusually accurate tree (Fig. 7.30).

Gini index attempts to separate classes by focusing on one class at a time. It will always favor working on the largest or, if you use costs or weights, the most "important" class in a node. While this approach might seem short-sighted, Gini index performance is frequently so good that you should always experiment with it to see how well it does. Gini index is the default rule in CART precisely because it is so often the best splitting rule (Fig. 7.31).

The philosophy of twoing rule is far different than that of Gini index. Rather than initially pulling out a single class, twoing rule first segments the classes into two groups, attempting to find groups that together add up to 50 % of the data. Twoing rule then searches for a split to separate the two subgroups. The diagram below shows the best possible split the twoing rule could find.

Again, this is an ideal split. It is unlikely that any real-world database would allow you to cleanly separate four important classes into two subgroups in this way. However, splits that approach this ideal might be possible, and these are the splits that twoing rule seeks to find. The entropy rule, which is very similar to twoing rule in practice, strives for similar splits. Table 7.3 shows the results which are applied for three types of criteria.

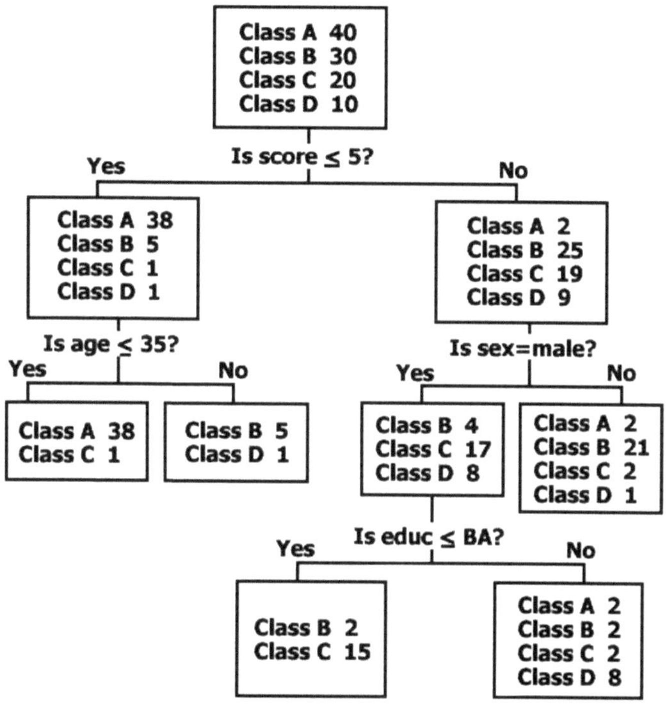

Fig. 7.30 Gini splitting rule

Fig. 7.31 Twoing splitting rule

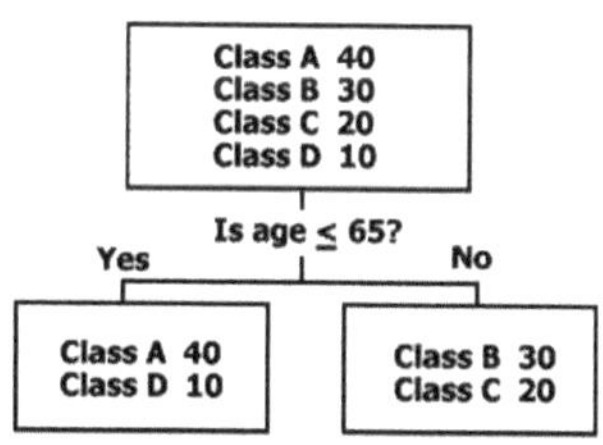

We concluded that entropy criterion is the best among them. So we constructed tree using entropy criterion which is based on the information theory and is proposed by Quinlan. We then use information entropy evaluation function based on the information theory.

This entropy evaluation function is calculated in the following way:

Step 1: Calculate info(T) necessary to identify the class in the training set T.

$$\text{info}(T) = -\sum_{j=1}^{k}\left\{\frac{\text{freq}(C_j, T)}{|T|} \times \log_2\left(\frac{\text{freq}(C_j, T)}{|T|}\right)\right\} \tag{7.93}$$

Table 7.3 The result of split criterion

Criterion	Type	Unknown attributes	Tree size	Error rate (%)
Gini index	Numeric	Ignored	21	9.5
Twoing			21	38
Entropy measure			21	0
	Nominal	Considered	25	3.8
			25	7.7

where $|T|$ is the number of cases in the training set. C_j is a class j, k is the number of classes, and freq(C_j, T) is the number of cases included in T.

Step 2: Calculate the expected information value, info$_X(T)$ for test X to divide into T.

$$\text{info}_X(T) = \sum_{i=1}^{n} \left\{ \frac{|T_i|}{|T|} \times \text{info}(T_i) \right\} \tag{7.94}$$

where n is the number of outputs for test X and T_i is a subset of T corresponding to output i.

Step 3: Calculate the mutual information value acquired from division according to test X.

$$\text{gain}(X) = \text{info}(T) - \text{info}_X(T) \tag{7.95}$$

Step 4: Calculate the dividing information value split info(X) acquired for T and divide into n subsets.

$$\text{split info}(T) = -\sum_{i=1}^{n} \left\{ \frac{|T_i|}{T} \times \log_2\left(\frac{|T_i|}{T}\right) \right\} \tag{7.96}$$

Step 5: Calculate the ratio of gain(X) over split info(X).

$$GR(X) = \frac{\text{gain}(X)}{\text{split info}(X)} \tag{7.97}$$

The $GR(X)$ compensates for the weak point of gain(X) which represents the quantity of information provided by X in the training set. Therefore, an attribute with the highest $GR(X)$ is taken as the root of the decision tree.

7.11.2 Pruning Decision Tree

Through the pruning step, we can construct a complete tree able to exactly classify all the instances, increase statistical reliability and predictability over the whole space of instances, and improve the understandability. There are many pruning methods to improve the tree's performance. Pruning methods are classified into two types according to those properties. One is overpruning method, and the other is underpruning method. Cost-complexity pruning (CCP) and reduced error pruning (REP) belong to the former. Error base pruning (EBP), pessimistic error pruning (PEP), and minimum error pruning (MEP) belong to the latter. Many pruning methods have been applied in decision trees and compared each other in performance of classification. EBP is powerful to classify the data with continuous attributes and used in C4.7. CCP is powerful which has discrete attributes. We can look for the empirical comparison among pruning methods proposed by Quinlan. Pessimistic pruning method is the best about many data in his test except EBP (Table 7.4).

The EBP is the best one among other methods through data group applied in Floriana. Consequently, we selected the EBP.

In this chapter, the examples of decision tree method for classification routine will be presented using MATLAB. The computational statistical toolbox is adopted.

Example 7.15 Build a tree using decision tree inducers CART and then prune it.

We see in Fig. 7.32 that the tree has partitioned the feature space into eight decision regions or eight terminal nodes.

Table 7.4 The comparison of pruning methods

Category	Properties
CCP, REP	Overpruning trendy, smaller size, less exact
EBP, PEP, MEP	Underpruning trendy, bigger size, better exact

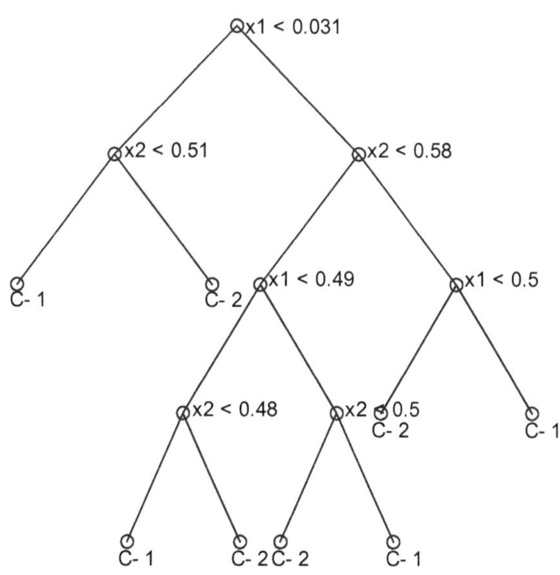

Fig. 7.32 The classification tree of CART

The input argument is a tree and the output argument is a cell array of subtrees, where the first tree corresponds to the tree T and the last tree corresponds to the root node.

The resulting sequence for complexity parameter (α) is

$$\text{alpha} = 0 \quad 0.0100 \quad 0.0300 \quad 0.0700 \quad 0.0800 \quad 0.1033$$

7.12 Random Forest

In recent years, a number of papers reported the classifiers to produce the fault diagnosis in the machine such as artificial neural networks (ANNs), support vector machines (SVMs), rule-based induction, and case-based reasoning. The proposed techniques and their extended research increase the intelligence, preciseness, and applicability in diagnosis domain. While the passion for developing fault diagnosis methods is increasing, there are a number of obstacles before researchers. That is, the correct diagnosis of faults is rather complicated. The reasons are listed as follows:

- Different kinds of faults may result in a certain symptom.
- Because of the background noise, some faults are difficult to be distinguished in the machine.
- There are a number of subassemblies with rotating machinery and a high-level internal interaction between these subassemblies such as bearings and rotor.

Hence, the machine fault diagnosis method which is employed to make hypotheses should be powerful enough to classify the malfunctions in a correct way. The critical issue in classification is how to integrate the classification power to achieve higher classification accuracy. To do that, improving the capability of diagnosis is the main motivation to inspire researchers synchronizing existent technologies and exploring new theories.

Ensemble classification methods train several classifiers and combine the decision of a set of classifiers by weighted or unweighted voting process to classify unknown examples. An ensemble classifier is generally found to be more accurate than any of the individual classifiers making up the ensemble (Dietterich 2002). The most widely used ensemble methods are *bagging* (an acronym of *bootstrap aggregating*) (Breiman 1996) and *boosting* (Bauer and Kohavi 1999). Bagging is based on training many classifiers on bootstrap samples from the training set, which has been shown to reduce the variance of the classification (Gislason et al. 2006). Boosting was introduced by Schapire (1990) as a method for boosting the performance of a weak learning algorithm. Boosting uses iterative retraining, where the incorrectly classified samples are given increased weighting as the iteration progresses. Therefore, it generally reduces both the variance and the bias of the classification and has been shown to be a very accurate classification method (Joelsson et al. 2005). However, it has various drawbacks: It is very slow, it can overtrain, and it is sensitive to noise (Briem et al. 2002).

Random forest algorithm (RF) which is introduced by Breiman (2001) is a general term for ensemble methods using tree-type classifiers. RF builds a large amount of decision trees (Quinlan 1986a, b; Yang et al. 2000a, b) out of sub-data set from a unique original training set by using bagging which is a meta-algorithm to improve classification and regression models according to stability and classification accuracy. Bagging reduces variance and helps to avoid overfitting synchronously. This procedure extracts cases randomly from original training data set, and the bootstrap sets are used to construct each of the decision trees in the RF. Each tree classifier is named component predictor. The RF makes decision by counting the votes of component predictors on each class and then selecting the winner class in terms of number of votes to it (Breiman 2006).

RF has been employed in various fields such as land cover (Pal 2005), drug discovery (Remlinger 2003), and geographic data (Gislason et al. 2006; Ham et al. 2005). The possibilities of using RF in machine fault diagnosis application are being considered only by authors (Di et al. 2006). RF provides good performance in applications in these fields. RF can be a competitor for rotating machinery fault diagnosis, because of these distinctive features:

- It is unexcelled in accuracy among current algorithms.
- It runs efficiently on large databases.
- It can estimate the importance of each variable in the classification.
- It has methods for estimating missing data and maintains accuracy when a large proportion of the data are missing.
- It computes proximities between pairs of cases that can be used in clustering, locating outliers, or five interesting views of the data.
- It generates an internal unbiased estimate of the generalization error as the forest building progresses.

In this section, we confirm the possibilities of using RF in machine fault diagnosis and propose an optimized RF method combined with genetic algorithm (GA) to improve the classification accuracy. To increase the diagnosis accuracy, we acquire the data of three-direction vibration signals as the original inputs of system. And a number of feature parameters in time and frequency domains and regression coefficients are calculated to extract helpful information and remove the background noise of the data. Then, RF diagnosis system detects the certain faulty-type bases on these features. It is an effective approach to promote the capability of diagnosis system (Yang et al. 2004a, b). Experimental result shows that the optimized RF-based method achieves a very high accuracy by combining RF with GA.

7.12.1 Random Forest

RF which is derived from decision tree classifier is an assembled method, and it grows tree using CART (classification and regression tree) methodology to

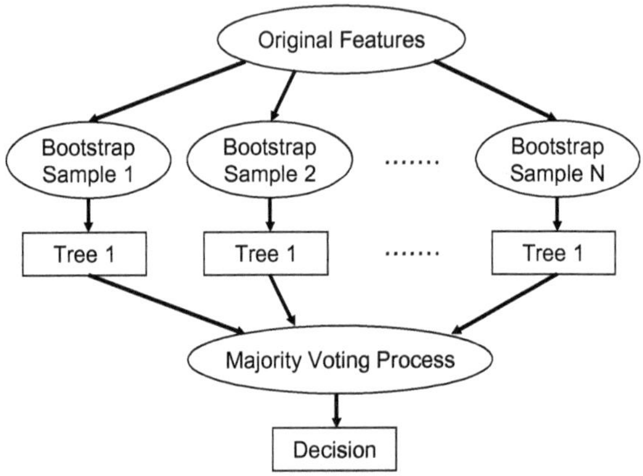

Fig. 7.33 Construction of random forest

maximum size and without pruning. Figure 7.33 shows the construction of a RF. Therefore, this of the building blocks for CART-based RF (RF-CART) explored in this section.

(1) *CART methodology*

CART (Breiman et al. 1984) grows classification and regression trees to predict continuous dependent variables (regression) and categorical predictor variables (classification). There are four basic steps in CART methodology. At first step, a tree is built by using recursive splitting of nodes. Each terminal node is assigned to a certain class, judged by class probability distribution of the dependent variable at terminal node. The assignment of a predicted class to each node occurs whether or not that node is subsequently split into child nodes. The second step consists of stopping the tree-growing process. And the final two steps, tree pruning and optimal tree selection, are ignored because RF grows trees freely without any pruning process.

(2) *Tree building*

The tree-building process begins with departing the root node into binary nodes by a very simple question of the form, $\mathbf{x} \leq d$? Here, $\mathbf{x}$ is a variable in the data set and d is a real number. Initially, all observations are located in the root node. CART implements a computer-intensive algorithm that searches for the best split at all possible split points for each variable. The methodology which CART uses for building trees is known as binary recursive partitioning. Adopting the Gini diversity index as a splitting rule, the tree-building process is as follows:

Step 1: CART splits the first variable at all of its possible split points, at all of the values the variable assumes in the sample. At each possible split point of a variable, the sample splits into binary or two child nodes. Cases with a "yes" response to the question posed are sent to the left node and those with "no" responses are sent to the right node.

Step 2: CART then applies its goodness-of-split criteria to each split point and evaluates the reduction in impurity that is achieved using the formula:

$$\Delta i(s,t) = i(t) - p_L[i(t_L)] - p_R[i(t_R)] \tag{7.98}$$

where s is a particular split, p_L is the proportion of the observations at node t which go into the left child node t_L, and p_R is for the right node t_R similar with p_L. $i(t_L)$ and $i(t_R)$ are impurity of left and right nodes, respectively.

Step 3: CART selects the best split of the variable as that split for which the reduction in impurity is highest. Three steps above are repeated for each of the remaining variables at the root node.

Step 4: CART then ranks all of the best splits on each variable according to the reduction in impurity achieved by each split and selects the variable and its split point that most reduced the impurity of the root or parent node.

Step 5: CART then assigns classes to these nodes according to the rule that minimizes misclassification costs. CART has a built-in algorithm that takes into account user-defined variable misclassification costs during the splitting process. The default is unit or equal to misclassification costs.

Because the CART procedure is recursive, steps 1–5 are repeatedly applied to each non-terminal child node at each successive stage.

(3) *Stopping tree building*

CART stops the splitting process when:

- There is only one observation in each of the child nodes.
- All observations within each child node have the identical distribution of predictor variables, making splitting impossible.
- The user sets an external limit on the number of levels in the maximal tree previously.

Stand by these steps, a CART algorithm-based decision tree without pruning and optimizing will be built.

7.12.2 Random Forest Algorithm (RF)

RF has greatly improved the classification accuracy resulting from growing an ensemble of trees and making them vote for the most promising class. A convenient method to build the ensembles is by random vectors which are generated via random selection procedure from integrated training set. The constituent in this method is that we prepare k random vectors, $\mathbf{\Theta}_k$, which are independent of the past random vectors $\mathbf{\Theta}_1$, $\mathbf{\Theta}_2$, $\mathbf{\Theta}_3$, …, $\mathbf{\Theta}_{k\text{-}1}$ but with the same distribution to build the trees among the RF. The corresponding individual classifier is noted by $C(\mathbf{X}, \mathbf{\Theta}_k)$. For example, in bagging processing, the random vector $\mathbf{\Theta}$ as the N observations randomly draws out from entire training data proportionally where N is the number of observations of training data. And then, they vote for the most popular class.

Breiman names these procedures as RFs. A definition drawn from the original paper is available here (Breiman 2001).

Definition 1 A RF is a classifier consisting of a collection of tree-structured classifiers $\{C(\mathbf{X}, \mathbf{\Theta}_k), k = 1, \ldots\}$ where the $\mathbf{\Theta}_k$ is identically distributed independent random vectors and each tree casts a unit vote for the most popular class at input $\mathbf{X}$.

(1) *Two randomized procedures in RF tree building*

As mentioned below, RF enhances the classification accuracy compared with decision tree classifier significantly. It is the reason that RF applies two randomized procedures when it builds trees. Each tree is built as follows. Firstly, assume that the number of cases in the training set is N and the number of variables in the classifier is M. Select the number of input variables that will be used to determine the decision at a node of the tree. This number, m, should be much less than M ($m \ll M$). Secondly, choose a training set by choosing N samples from the training set with replacement. And then, for each node of the tree, randomly select m of the M variables on which to base the decision at that node. Calculate the best split based on these m variables in the training set. Finally, each tree is fully grown and not pruned.

The two distinctive randomized procedures exist among the four steps below. That is, RF extracts a fixed quantity from training set randomly, or names it a bagging processing (Breiman 1996). Each base classifier in the ensemble is trained on a bootstrap from the entirety of available data. However, each of these bootstrap replicates tends to leave out roughly one-third of the sample. Thus, each classifier in the ensemble is trained on roughly two-thirds of the original data. Consequently, each element in the sample of size n trains roughly $(2/3)k$ of all classifiers in the ensemble so that it can be used to validate the remaining $k/3$ classifiers (Fig. 7.34) where n is the number of training data and k is the total number of single-tree classifier. This part of data is named *out-of-bag data* to get an unbiased estimate of

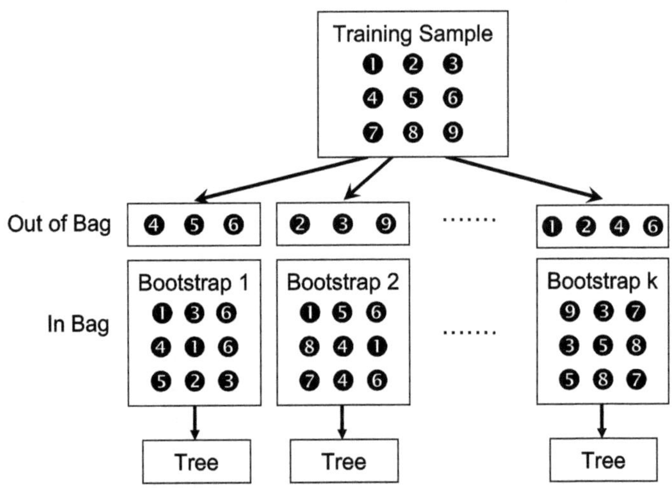

Fig. 7.34 Schematic of bagging using the decision tree as the base classifier

the test set error of an individual tree. The rest of the data are used to build the single-tree classifier.

After the bagging processing, the other randomized procedure is emerged in node splitting during tree classifier built. Different from the normal CART-like decision tree-splitting algorithm, CART within RF algorithm searches only in n variables which are small amount and drawn at random from all M variables instead of entire variables.

The research of Breiman states why these two randomized procedures make classification accuracy increase effectively: Improvement will occur for unstable procedures where a small change in training set can result in large change between component classifiers and classifier trained by entire training set. In RF, whatever the bagging processing or the random selection of variables to split the node both makes difference in individual tree and forests. Therefore, these two sources of randomness are most important features of RF.

(2) *Convergence of RF*

RF adopts an ensemble of decision trees and determines the categorical classes by majority vote algorithm. Thus, a serious consideration of overfitting is necessary for testing RF performance. Normally, an overfitting will occur where learning is performed too long or where training examples are rare, and the learner may be limited in very specific random features of the training data that have no causal relation with the target function. But RF can avoid the overfitting completely (Breiman 1996). To affirm this point, we define a margin function first.

Given an ensemble of a series of classifiers $C_1(\mathbf{X})$, $C_2(\mathbf{X})$, …, $C_k(\mathbf{X})$, and with the training set drawn at random from the distribution of the random vector Y, $\mathbf{X}$, define the margin function as follows:

$$mg(\mathbf{X}, Y) = av_k I(C_k(\mathbf{X}) = Y) - \max_{j \neq Y} av_k I(C_k(\mathbf{X}) = j) \tag{7.99}$$

where $\mathbf{X}$ is the input metric, avk is the average number of votes at $\mathbf{X}$, Y for the corresponding class, and $I(\cdot)$ is the indicator function. The margin measures the extent to which the average number of votes at $\mathbf{X}$, Y for the right class exceeds the average vote for any other class. The larger the margin, the more confidence the classification.

According to this function, the generalization error is given by:

$$PE^* = P_{\mathbf{X},Y}(mg(\mathbf{X}, Y) < 0) \tag{7.100}$$

where $P\mathbf{X},Y$ indicates the probability which is over the $\mathbf{X}$, Y space.

Theorem 7.1 *As the number of trees increases, for almost surely all sequences $\Theta 1$, PE^* converges to:*

$$P_{\mathbf{X},Y}\left(P_\Theta(C(\mathbf{X}, \Theta) = Y) - \max_{j \neq Y} P_\Theta(C(\mathbf{X}, \Theta) = j) < 0\right) \tag{7.101}$$

Theorem 7.1 is proved with the strong law of large numbers and the tree structure. It indicates that it is unnecessary for RF to employ common anti-overfitting methods for instance, cross-validation, early stopping, etc. RF does not overfit when more trees are added; meanwhile, it results in a limiting value of the generalization error. This is another important feature of RF besides the two randomized procedures mentioned above.

(3) *Accuracy of RF depending on strength and correlation*

In the previous section, the anti-overfitting characteristic of RF is proved. But we concern more about its accuracy. According to the analysis built in references, an upper bound of RF can be derived for the generalization error in terms of two parameters that are measures of how accurate the individual classifiers are and of the dependence between them. These also lead an in-depth view of how RF works. Firstly, we define a margin function and raw margin function for RF.

The margin function for a RF is:

$$mr(\mathbf{X}, Y) = P_{\Theta}(C(\mathbf{X}, \Theta) = Y) - \max_{j \neq Y} P_{\Theta}(C(\mathbf{X}, \Theta) = j) \tag{7.102}$$

The raw margin function is:

$$rmg(\Theta, \mathbf{X}, Y) = I(C(\mathbf{X}, \Theta) = Y) - I(C(\mathbf{X}, \Theta) = \hat{j}(\mathbf{X}, Y)) \tag{7.103}$$

Distinctively, $mr(\mathbf{X}, Y)$ is the expectation of $rmg(\boldsymbol{\Theta}, \mathbf{X}, Y)$ with respect to $\boldsymbol{\Theta}$. And the strength of the number of individual classifiers $C(\mathbf{X}, \boldsymbol{\Theta})$ is:

$$S = E_{\mathbf{X},Y} mr(\mathbf{X}, Y) \tag{7.104}$$

where $E\mathbf{X},Y$ is the expected value of margin function over $\mathbf{X}$, Y space.

Then, we compute the variance of margin function:

$$\text{var}(mr) = \bar{\rho}(E_{\Theta} sd(\Theta))^2 \leq \bar{\rho} E_{\Theta} \text{var}(\Theta) \tag{7.105}$$

Write

$$E_{\Theta} \text{var}(\Theta) \leq E_{\Theta}(E_{\mathbf{X},Y} rmg(\Theta, \mathbf{X}, Y))^2 - S^2 \leq 1 - S^2 \tag{7.106}$$

where $\bar{\rho}$ is the mean value of the correlation and $sd(\cdot)$ is the standard deviation of $rmg(\boldsymbol{\Theta},\mathbf{X},Y)$. Considering Eqs. (7.104) and (7.105) and Chebyshev inequality, Theorem 7.2 can be concluded.

Theorem 7.2 *An upper bound for the generalization error is given by:*

$$PE^* \leq \bar{\rho}(1 - S^2)/S^2 \tag{7.107}$$

Although the bound is likely to be loose, it fulfills the same suggestive function for RF as VC-type bounds do for other types of classifiers. It shows that the two ingredients involved in the generalization error for RFs are the strength of the individual classifiers in the forest and the correlation between them in terms of the raw margin functions. There is a conclusion drawn from this upper bound, that is, the smaller this ratio is, the better performance RF provided.

Example 7.16 Demonstrate the random forest algorithm for classification of artificial data.

Number of trees: 50

No. of variables tried at each split: 2
OOB estimate error rate for training data: 9.8083 %

Confusion Matrix For Training Set

	1	2	3	4	5	6	err %
1	1050	2	15	0	5	0	2.522
2	1	468	1	2	3	4	2.2965
3	6	1	921	22	0	11	4.1623
4	7	4	84	237	3	80	42.8916
5	32	4	1	4	393	36	16.3830
6	1	0	29	60	17	931	10.3083

See Fig. 7.35.

7.12.3 Genetic Algorithm

RF is strengthened by a standard genetic algorithm (GA) (Goldberg 1989) in this paper. GA is a simulation of evolution where the rule of survival of the fittest is applied to a population of individuals, or it can be considered as a parallel search procedure that simulates the evolutionary process by applying genetic operators.

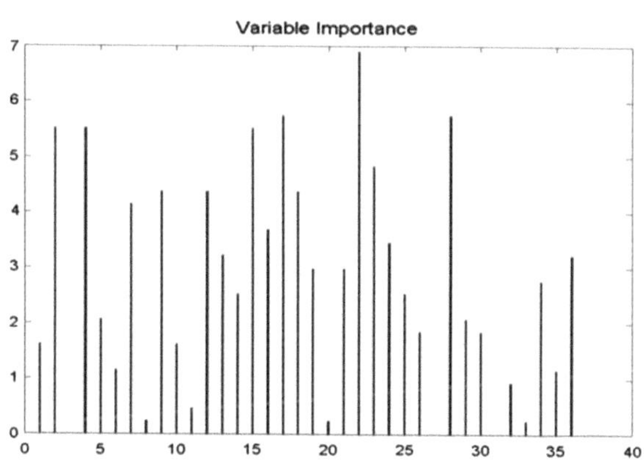

Fig. 7.35 Variable importance

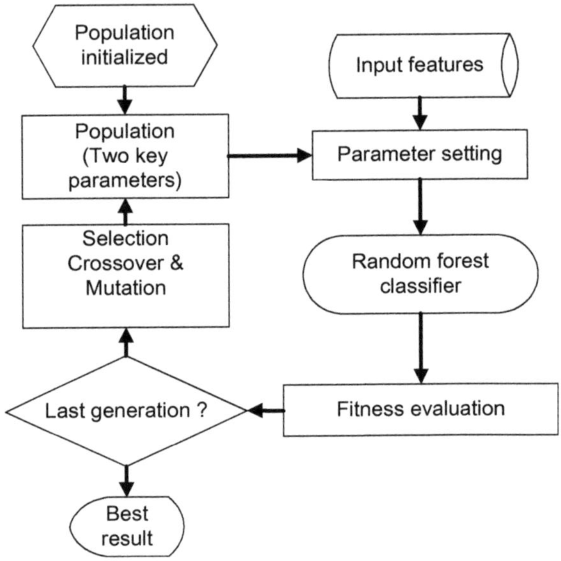

Fig. 7.36 Flowchart of genetic algorithm

Comparing with other search algorithms, GA has been well known for its superior performance. And the most powerful feature of GAs is its great simplicity. They do not need too much code and no differentiability or continuity requirements to be satisfied. The usual GA flowchart (Fig. 7.36) and steps are shown as follows:

Step 1: Coding. Generate an initial population (usually a random string).
Step 2: Fitness evaluation. Apply some function or formula to the individuals to get the fitness of each individual.
Step 3: Selection. According to the fitness, individuals are selected to be the parents of next generation.
Step 4: Crossover. It is used to create two child individuals from the parent which pass the selection successfully via exchanging their chromosomes.
Step 5: Mutation. It assigns a new value to a randomly chosen gene and is controlled by a mutation probability.
Step 6: Repeat steps 3 to 5 until the evolved result satisfies the termination criteria, or a certain fixed number of generations are achieved.

The function of GA is to evaluate the best parameters of RF. Fitness is the criterion which indicates the capacity of each individual. In RF, the diagnosis accuracy rate value is assigned to fitness which represents the performance of certain parameters. After generating the initial population, fitness values are calculated and assigned to individuals which include two key parameters of RF. The GA proceeds to the next generation through three genetic operators: selection, crossover, and mutation.

Selection is the most important part of GA. This operator impacts on the trend of GA and makes GA's running time shorten. It picks up the excellent parents to reproduce the individuals within the limitation. The normalization probability for individuals to be selected, N_p, is described as follows:

$$N_p(i) = \frac{B_s(i)}{1 - (1 - B_s(i))^{N_g(i)}} \tag{7.108}$$

where i is an individual and N_g is the number of generations. B_s is the probability of selecting best individual from the current population.

The selection probability of each individual is:

$$P_s(i) = N(i)(1 - B_s(i))^{I(i)} . \tag{7.109}$$

where $I(i)$ is the sorted index of individuals according to the fitness.

The selection probability stands for the opportunity of individuals to be chosen as parents of the next generation. The new individuals are reproduced by the survivals from selection by crossover and mutation procedure.

7.13 Adaptive Neurofuzzy Integrated System (ANFIS)

Artificial neural networks (ANNs) have been proven as a reliable technique to diagnose the condition of a motor and have a good learning capability. However, ANNs are not highly interpretable and understandable, i.e., they are incapable of explaining a particular decision to the user in a human-comprehensible form. Fuzzy logic is another method, which has been used for fault detection and diagnosis (Benbouzid and Nejjari 2001). It has the ability of modeling human knowledge in the form of if-then rules by easily understandable linguistic term. It has the capability of transforming linguistic and heuristic term into numerical values for use in complex machine computation via fuzzy rules and membership functions. The if-then rules and the initial parameters of membership functions are prepared by an expert. Thus, fuzzy logic requires fine-turning in order to get acceptable rule base and optimize parameters for available data (Shukri et al. 2004). The problems arise from fuzzy logic or ANN alone can be solved by the integration of both methods and is proven for motor fault diagnosis (Goode and Chow 1995).

The adaptive neurofuzzy inference system (ANFIS) (Jang 1993) is a specific kind of neurofuzzy classifier approach which integrates the ANNs' adaptive capability and the fuzzy logic qualitative approach. ANFIS has been successfully applied to automated fault detection and diagnosis of induction machines (Shukri et al. 2004; Altug et al. 1999). Recently, ANFIS and its combination with other methods were also employed as an enhanced tool for fault classification such as ANFIS and genetic algorithms (Lei et al. 2007), and ANFIS and wavelet transform (Lou and Loparo 2004) for bearing fault diagnosis. ANFIS has been also applied in classifying the faults of induction motor with variable driving speed (Ye et al. 2006).

The data obtained from measurements are normally high dimension and have a large amount of redundant features. If it is directly inputted into classifier, the performance will be significantly decreased. Feature extraction and feature selection have been utilized for reducing dimension of data and selecting the important features wherein feature extraction means transforming the existing features into a lower dimensional space (Yang et al. 2006). Nevertheless, each feature set contains redundant or irrelevant features as well as salient features in feature space after the feature extraction has been done. Consequently, we need feature selection procedure for selecting a few features which obviously characterize the machine conditions from the whole feature set (Lei et al. 2007). In this study, decision tree is utilized as feature selection procedure to remove irrelevant features for the purpose of reducing the amount of data needed to achieve good learning, improving classification accuracy, attaining compact and easily understanding knowledge base, and reducing computational time (Kumar et al. 2005).

In this study, an integrated method which combines classification and regression tree (CART) and ANFIS is used for the fault diagnosis of induction motors. The proposed approach includes the two following procedures. First, the CART is performed as a feature selection tool to get the valuable features and identifies the structure of classifier in the next step. Second, the ANFIS classifier is used to diagnose the faults of induction motors wherein the parameters of membership functions are tuned throughout the learning process.

7.13.1 Classification and Regression Tree (CART)

CART algorithm (Breiman et al. 1984) is similar to other seminal ones used in decision tree induction such as ID3 and C4.5 (Quinlan 1986a, b). One of the major distinctions is that CART induces strictly binary trees through a process of binary recursive partitioning of feature space of a data set (Jang et al. 1996). The trees produced by CART also consist of internal nodes (with two children) and terminal nodes or leaf nodes (without children). Each internal node is associated with a decision function to indicate which node to visit next, while each terminal node shows the output of a given input vector that leads the visit to this node (Jang 1994). The decision tree shown in Fig. 7.37 evidently classifies the input space into four mutually exclusive rectangular regions which are assigned a labeled class. As ID3 or C4.5, CART extensively builds the tree by using the data set of already classified instances which is called training set and then prunes the tree back based on a minimum cost-complexity principle. The first phase is the so-called *tree building*, and the other is *tree pruning*.

(1) *Tree building*

The initial state of a decision tree is the root node (the first internal node) that is assigned all the examples of training set. If it is the case that all examples belong to

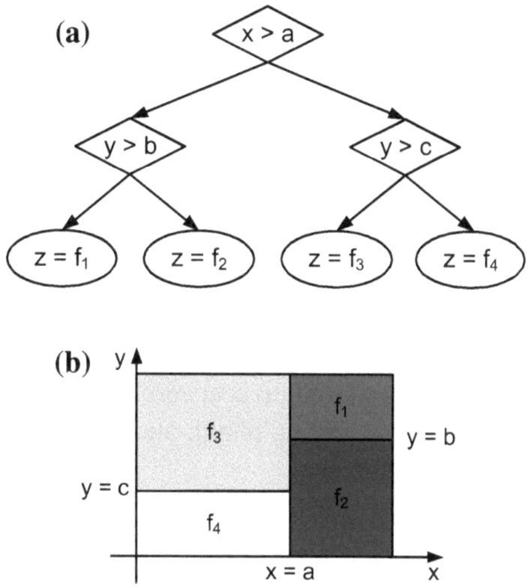

Fig. 7.37 Decision tree (Jang 1994). **a** Binary decision tree. **b** Feature space partitioning

the same class, no further decisions need to be made to partition the examples, and the solution is completed. Conversely, if the examples at this node belong to two or more classes, a test is made at the node that will result in a split and the training set is then divided into two subspaces. The process is recursively repeated for each of the new terminal nodes until a completely discriminating tree is obtained.

The test at internal nodes is determined based on a measure of impurity to choose which feature to be selected and which threshold value to be chosen. The best-known measure of impurity for CART is entropy impurity given by:

$$E(t) = -\sum_{j=1}^{\#\text{class}} p(w_j|t) \log p(w_j|t) \tag{7.110}$$

where $E(t)$ is the entropy impurity at node t and $p(w_j|t)$ is the fraction of patterns at node t that belongs to the class w_j.

The optimal splitting value s^* at node t is chosen from a set of all splitting candidates S so that the drop of impurity is maximized as follows:

$$\Delta E(s^*, t) = \max_{s \in S} \Delta E(s, t) \tag{7.111}$$

where $\Delta E(s,t)$ is the drop of impurity given by:

$$\Delta E(s, t) = E(t) - p_L E(t_L) - p_R E(t_R) \tag{7.112}$$

where t_L and t_R are left and right branch nodes, $E(t_L)$ and $E(t_R)$ are the impurities of the left and right branch nodes, and p_L and p_R are the fraction of patterns at node t, respectively.

(2) *Tree pruning*

The tree obtained by preceding building phase is biased the training set and may have a large number of branches which substantially increase the tree's complexity, while they do not yield higher accuracy if resulting from noisy data. It is therefore necessary to prune the tree to improve the accuracy of the classifier and overcome the familiar overfitting problem. The method for pruning in CART is based on the principle of minimum cost complexity. Let $T_{\max}$ denote a wholly expanded tree that is grown in building phase, and the cost-complexity measure $E_\alpha(T)$ of subtree $T \subset T_{\max}$ is defined by:

$$E_\alpha(T) = E(T) - \alpha\left|\widetilde{T}\right| \tag{7.113}$$

where $\left|\widetilde{T}\right|$ is the number of terminal nodes in T and α is a complexity parameter.

The general process for pruning tree is executed as follows:

Step 1: Beginning at the internal node t that is upward terminal node of a tree T.
Step 2: Calculating the value of α, denote by α_t, that makes $T - T_t$ as the next minimizing tree for each internal node t. The α_t is given by:

$$\alpha_t = \frac{E(t) - E(T_t)}{\left|\widetilde{T}_t\right| - 1} \tag{7.114}$$

Step 3: Finding the minimal α_t and choosing $T - T_t$ as the next minimizing tree.
Step 4: Repeating the process until getting the optimum-sized tree by using an independent testing data set or performing cross-validation.

The resulting decision tree is an easy to interpret representation of the nonlinear input–output mapping. Furthermore, it is also easy for generating decision rules. For instance, the decision tree shown in Fig. 7.37 is equivalent to a set of crisp rules

$$\begin{aligned} &\text{If } x > a \text{ and } y > b \text{ then } z = f_1 \\ &\text{If } x > a \text{ and } y \le b \text{ then } z = f_2 \\ &\text{If } x \le a \text{ and } y > c \text{ then } z = f_3 \\ &\text{If } x \le a \text{ and } y \le c \text{ then } z = f_4 \end{aligned} \tag{7.115}$$

Those crisp rules and thresholds are utilized to define the structure of neurofuzzy classifier that will be briefly described in the next section. However, the discontinuities at the decision boundaries are crisp and lead to large output variations for small changes in input features when such features are closed to decision boundaries.

7.13.2 Adaptive Neurofuzzy Inference System (ANFIS)

(1) *Architecture of ANFIS based on CART*

The ANFIS architecture is an integration of fuzzy logic and neural network algorithm (Jang 1994; Jang et al. 1996) so as to use learning abilities of neural networks with human knowledge representation abilities of fuzzy systems.

In order to present the ANFIS architecture based on CART, the crisp rules (6) in the previous section are considered. Assuming that any input vector (x, y) is given, only one rule out of four will be fired at full strength, while the other three rules will not be activated and the output is solely determined by the fired rule. Furthermore, the crisp sets reduce the computation burden in constructing the tree using CART, but it also gives undesired discontinuous boundaries as mentioned above. This problem can be solved by using fuzzy sets that can smooth out the discontinuities at each split. Fuzzy sets, therefore, are used to represent the premise part of the rule set in Eq. (7.115). Equation (7.115) is converted into a set of fuzzy rules of either zero order (when f_i's are constant) or first order (when f_i's are linear equations). Assuming that a first-order Sugeno fuzzy model (Sugeno and Kang 1988) is considered, the crisp rules in Eq. (7.115) can be expressed as follows:

$$\begin{aligned}
&\text{Rule 1 : If } x > a \text{ and } y > b \text{ then } z = f_1 = p_1x + q_1y + r_1\\
&\text{Rule 2 : If } x > a \text{ and } y \leq b \text{ then } z = f_2 = p_2x + q_2y + r_2\\
&\text{Rule 3 : If } x \leq a \text{ and } y > c \text{ then } z = f_3 = p_3x + q_3y + r_3\\
&\text{Rule 4 : If } x \leq a \text{ and } y \leq c \text{ then } z = f_4 = p_4x + q_4y + r_4
\end{aligned} \tag{7.116}$$

where x and y are the inputs, f_i are the outputs within the fuzzy region specified by the fuzzy rule, and p_i, q_i, and r_i are the design parameters that are determined during the learning process.

The ANFIS architecture to implement these rules consists of five layers as shown in Fig. 7.38. In this architecture, circles indicate fixed nodes, while squares indicate adaptive nodes. Nodes within the same layer perform identical functions as detailed below.

Layer 1: All the nodes are adaptive nodes. The outputs of this layer are the fuzzy membership grade of the inputs, which are given by:

$$O_i^1 = \mu_{x*d}(x), \quad O_i^1 = \mu_{x*d}(y) \tag{7.117}$$

where d is the decision boundaries and $\mu_{x*d}(x), \mu_{x*d}(y)$ can adopt any fuzzy membership function. For instance, the statement $y > c$ can be represented as a fuzzy set characterized by either the sigmoid membership function with one parameter α (Jang et al. 1996):

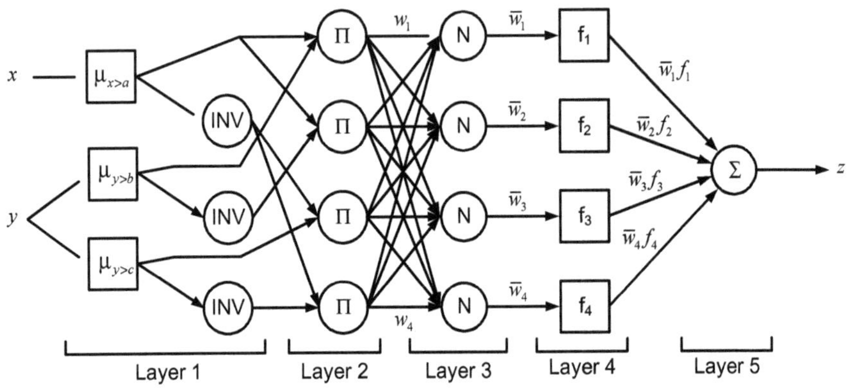

Fig. 7.38 ANFIS architecture of first-order Sugeno fuzzy model

$$\mu_{y>c}(y;\alpha) = sig(y;\alpha,c) = \frac{1}{1+\exp[-\alpha(y-c)]} \tag{7.118}$$

or the extended sigmoid membership function with two parameters: α and γ

$$\mu_{y>c}(y;\alpha,\gamma) = sig(y;\alpha,c,\gamma) = \begin{cases} 0 & \text{if } \; y \le c-\alpha \\ \frac{1}{2}\left[\frac{y-(c-\alpha)}{\alpha}\right]^{2\gamma} & \text{if } \; c-\alpha < y \le c \\ 1-\frac{1}{2}\left[\frac{c+\alpha-y}{\alpha}\right]^{2\gamma} & \text{if } \; c<y\le c+\alpha \\ 1 & \text{if } \; c+\alpha<y \end{cases} \tag{7.119}$$

where α, γ, and c are the modifiable parameters governing the shape of the membership functions. Parameters in this layer are referred to as premise parameters.

Layer 2: The nodes are fixed nodes labeled with Π, indicating that they perform as a simple multiplier. Each node in this layer calculates the firing strengths of each rule via multiplying the incoming signals and sends the product out. The outputs of this layer can be represented as follows:

$$O_i^2 = w_i = \mu_i(x)\mu_i(y), \quad i = 1,2,3,4 \tag{7.120}$$

Layer 3: The nodes are also fixed nodes. They are labeled with N, indicating that they play a normalization role to the firing strengths from the previous layer. The ith node of this layer calculates the ratio of the ith rule's firing strength to the sum of all rules' firing strengths:

$$O_i^3 = \bar{w}_i = \frac{w_i}{\sum_{i=1}^{4} w_i} = \frac{w_i}{w_1+w_2+w_3+w_4} \tag{7.121}$$

Note that this layer is not needed if the constraints: (a) $\mu_{x>a}(x) + \mu_{x\leq a}(x) = 1$ and (b) multiplication is used as the T-norm operator to calculate each rule's firing strength, then the summation over each rule's firing strength is always equal to one.

Layer 4: The nodes are adaptive nodes. The output of each node in this layer is simply the product of the normalized firing strength and a first-order polynomial. Thus, the outputs of this layer are given by:

$$O_i^4 = \bar{w}_i f_i = \bar{w}_i(p_i x + q_i y + r_i), \quad i = 1, 2, 3, 4 \tag{7.122}$$

Layer 5: There is only single fixed node labeled with Σ. This node performs the summation of all incoming signals. Hence, the overall output of the model is given by:

$$O_i^5 = \sum_{i=1}^{4} \bar{w}_i f_i = \frac{\sum_i w_i f_i}{\sum_i w_i} \tag{7.123}$$

Thus, we have constructed an adaptive network that has exactly the same function as a Sugeno fuzzy model.

(2) *Learning algorithm of ANFIS*

The task of learning algorithm for ANFIS architecture is to tune all the modifiable parameters, namely premise parameters $\{\alpha, \gamma, c\}$ and consequent parameters $\{p_i, q_i, r_i\}$, to make the ANFIS output match the training data. The least squares method can be used to identify the optimal values of these parameters easily. When the premise parameters are not fixed, the search space becomes larger and convergence of the training becomes slower. A hybrid algorithm combining the least squares method and the gradient descent method is adopted to solve the problem. The hybrid algorithm is composed of a forward pass and a backward pass. In the forward pass, the least squares method is used to optimize the consequent parameters with the fixed premise parameters. Once the optimal consequent parameters are found, the backward pass commences immediately. In the backward pass, the gradient descent method is used to adjust the premise parameters corresponding to the fuzzy sets in the input domain, while the consequent parameters remain fixed. This procedure is repeated until either the squared error is less than a specified value or the maximum number of training epochs is encountered.

Example 7.17 Classify the iris data using ANFIS algorithm. The membership functions before and after training are presented in Figs. 7.39 and 7.40. The result of classification using ANFIS is presented in Fig. 7.41.

Example 7.18 Classify the riply data using ANFIS algorithm (Fig. 7.42).

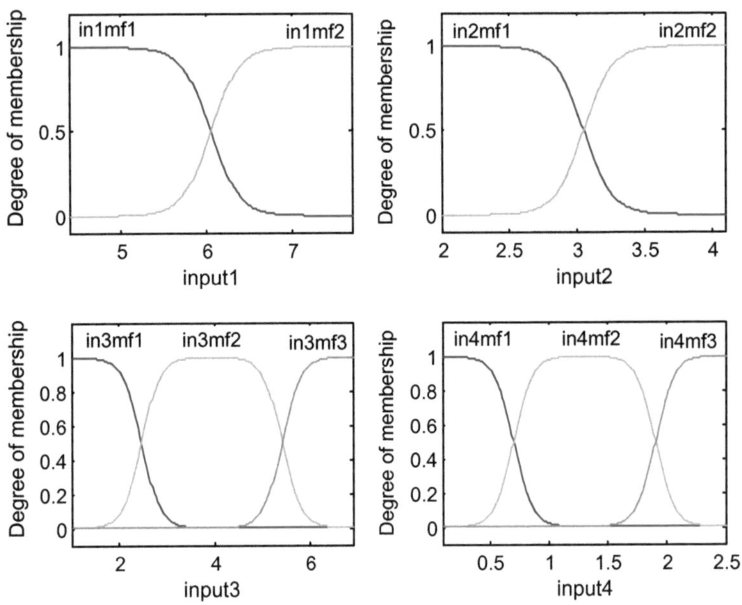

Fig. 7.39 Initial membership functions

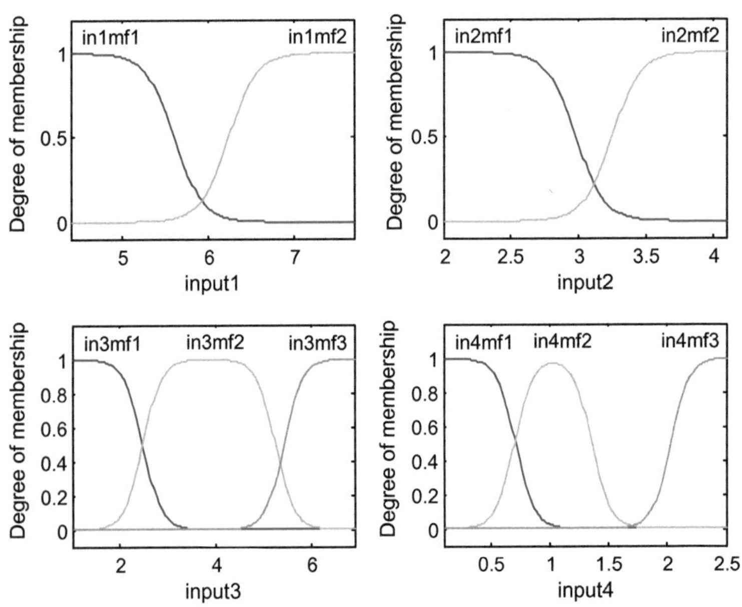

Fig. 7.40 Final membership functions

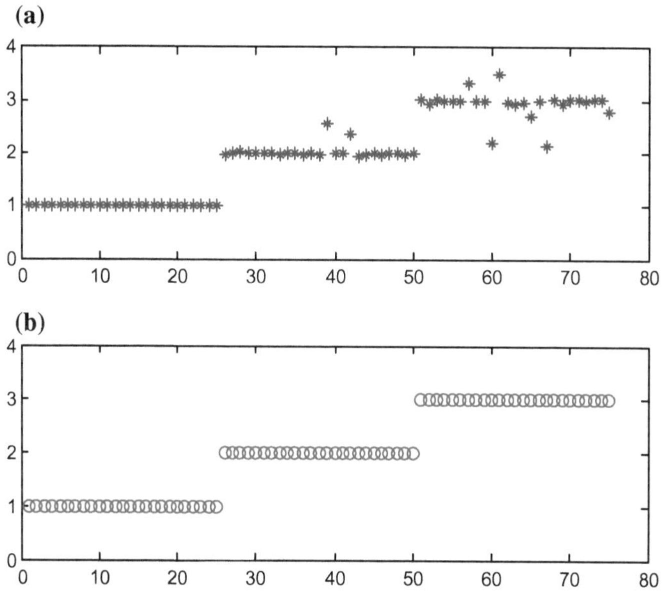

Fig. 7.41 Classification by ANFIS. **a** Result of testing. **b** Testing data

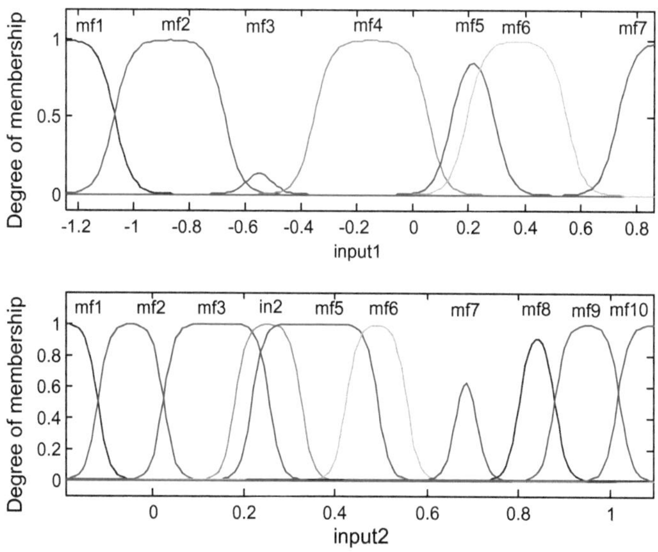

Fig. 7.42 Membership function plot

7.14 Case Studies: Fault Diagnosis of Induction Motors

Several case studies are presented for giving better understanding of the classification process. The selected methods for case studies are wavelet SVM (W-SVM), decision tree, random forest algorithm, and CART-ANFIS. Those classifier algorithms are applied to induction motor fault diagnosis using vibration and current signals.

7.14.1 W-SVM

(1) *Experiment and data acquisition*

Data acquisition was conducted on induction motor of 160 kW, 440 V, 2 poles as shown in Fig. 7.43. Six accelerometers were used to pick up vibration signal at drive-end and non-drive-end on vertical, horizontal, and axial directions, respectively. The maximum frequency of the used signals and the number of sampled data were 60 Hz and 16,384, respectively (Table 7.5).

(2) *Feature calculation*

The condition of the induction motor is briefly summarized in Table 7.7. Each condition was labeled as class from 1 to 7. There are totally 126 features calculated

Fig. 7.43 Data acquisition of induction motor

Table 7.5 Condition of induction motor

Class No.	Condition	Description	Others
1	Bent rotor	Maximum shaft deflection	1.45 mm
2	Eccentricity	Static eccentricity (30 %)	Air gap: 0.25 mm
3	MCDE	Magnetic center moved (DE)	6 mm
4	MCNDE	Magnetic center moved (NDE)	6 mm
5	Normal	No faults	–
6	Unbalance	Unbalance mass on the rotor	10 gr
7	Weak-end shield	Stiffness of the end cover	–

from 6 signals, 21 features, and 98 data calculated from 7 conditions and 14 measurements.

(3) *Feature extraction and reduction*

Basically, feature extraction is the mapping process of data from high-dimension into low-dimension space. This step is intended to avoid the curse dimensionality phenomenon. Structures of three first original features, which are mean, RMS, and shape factor, are plotted in Fig. 7.44. This figure shows the performance of original features that are containing overlap in some conditions. Then, applying component analysis is suggested to make original features well clustered.

Component analysis via ICA, PCA, and their kernel is then used to extract and reduce the feature dimensionality based on the eigenvalue of covariance matrix as described in Fig. 7.45. After performing the component analysis, the features have been changed into independent and principal components, respectively. The first three independent and principal components from PCA, ICA, and their kernel are plotted in Fig. 7.46. It can be observed that the clusters for seven conditions are separated well. It indicates that component analysis can perform feature extraction and all at once do clustering of each condition of induction motors.

Fig. 7.44 Original features

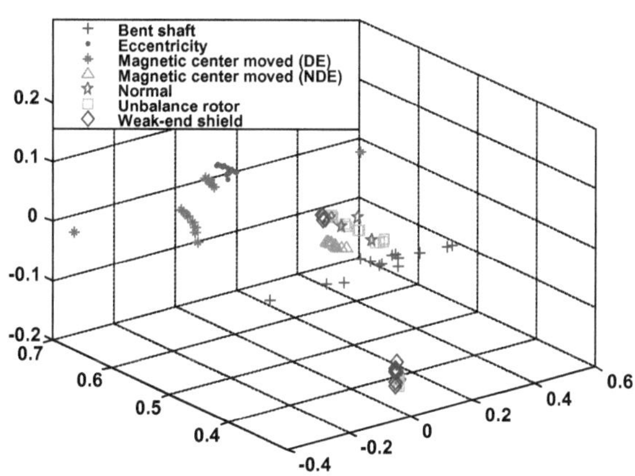

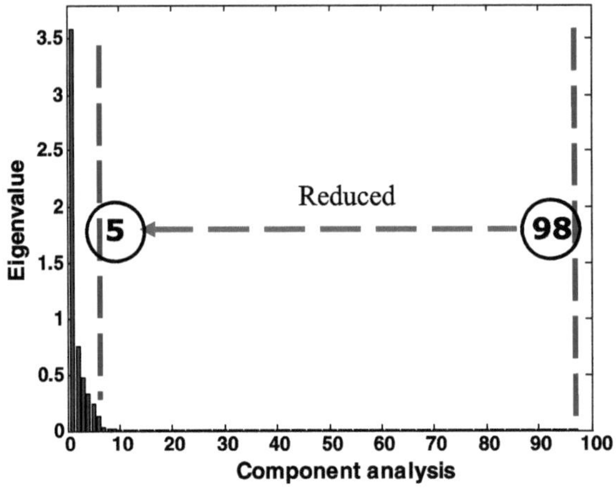

Fig. 7.45 Feature reduction using component analysis

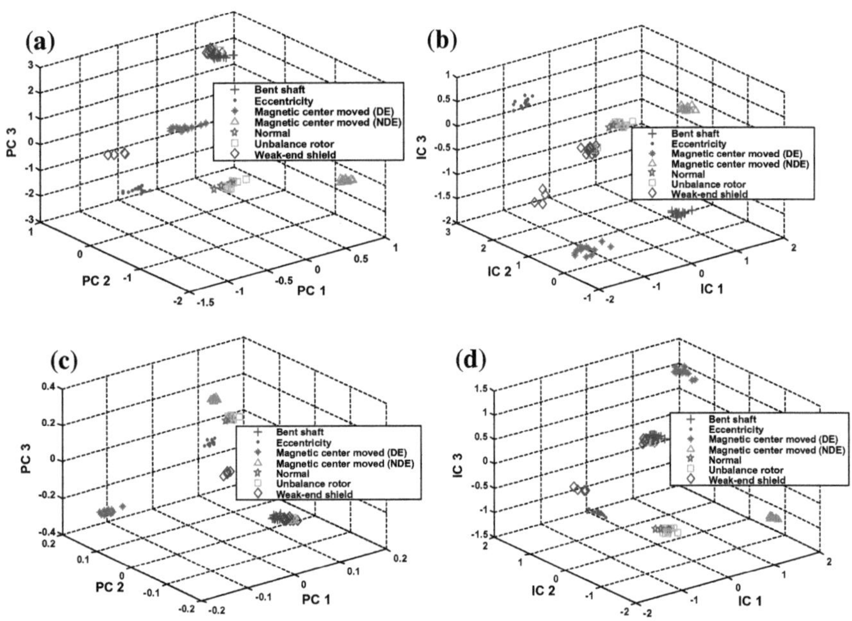

Fig. 7.46 The first three principal and independent components. **a** Principal components. **b** Independent components. **c** Kernel principal components. **d** Kernel independent components

According to the eigenvalue of covariance matrix, the features were changed into component analysis and reduced only 5-component analysis needed for classification process. The other features are discarded due to small eigenvalue of covariance matrix. The selected component analysis is then used by W-SVM classifier as input vectors for fault diagnosis using classification routine.

(4) *Training and classification*

The SVM-based multi-class classification is applied to perform the classification process using one-against-all methods. To solve the SVM problem, Vapnik (1982) describes a method which used the projected conjugate gradient algorithm to solve the SVM-QP problem. In this study, SVM-QP was performed to solve the classification problem of SVM. The parameters C (bound of the Lagrange multiplier) and A (condition parameter for OP method) were 1 and 10^{-7}, respectively.

Wavelet kernel function using Daubechies series was performed in this study. The parameter δ in wavelet kernel refers to the number of vanishing moments and is set 4. In the training process, the data set was also trained using RBF kernel function as comparison. The parameter γ for bandwidth RBF kernel was user-defined which is equal to 0.7.

(5) *Result and discussion*

The complex separation boundaries are presented in Fig. 7.47 from which the separation of W-SVM can be shown. In these figures, the circle refers to the support vector that states the correct recognition in W-SVM. Each condition of induction motor is well recognized using Daubechies wavelet kernel. In the classification

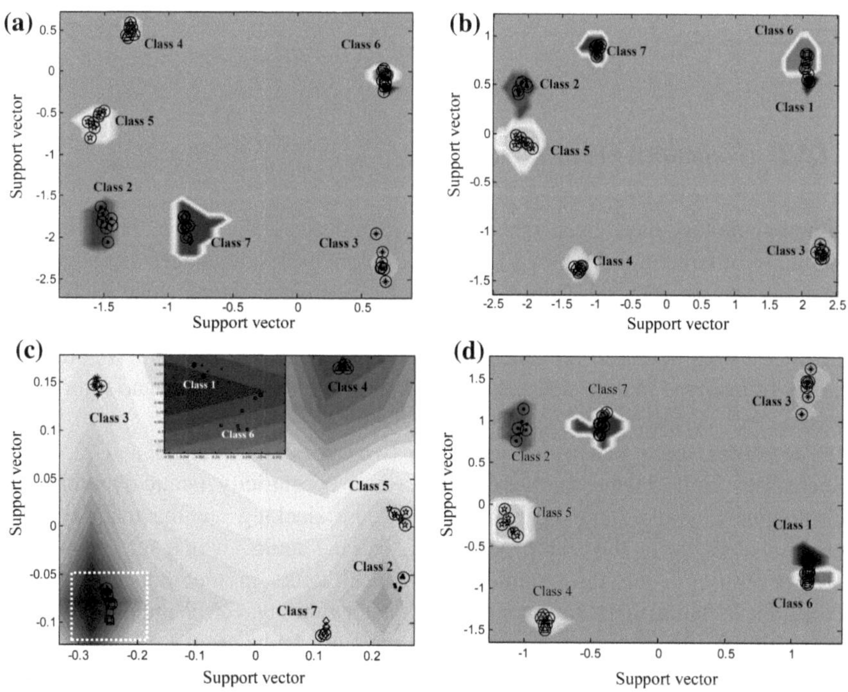

Fig. 7.47 Separation boundaries of W-SVM. **a** Daubechies kernel with PC data. **b** Daubechies kernel with IC data. **c** Daubechies kernel with kernel PC data. **d** Daubechies kernel with kernel IC data

Table 7.6 Results of classification

Kernel	Accuracy (train/test), %				Number of SVs			
	IC	PC	Kernel IC	Kernel PC	IC	PC	Kernel IC	Kernel PC
Wavelet Daubechies	100/100	100/100	100/100	100/100	35	39	39	17
RBF Gaussian ($\gamma = 0.5$)	100/100	100/100	100/100	100/100	22	22	25	33

process using W-SVM, each condition of induction motors can be clustered well. The good separation among conditions shows the performance of W-SVM doing recognition of component analysis from vibration signal features.

The performance of classification process is summarized in Table 7.6. All data sets come from the component analysis are accurately classified using Daubechies wavelet kernel and SVM and reached 100 % accuracy in training and testing, respectively. SVM using RBF kernel function with kernel width $\gamma = 0.5$ is also performed in classification for comparison with Daubechies wavelet kernel. The results show that the performance of W-SVM is similar to SVM using RBF kernel function, which are 100 % in accuracy of training and testing, respectively. In the case of number support vectors, SVM with RBF kernel function needs lower than W-SVM except kernel PCA.

7.14.2 Decision Tree

Induction motors have been widely used in industries due to its reliability and simplicity of structure. Although induction motors have advantages as above, they are subject to some faults. These faults may be inherent to the machine itself or due to operation conditions. The cause of inherent faults results from the mechanical or electrical forces acting in the machine enclosure. Many researchers have studied a variety of machine faults, such as winding faults, unbalanced stator and rotor faults, broken rotor bar, eccentricity, and bearing faults.

The motor faults can be divided into two types: mechanical and electrical. Mechanical faults in the rotor are identified as eccentricity (static/dynamic) and misalignment, while stator eccentricity and core slacking are the main types of mechanical faults in the stator. Moreover, bearing faults, which may also cause rotor eccentricity, are the common mechanical faults in the induction motors. Except these faults, there are rotor rubbing, stator rubbing, etc. Winding faults such as turn-to-turn, phase-to-phase, and winding to earth faults are the root of electrical faults in the rotor of slip ring induction motor. The roots for electrical faults in the squirrel cage rotor are bars crack, bars slack, and bad connection with the end rings.

When these faults happened, we can see some features from vibration or current spectrum. For example, if induction motor has a fault in rotor bar, we can find many

Table 7.7 Example of cause–result matrix

Class		1X	m_1X	s_1X	$1X \pm f_p$
Broken rotor bar	Initial	O			O
	Bad	O	O		O
Rotor shorting rings		O			O
Rotor bar crack		O	O		O
Rotor rubbing		O	O	O	
Rotor bar looseness		O			

symptoms and properties. Broken or cracked rotor bars will produce high 1X (running speed) vibration with pole pass frequency sidebands F_p. Severity of broken rotor bar is difficult to diagnose through the vibration spectrum. So we need to utilize the current signal. And loose rotor bars are indicated by 2X line frequency ($2F_L$) sidebands surrounding the rotor bar pass frequency (RBPF) and/or its harmonics. Besides, there are many attributes to indicate a feature of fault, such as whirling motion, phase angle variation, and temperature.

This is considered as the most applicable and effective tool for diagnosis of broken rotor bar. And this is needed zooming for detecting the sidebands of the fundamental frequency. The spectrum of current signal is different according to the load condition. This is very important method to judge the state of machine.

As mentioned before, we made a cause–result matrix which consists of the relations between attributes and classes as shown in Table 7.7. Attributes means the feature frequencies incurrent and vibration signal and other symptoms when the induction motors are under operation. The relations between class and attribute are collected from general knowledge of induction motor faults and the fault histories of some motors. Especially, fault histories of motors have been reported primarily about the symptom and diagnosis result when the field experts observed the state of induction motor using the analysis techniques through the world industry.

First, we made two data. One is motor fault data, and the other is bearing data. There are some reasons to divide data like this. The attributes of motor fault are different from those of bearing. The size of the data becomes bigger when we combine the two data. And it demands more cases to construct a complete decision tree (Fig. 7.48).

We tested the training data which are constructed through the cause–result matrix, using C7.0 program. C7.0 uses C4.5 criterion and error-based pruning method. The purpose of this test is to find the best training set for classification rate. Test data were mixed with many random unknown values. Table 7.8 shows the results for applied three types of data.

Data type I is constructed with 280 instances, and each class has instances of the same number. Type II is done with the same instances, and each class has instances of the random number. Type III complies with the survey ratio of EPRI and IEEE. All types of data have 280 instances and same unknown attributes in common. We selected data type III from this result. The effect of pruning revealed is to be

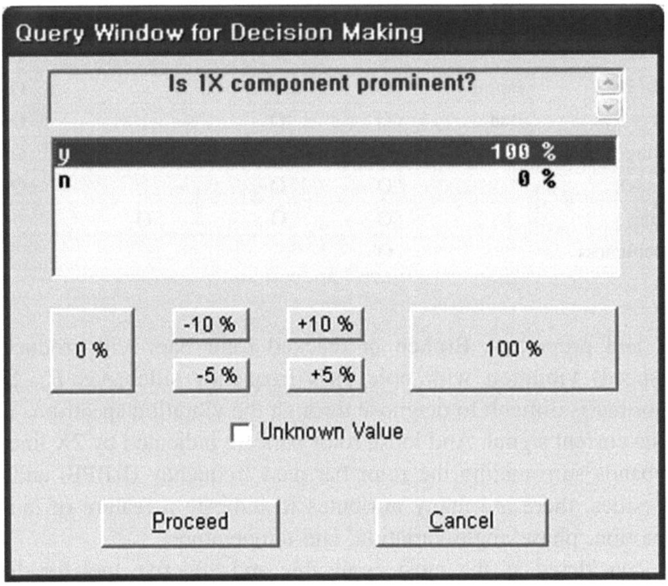

Fig. 7.48 Consultant window for diagnosis

Table 7.8 Test results of data type using C7.0

Data type	Type I (made with same number)				Type II (random weighting)				Type III (weighting for standards)			
	Training		Test		Training		Test		Training		Test	
Degree of pruning												
Degree	Size	Error (%)	Size	Error (%)	Size	Error (%)	Size	Error (%)	Size	Error (%)	Size	Error (%)
1	31	0	31	32.3	19	3.3	19	54.8	31	0	31	16.1
2	31	0	31	32.3	19	3.3	19	54.8	31	0	31	16.1
5	31	0	31	32.3	19	3.3	19	54.8	31	0	31	16.1
10	31	0	31	32.3	21	2.8	21	48.4	31	0	31	16.1
15	31	0	31	32.3	21	2.8	21	48.4	31	0	31	16.1
20	31	0	31	32.3	21	2.8	21	48.4	31	0	31	16.1
25	31	0	31	32.3	21	2.8	21	48.4	31	0	31	16.1

improved for data type II. The others are the same according to the degree of pruning without changing the error rate.

Figure 7.48 expresses the consultant window of system which asks users for what they have measured about the state of induction motor. Questions for attributes are seen and proceeded step by step. When the users do not know or have not measure any attributes, they can check in unknown value term. This means that most of the users do not have all the information about the state or symptom of

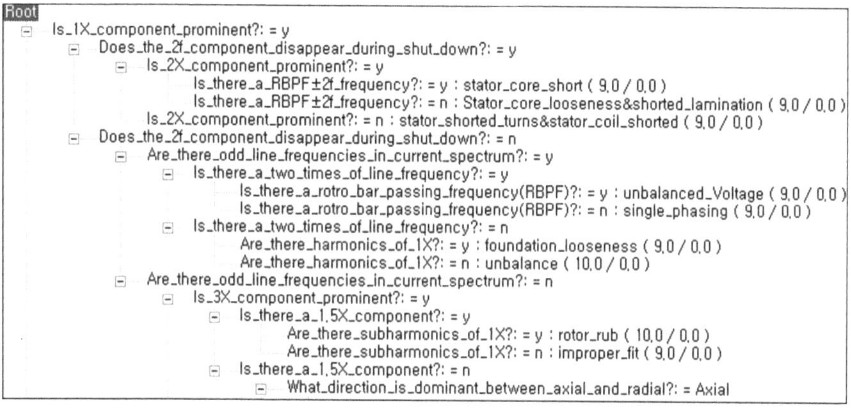

Fig. 7.49 Decision tree of induction motor

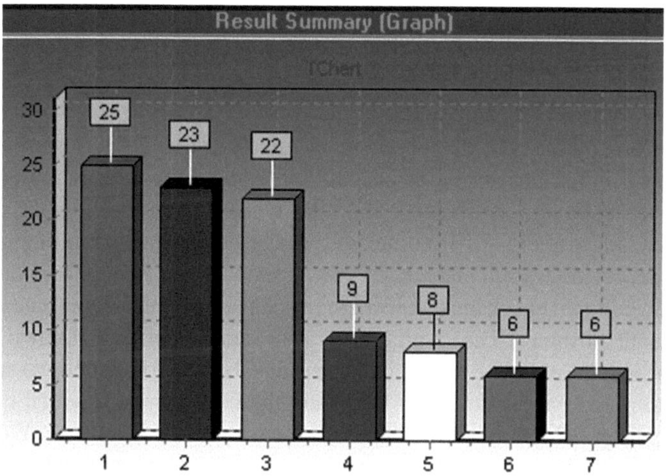

Fig. 7.50 Diagnosis result of decision tree

induction motor. Answer "y" or "n" means that what percents users have confidence for some attributes they have measured. We made an expert system with visual basic programming using the C4.5 inducer of Quinlan. We used the input data decided as above. Figure 7.49 stands for the decision tree of induction motor which is made from input data. We have difficulty finding the path to arrive from the root to special fault due to their complex constructions. So tree needs to be automated to look for the paths.

Figure 7.50 shows the diagnosis result of decision tree after some questions through interfacing with consultant window. This is expressed as charts which are easy for users to find the type of faults.

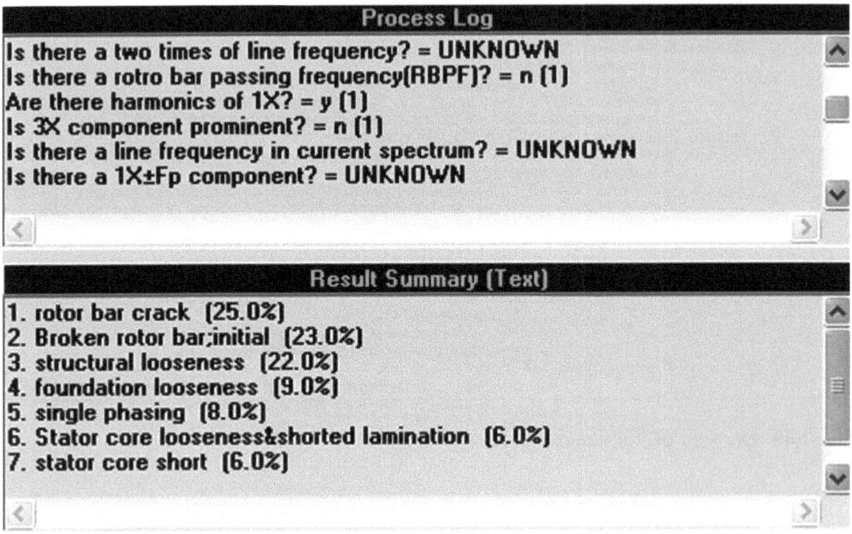

Fig. 7.51 Description of consultant process

We can look for the consultant process in window of Fig. 7.51. Some faults can be detected with minimum attributes and the others be done with many ones. If we cannot observe enough to diagnose the fault type, we can look for the consultant process in window of Fig. 7.51. Some faults can be detected with minimum attributes and the others be done with many ones. If we cannot observe enough to diagnose the fault type, we should check the unknown values. In general, there are some difficulties to measure all attributes. The result of Fig. 7.48 is due to many unknown values.

Fault of induction motor results in serious economical loss when we cannot find the symptom and anticipate them. We made the expert system which can have a role of expert in field and is based on the inducer called decision tree. We arranged the relations between attributes and classes and symptoms and faults as in Table 7.8. Then, we made the cause–result matrix as described above. Input data for expert system was selected with criterion test and data construction which can choose the best data for classification rate. Expert system is applied to induction motors for fault diagnosis through the consultant window, using real data for inspection.

7.14.3 Random Forest

(1) *Data acquisition and feature calculation*

The experiment is designed to simulate six most universal categories of induction motor faults which are broken rotor bar, bowed rotor, bearing outer race fault, rotor

Fig. 7.52 Faults on introduction motor

unbalance, and adjustable eccentricity motor (misalignment) and phase unbalance (Yang et al. 2006). First, four motor faults are shown in Fig. 7.52 as an example. The platform of this experiment consists of six 0.5 kW, 60 Hz, 4-pole induction motors, pulleys, belt, shaft, and fan. The detailed information of faults is listed in Table 7.9.

Three AC current probes and another three accelerometers were used to measure the stator current of three-phase power supply and vibration signals of horizontal, vertical, and axial directions for evaluating the RF-based fault diagnosis system.

After measuring the raw data, a preprocessing is implemented on the data to obtain the most important features for the RF-based diagnosis methodology. Finally, there are 63 features left which are prepared for the next procedure, induction motor fault diagnosis by RF. The description of features is shown in Table 7.10.

Figure 7.53 marks the eight conditions of faulty motor and one normal condition via selecting the two features from the all randomly for three times. The purpose is to explain the relation between features and fault categories. It can be seen that although selected features are calculated for standing difference out among diverse fault categories, but overlap cannot be avoided, in some samples it is even serious based on two randomly selected features.

Table 7.9 Fault category of induction motors

Fault condition	Fault description	Others
Broken rotor bar	Number of broken bars: 12 ea	Total number of 34 bars
Bowed rotor	Max. bowed shaft deflection: 0.075 mm	Air gap: 0.25 mm
Faulty bearing	A spalling on outer raceway	#6203
Rotor unbalance	Unbalance mass on the rotor	8.4 g
Eccentricity	Parallel and angular misalignments	Adjusting the bearing pedestal
Phase unbalance	Add resistance on one phase	8.4 %

Table 7.10 Representation of input features for RF

Time domain	Frequency domain	Autoregression
Mean	Root-mean-squared frequency	AR coefficients
RMS	Frequency center	($a1 \sim a8$)
Shape factor	Root variance frequency	
Skewness		
Kurtosis		
Crest factor		
Entropy error		
Entropy estimation		
Histogram lower		
Histogram upper		

Generally, some statistical techniques, such as principal component analysis (PCA), kernel PCA, and linear discriminant analysis, are employed to compress the data again by reducing the number of dimensions of data. But for RF-based system, it is wised to do diagnosis without such techniques because of the two reasons. First, information of data has been extracted when features are calculated. If dimension-reducing technique is employed, it may cause overcompressing problem of data. Second, according to Breiman's test (Breiman 2001), RF always gives good performance when the data scale is large. Therefore, RF is more sensitive to the overcompressing problem. Thus, RF finished diagnostic task without any feature extraction methods, and excellent experiment result shows that it is unnecessary at all.

(2) *Fault diagnosis result and discussion*

In this section, RF was run on the induction motor fault data. Number and faulty sorts of training and testing data are shown in Table 7.11. The experimental results for RF-based method are given in Table 7.12. Confusion matrices for the training data in RF are given by Tables 7.13 and 7.14 which indicate the accuracies of each fault class for training and testing data with 907 trees and selecting 1 variable in every split.

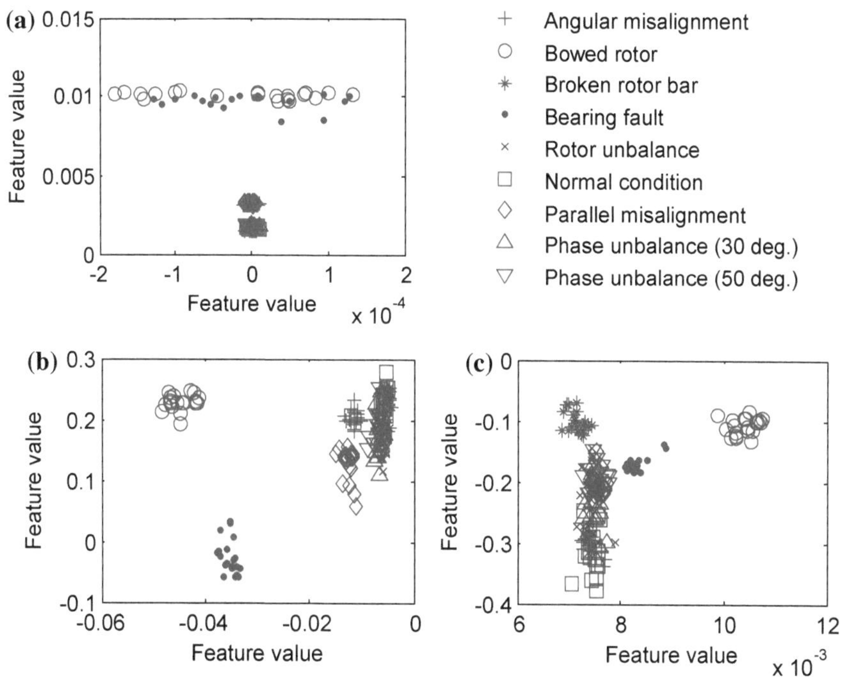

Fig. 7.53 Connection between features and fault categories. **a** 1st time. **b** 2nd time. **c** 3rd time

Table 7.11 Information of class and samples

Class No.	Class	Training samples	Test samples
1	Angular misalignment	20	10
2	Bowed rotor	20	10
3	Broken rotor bar	20	10
4	Bearing outer race fault	20	10
5	Mechanical unbalance	20	10
6	Normal condition	20	10
7	Parallel misalignment	20	10
8	Phase unbalance (30°)	20	10
9	Phase unbalance (50°)	20	10
Total samples		180	90

Figure 7.54 shows the classification rate according to the experiment which represents three characteristics of RF very clearly. First, compared with the number of component classification trees, the parameter, random split number at each node, is more sensitive to the classification accuracy. Hence, a prudential searching procedure is necessary to find the best split variable number by an experimental way. Second, if the split variable number is decided, the sum of individual tree

Table 7.12 Fault diagnosis accuracies (%) based on RF

No. of trees	Split variables		
	1	5	8
200	88.89	71.11	81.22
500	94.44	77.56	83.33
1200	97.56	72.23	82.34
2000	93.33	73.34	83.33
5000	92.23	72.23	78.89
10,000	92.25	74.44	77.78

Table 7.13 Accuracy of each fault class for training data with 907 trees and selecting 1 variable every split

Class No.	1	2	3	4	5	6	7	8	9	Accuracy (%)
1	20	0	0	0	0	0	0	0	0	100
2	0	20	0	0	0	0	0	0	0	100
3	0	0	20	0	0	0	0	0	0	100
4	0	0	0	20	0	0	0	0	0	100
5	0	0	0	0	20	0	0	0	0	100
6	0	0	0	0	0	20	0	0	0	100
7	0	0	0	0	0	0	20	0	0	100
8	0	0	0	0	0	0	0	20	0	100
9	0	0	0	0	0	0	0	0	20	100

Table 7.14 Accuracy of each fault class with 907 trees and selecting 1 variable every split

Class No.	1	2	3	4	5	6	7	8	9	Accuracy (%)
1	10	0	0	0	0	0	0	0	0	100
2	0	10	0	0	0	0	0	0	0	100
3	0	0	10	0	0	0	0	0	0	100
4	0	0	0	10	0	0	0	0	0	100
5	0	0	0	0	9	1	0	0	0	90
6	0	0	0	0	0	10	0	0	0	100
7	0	0	0	0	0	0	10	0	0	100
8	0	0	0	0	0	0	0	10	0	100
9	0	0	0	0	0	0	0	0	10	100

classifier should achieve an appropriate quantity to get a better performance. Last one, when we increase trees to a high number, for example 5000 or 10,000, there is no overfitting occurred, but a little undulating exists.

Table 7.14 indicates that incorrect diagnosis of RF-based methodology often occurs at certain fault category. So we can apply some assistant diagnosis method which is function in that specific kind of fault to improve the diagnosis precision.

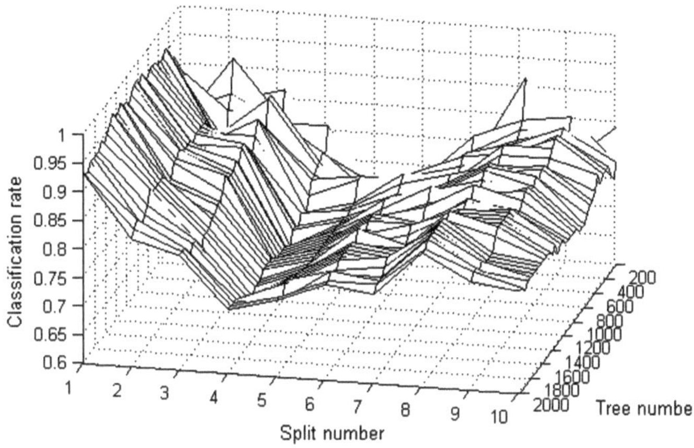

Fig. 7.54 Classification rate against random split number and tree number

In general, the normal RF has achieved the satisfied fault diagnosis accuracy. But it should be noticed that two parameters, the number of trees and random split number, which greatly affect classification result are set manually. It means accuracy of normal RF depends on researcher's experience. This situation exists at almost all the applications of RF. So that applying the GA to do the parameter optimization is exigent. The effect of this cooperation is proved by using the same data. According to the pervious research, in order to reduce executed time of GA program and find the optimized point synchronously, the number of trees and random split number are limited in the range from 500 to 1500 and from 1 to 10, respectively.

Figure 7.55 shows the trace information of every generation. Fitness adopts the classification accuracy of the test data set. The solid line is the best fitness value, and the other one is mean fitness value of each generation. The risen and convergent trend of mean fitness value indicates that GA well cooperates with RF-based methodology on the motor fault diagnosis, and the best fitness value lays out of the optimization point which is 907 trees and 1 random split created by 9th generation. The classification accuracy at this point achieves the 98.89, 3.33 % higher than the best value of normal RF. It means GA can enhance the capability of RF distinctly.

Finally, we have also investigated the accuracy of other classifiers, adaptive resonance theory Kohonen neural network (ART-KNN) (Yang et al. 2004a, b), SVM (Widodo et al. 2007), and CART (Breiman et al. 1984). Classification accuracies (Table 7.15) were obtained as 97.56 % for RF only and 98.89 % for RF optimized by GA (RFOGA), while the classification accuracies were 86.67 % for ART-KNN, 87.15 % for SVM, and 77.78 % for CART. It can be seen from the results that RFOGA and RF only achieve higher classification rates than ART-KNN, SVM, and CART. RF has greatly increased the capability of tree classification method, to 97.56 % from 77.78 % for CART.

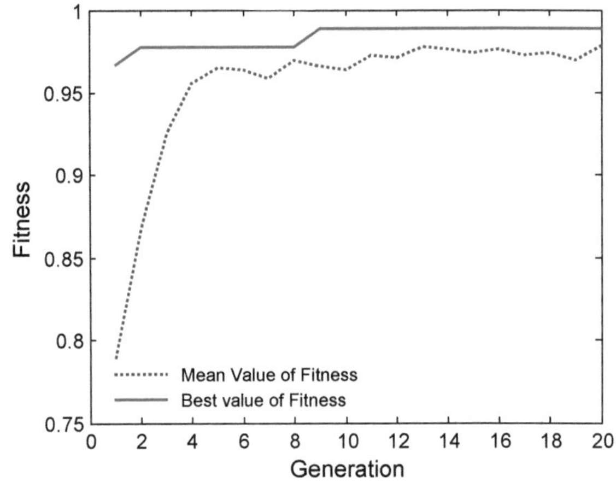

Fig. 7.55 Optimization trace within 40th generation

Table 7.15 Overall classification accuracy of each classifier

Classifier	Overall accuracy (%)
ART-KNN	86.67
SVM	87.15
CART	77.78
RF only	97.56
RFOGA	98.89

This comparative result means all we have done is significant and the farther research is important and necessary.

The purpose of this paper is to confirm the possibilities of using random forest algorithm (RF) in machine fault diagnosis and propose a hybrid method combined with genetic algorithm to improve the classification accuracy. The proposed method is based on RF, a novel ensemble classifier which builds a large amount of decision trees to improve the single-tree classifier. Although there are several existed techniques for fault diagnosis, the research on RF is meaningful and necessary because of its fast execution speed, the characteristic of tree classifier, and high performance in machine fault diagnosis. Evaluation of the RF-based method has been demonstrated by a case study on induction motor fault diagnosis. Experimental results indicate the validity and reliability of RF-based fault diagnosis method. In this paper, the RF and optimized RF-based fault diagnosis method of rotating machinery were investigated. The performance of two methods was proved by the fault diagnosis test of an induction motor. The optimized approach attains a high accuracy rate of diagnosis, 98.89 %. The comparison result also shows that the optimized RF-based method is competitive with other classification methods. The extended research will focus on two parts. First part is to improve this hybrid method RFOGA: Not only GA is used for the parameter optimization, but also it can be used to select the best combination of subclassification trees from the forest to get the more accurate result. As the second part, we will decrease the redundancy of the RF and try other optimization algorithms or more effective voting principle. The extended research will focus on decreasing the redundancy of the RF and try other optimization algorithms or more effective voting principle.

7.14.4 CART-ANFIS

(1) *Data acquisition*

To validate the CART-ANFIS model, the experiment was carried out by using test rig that consists of a motor, pulleys, belt, shaft, and fan with changeable blade pitch angle that represents the load. The load can be changed by adjusting blade pitch angle or the number of blades. Six induction motors of 0.5 kW, 60 Hz, 4 poles were used to create data. One of the motors is good condition (healthy), which is considered as a benchmark in comparison with faulty motors. The others are faulty motors, which are rotor unbalance, broken rotor bar, phase unbalance, bearing outer race fault, bowed rotor, and adjustable eccentricity motor, as shown in Fig. 7.52. The conditions of faulty motors are described in Table 7.9.

For acquiring data from test rig, three AC current probes and three accelerometers were used to measure the stator current of three-phase power supply and vibration signal of horizontal, vertical, and axial directions for evaluating the fault diagnosis system, respectively. The maximum frequency of the signal was 3 kHz, the number of sampled data was 16,384, and the measured time was 2.133 s. The time waveform of vibration and stator current signals is shown in Fig. 7.56. From vibration signals, we can see that there is much difference between the waveforms of normal, rotor unbalance, rotor bar broken, and phase unbalance, which are approximate sine waves same with running speed; the characteristics of misalignment waveforms are sinusoidal with one or two clear cycles per revolution; and many impacts are in bowed rotor and faulty bearing waveforms. For stator current signals, much difference cannot be found among these faults from time waveforms since the main component is line frequency and fault signals are modulated or riding on the sine wave of line frequency (60 Hz).

(2) *Feature calculation*

The measured signals after being obtained from the experiment were calculated to get the most significant features by feature calculation. The accuracy of feature calculation is of substantial importance since it directly affects the final diagnosis results. In this paper, the feature calculation using statistical feature parameters from time domain and frequency domain was used. Sixty-three features in total are calculated from 10 feature parameters of time domain. These parameters are mean, RMS, shape factor, skewness, kurtosis, crest factor, entropy error, entropy estimation, histogram upper, and histogram lower. And three parameters are from frequency domain (RMS frequency, frequency center, and root variance frequency) using three-direction vibration signals and three-phase current signals. The total of

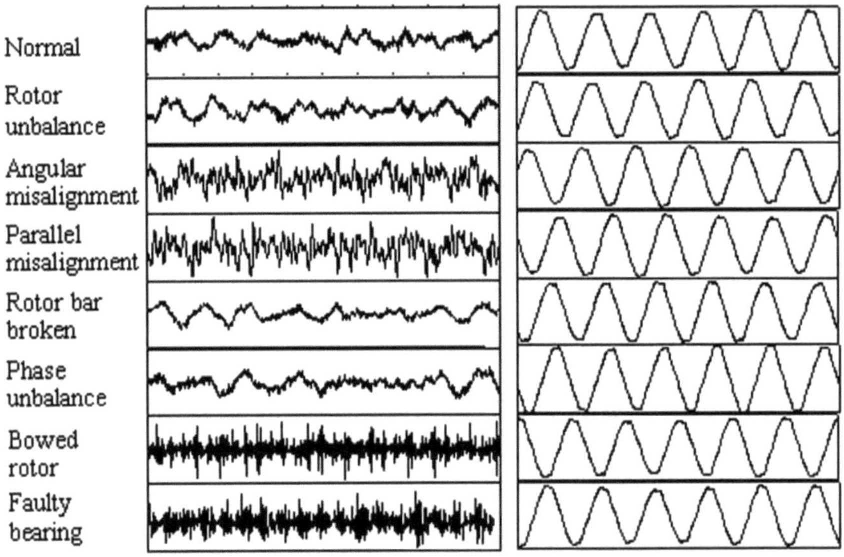

Fig. 7.56 Vibration and current signals of each condition

feature parameters is shown in Table 7.16. The data sets of the features include 270 samples. For each operating condition, 20 samples are employed for training process and 10 samples for testing. The detailed descriptions of those data sets are shown in Table 7.17.

(3) *Feature selection and classification*

A decision tree grows wholly based on training data sets and then prunes the tree back to give the final tree. Figures 7.57 and 7.58 depict the trees corresponding to the data set of features obtained from vibration signal and current signal, respectively. Obviously, a number of features are strikingly diminished and only 4 features (x_2, x_5, x_{15}, and x_{23}) of vibration signal and 7 features (x_2, x_5, x_6, x_8, x_{11}, x_{15}, and x_{19}) of current signal are remained. The reduction of features will decrease burden computation for ANFIS classifier in the next step. Furthermore, the structure of ANFIS classifier can be defined based on the crisp rules and boundary values of the decision trees.

In order to implement the fault diagnosis of induction motors by using ANFIS classifier, the structure identification for classifier is antecedently defined. This structure includes fuzzy rule set and membership functions. The fuzzy rule set is also crisp rule set of decision tree that has been fuzzified. Bell-shaped membership functions, whose initial parameters are determined based on boundary values, are chosen for our classifiers. For instance, the topology of ANFIS architecture designed by using MATLAB software package with fifteen fuzzy rules for vibration signal data set is shown in Fig. 7.59, from where the number of nodes in each layer, the number of fuzzy rules, and other meaningful information can be seen.

Table 7.16 Feature parameters

Signals	Position	Feature parameters	
		Time domain	Frequency domain
Vibration	Vertical	Mean	RMS variance frequency
	Horizontal	RMS	Frequency center
	Axial	Shape factor	Root variance frequency
Current	Phase A	Skewness	
	Phase B	Kurtosis	
	Phase C	Crest factor	
		Entropy error	
		Entropy estimation	
		Histogram lower	
		Histogram upper	

Table 7.17 Descriptions of data sets

Label of classification	Condition	Number of training samples	Number of testing samples
C1	Angular misalignment	20	10
C2	Bowed rotor	20	10
C3	Broken rotor bar	20	10
C4	Bearing outer race fault	20	10
C5	Mechanical unbalance	20	10
C6	Normal condition	20	10
C7	Parallel misalignment	20	10
C8	Phase unbalance (30°)	20	10
C9	Phase unbalance (50°)	20	10
Total samples		180	90

The system parameters and the chosen membership functions are automatically adjusted during the learning process. The convergence of the root-mean-squared (RMS) error is utilized to evaluate the learning process. If the decreasing rate of the RMS error and the performance are not significant, the learning process can be terminated.

In Fig. 7.60, the RMS error decreased to 0.087 after 800 training epochs for vibration signal data set which meant the network had learned the training data very well. In other words, the premise parameters of the membership functions corresponding to the inputs were changed for the sake of network convergence according

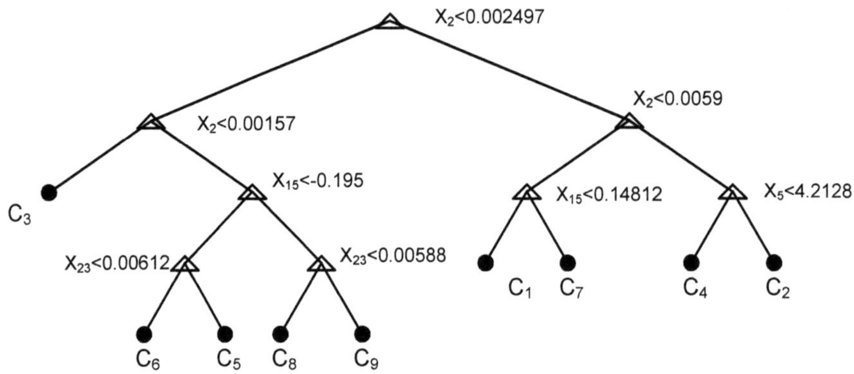

Fig. 7.57 Decision tree of features obtained from vibration signals

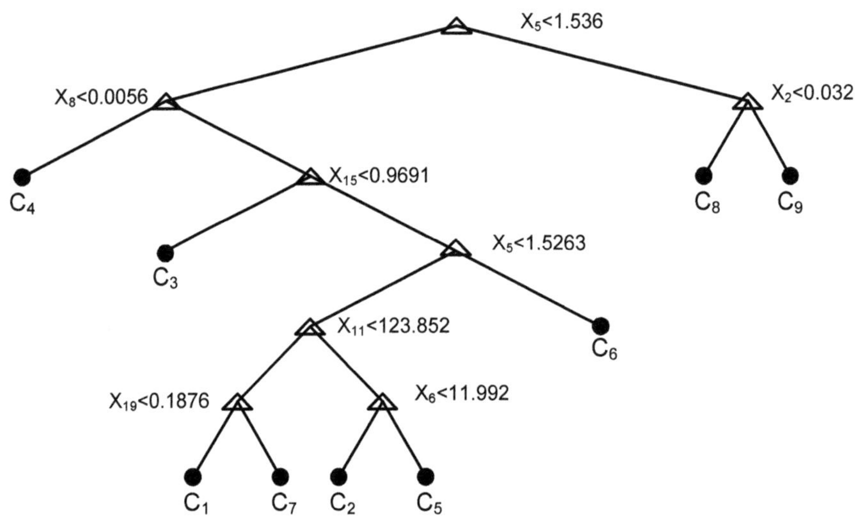

Fig. 7.58 Decision tree of features obtained from current signals

to the given training samples. The membership function of each input parameter was divided into three regions, namely small, medium, and large.

Figure 7.61 shows the initial (before training) and final (after training) membership functions of the four input parameters, using the generalized bell-shaped membership function. From this figure, one can see that changes of the final membership functions of input 2 (x_5) and input 3 (x_{15}) are similar, while input 1 (x_2) and input 4 (x_{23}) are changed.

The classification results are calculated using a tenfold cross-validation evaluation where the data set to be evaluated is randomly partitioned so that 180 samples are used for training and 90 samples are used for testing. The process is iterated

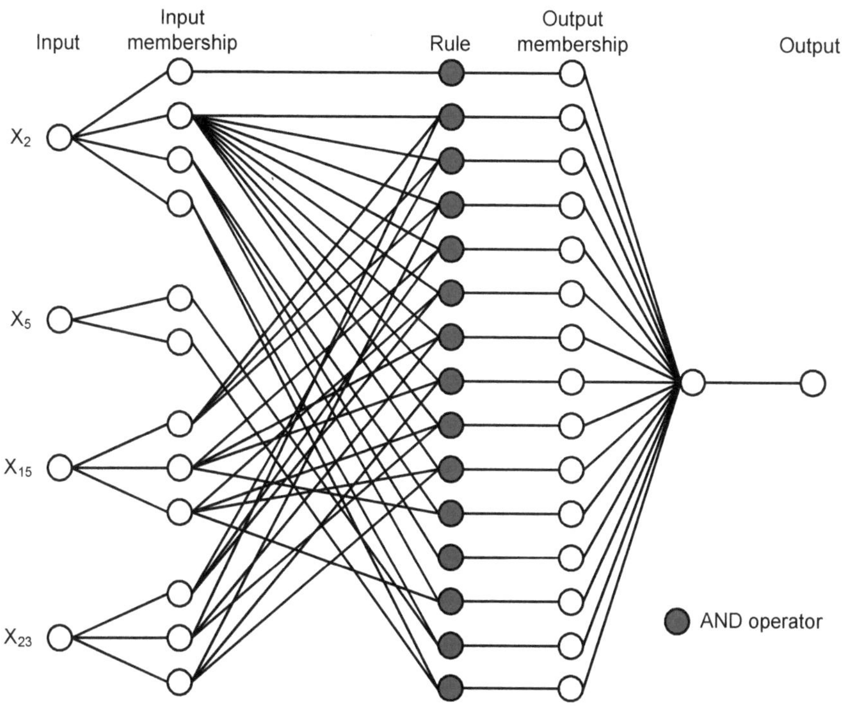

Fig. 7.59 Topology of ANFIS architecture for vibration signals

Fig. 7.60 The network RMS error convergence curve

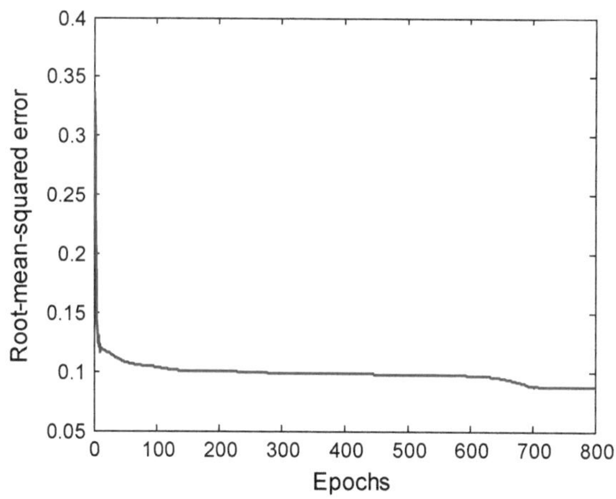

with different random partitions, and the results are averaged. The CART-ANFIS achieved 100 % classification accuracy without misclassification out of 180 training data for vibration and current signals. After training, the CART-ANFIS was tested against the testing data. The confusion matrix showing the classification results of

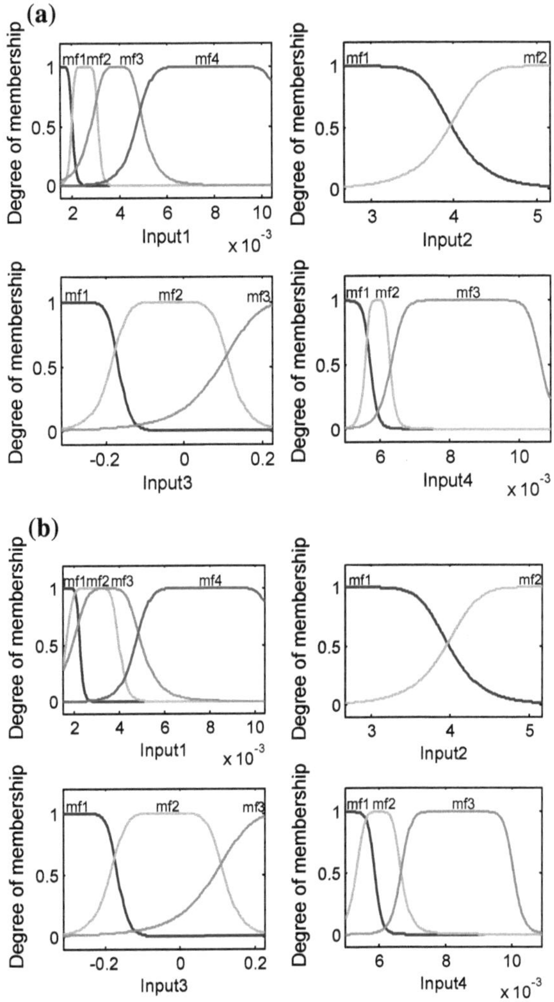

Fig. 7.61 Bell-shaped membership function for vibration signals. **a** Initial (before training). **b** Final (after training)

the CART-ANFIS created with 800 epochs of training cycle is given in Table 7.18. In confusion matrix, each cell contains the raw number of samples classified for the corresponding combination of desired and actual network outputs. The total classification accuracy for the test data was found as 91.11 % with 8 misclassifications out of 90 test samples for vibration signal, while 76.67 % with 21 misclassifications out for current signal. For example, according to the confusion matrix for current signals, one subject from C1 was incorrectly classified as subjects from C3, and two subjects from C1 were classified as subjects from C8. Two subjects from C2 were classified as subjects from C8, while one subject from C3 was classified as a subject from C8.

Table 7.18 The confusion matrix for CART-ANFIS of 800 epochs

Output/ desired	Confusion matrix (vibration/current signals)								
	C1	C2	C3	C4	C5	C6	C7	C8	C9
C1	10/7	0/0	0/0	0/0	0/0	0/0	0/0	0/0	0/0
C2	0/0	10/8	0/0	1/0	1/0	0/0	1/1	0/1	0/0
C3	0/1	0/0	7/9	0/1	0/1	0/2	0/0	0/0	1/2
C4	0/0	0/0	0/0	9/8	0/1	0/0	0/1	0/1	0/0
C5	0/0	0/0	0/0	0/0	8/7	0/0	0/0	0/0	0/0
C6	0/0	0/0	0/0	0/0	0/0	10/7	0/0	0/0	0/0
C7	0/0	0/0	0/0	0/0	0/0	0/1	9/7	0/0	0/0
C8	0/2	0/2	1/1	0/1	1/0	0/0	0/0	10/8	0/0
C9	0/0	0/0	2/0	0/0	0/1	0/0	0/1	0/0	9/8

Table 7.19 The value of statistical parameters

Data set label	Statistical parameters (vibration/current signals)		
	Sensitivity (%)	Specificity (%)	Total classification accuracy (%)
C1	100/70	100/100	91.11/76.67
C2	100/80	96.5/97.5	
C3	70/90	98.75/91.25	
C4	90/80	100/96.25	
C5	80/70	100/100	
C6	100/70	100/100	
C7	90/70	100/98.75	
C8	100/80	97.5/92.5	
C9	90/80	97.5/97.5	

The test performance of the classifier can be determined by the computation of statistical parameters such as sensitivity, specificity, and total classification accuracy defined by:

- *Sensitivity:* number of true-positive decisions/number of actually positive cases.
- *Specificity:* number of true-negative decisions/number of actually negative cases.
- *Total classification accuracy:* number of correct decisions/total number of cases.

The values of statistical parameters are given in Table 7.19. The CART-ANFIS model classified C1 to C9 subject with the accuracy of 100/70, 100/80, 70/90, 90/80, 80/70, 100/70, 90/70, 100/80, and 90/80 % for vibration and current signals, respectively. All of the data sets were classified with the accuracy of 91.11 %/ 76.67 % (total classification accuracy).

A combination of classification and regression tree (CART) algorithm and adaptive neurofuzzy inference system (ANFIS) have been presented to perform the fault diagnosis of induction motors. The implementation of CART-ANFIS-based classifier requires two consecutive steps. Firstly, CART is utilized to select the relevant features in data set obtained from feature calculation part. The output of CART is decision tree that is employed to product the crisp if-then rule set. Secondly, the structure of ANFIS classifier is defined based on the obtained rules, which are fuzzified in order to avoid classification surface discontinuity; a hybrid algorithm is further used to tune the parameters' fuzzy memberships. The classification results and statistical measures were used for evaluating the CART-ANFIS model. The total classification accuracy was 91.11 and 76.67 % for vibration and current signals, respectively. The results indicate that the proposed CART-ANFIS model can be used in diagnosing induction motor faults.

References

Altug S, Chow MY, Trussell HJ (1999) Fuzzy inference systems implemented on neural architectures for motor fault detection and diagnosis. IEEE Trans Indus Electron 46(6):1069–1079

Bauer E, Kohavi R (1999) An empirical comparison of voting classification algorithms: bagging, boosting, and variants. Mach Learn 36:105–139

Benbouzid MEH, Nejjari H (2001) A simple fuzzy logic approach for induction motors stator condition monitoring. In: Proceedings of the IEEE IEMDC2001, pp 634–639

Breiman L (1996) Bagging predictors. Mach Learn 26(2):123–140

Breiman L, Friedman JH, Olshen RA, Stone CJ (1984) Classification and regression trees. Wadsworth Inc, Belmont, California

Breiman L (2001) Random forests. Mach Learn 45(1):5–32

Breiman L (2006) Random forest user notes, Statistics Department, Univ. of California, Berkeley, ftp://stat.berkeley.edu/pub/users/breiman/notes_on_random_forests_v2.pdf

Briem GJ, Benediktsson JA, Sveinsson JR (2002) Multiple classifiers applied to multisource remote sensing data. IEEE Trans Geosci Remote Sens 40:2291–2299

Carpenter GA, Grossberg S (1988) The ART of adaptive pattern recognition by a self-organizing neural network. IEEE Trans Computer 21(3):77–88

Carpenter GA, Grossberg S (1987) ART2: self-organization of stable category recognition codes for analog input patterns. Appl Opt 26(23):217–231

Carpenter GA, Grossberg S (1992) A self-organizing neural network. IEEE Commun Mag 30 (9):38–49

Cristianini N, Shawe-Taylor J (2000) An introduction to support vector machines. Cambridge University Press, Cambridge

Daubechies I (1990) The wavelet transform, time-frequency localization and signal analysis. IEEE Trans Inform Theory 36:961–1007

Di X, Han T, Yang BS (2006) Application of random forest algorithm in machine fault diagnosis. Inaugural World Congress on Engineering Asset Management, Gold Coast, Australia

Dietterich TG (2002) Ensemble learning. In: Arbib MA (ed) The handbook of brain theory and neural networks. The MIT Press, Cambridge, MA

Duda RO, Hart PE, Stork DG (2001) Pattern classification. Wiley-Interscience, New York

Duffin RJ, Schaeffer A (1952) A class of nonharmonic Fourier series. Trans Am Math Soc 72:341–366

Gislason PO, Benediktsson JA, Sveinsson JR (2006) Random forests for land cover classification. Pattern Recogn Lett 27(4):294–300

Goldberg GE (1989) Genetic algorithm in search, optimization and machine learning. Addison Wesley, New York

Goode P, Chow MY (1995) Using a neural/fuzzy to extract knowledge of incipient fault in induction motor: part 1-methodology. IEEE Trans Indus Electron 42(2):131–138
Ham JS, Chen Y, Crawford MM, Ghosh J (2005) Investigation of the random forest framework for classification of hyperspectral data. IEEE Trans Geosci Remote Sens 43(3):492–501
Jang JSR, Sun CT, Mizutani E (1996) Neuro-fuzzy and soft computing: a computational approach to learning and machine intelligent. Prentice Hall
Jang JSR (1994) Structure determination in fuzzy modeling: a fuzzy CART approach. In: Proceedings of the IEEE international conference on fuzzy systems, pp 1–6
Jang JSR (1993) ANFIS: adaptive-network-based fuzzy inference system. IEEE Trans Syst Man Cybernet 23(3):665–687
Joelsson SL, Benediktsson JA, Sveinsson JR (2005) Random forest classifiers for hyperspectral data. IEEE Publications, pp 160–163
Kohonen T (1992) LVQ_PAK—The learning vector quantization program package. Department of Technical, Physics, Helsinki University of Technology
Kohonen T (1995) Self-organizing maps. Springer, Berlin
Kumar R, Jayaraman VK, Kulkarni RD (2005) An SVM classifier incorporating simultaneous noise reduction and feature selection: illustrative case examples. Pattern Recogn 38:41–49
Lei Y, He Z, Zi Y, Hu Q (2007) Fault diagnosis of rotating machinery based on multiple ANFIS combination with GAs, Mechanical Systems and Signal Processing (in press)
Liobet E, Hines EL, Gardner JW, Bartlett PN, Mottram TT (1999) Fuzzy ARTMAP based electronic nose data analysis. Sens Actuators B 61:183–190
Lou X, Loparo KA (2004) Bearing fault diagnosis based on wavelet transform and fuzzy inference. Mech Syst Signal Process 18:1077–1097
Novikoff AB (1962) On convergence proofs on perceptron. In: Symposium on the mathematical theory of automata, vol 12, pp 615–622
Osuna E, Freund RR, Girosi FF (1997) Improved training algorithm for support vector machines. In: Proceeding of IEEE neural networks for signal processing, pp 276–287
Pal M (2005) Random forest classifier for remote sensing classification. Int J Remote Sens 26 (1):217–222
Platt J (1999) Fast training of support vector machines using sequential minimal optimization. In: Scholkopf B et al (eds) Advances in kernel methods − support vector learning. MIT Press, Cambridge, pp 185–208
Quinlan JR (1986) C4.5: programs for machine learning. Morgan Kaufmann, San Mateo, CA
Quinlan JR (1986b) Induction of decision trees. Mach Learn 1:81–106
Remlinger K (2003) Introduction and application of random forest on high throughput screening data from drug discovery. In: proceedings of workshop for the SAMSI program on data mining and machine learning, pp 1–6
Schapire RE (1990) The strength of weak learnability. Mach Learn 5(2):197–227
Shukri M, Khalid M, Yusuf R, Shafawi M (2004) Induction machine diagnostic using adaptive neuron fuzzy inference system. In: Negoita M Gh et al (eds) KES 2004, pp 380–387
Smola A, Scholköpf B, Müller KR (1998) The connection between regularization operators and support vector kernels. Neural Network 11:637–649
Sugeno M, Kang GT (1988) Structure identification of fuzzy model. Fuzzy Sets Syst 28:15–33
Vapnik VN (1982) Estimation dependences based on empirical data. Springer, Berlin
Vapnik VN (1995) The nature of statistical learning theory. Springer, New York
Vapnik VN (1999) The nature of statistical learning theory. Springer, New York
Widodo A, Han T, Yang BS (2007) Combination of independent component analysis and multi-class support vector machines for fault diagnosis of induction motors. Expert Syst Appl 32(2):299–312
Yang BS, Park CH, Kim HJ (2000a) An efficient method of vibration diagnostics for rotating machinery using a decision tree. Int J Rotating Mach 6(1):19–27
Yang BS, Lim DS, Seo SY, Kim MH (2000) Defect diagnostics of rotating machinery using SOFM and LVQ. In: Proceedings of 7th international congress on sound and vibration, pp 567–574

Yang BS, Kim K, Rao Raj BKN (2002) Condition classification of reciprocating compressors using RBF neural network. Int J COMADEM 5(4):12–20

Yang BS, Han T, An JL (2004a) ART-Kohonen neural network for fault diagnosis of rotating machinery. Mech Syst Signal Process 18(3):645–657

Yang BS, Jeong SK, Oh YM, Tan ACC (2004b) Case-based reasoning with Petri nets for induction motors fault diagnosis. Expert Syst Appl 27(2):301–311

Yang BS, Han T, Yin ZJ (2006) Fault diagnosis system of induction motors using feature extraction, feature selection and classification algorithm. JSME Int J (C) 49(3):734–741

Ye Z, Sadeghian A, Wu B (2006) Mechanical fault diagnosis for induction motor with variable speed drives using adaptive neuro-fuzzy inference system. Electr Power Syst Res 76(9–10):742–752

Zhang QH, Benveniste A (1992) Wavelet networks. IEEE Trans Neural Netw 3:889–898

Chapter 8
Science of Prognostics

8.1 Introduction

The word *prognosis* comes from the Greek *prognostikos* (of knowledge beforehand). It combines *pro* (before) and *gnosis* (a knowing). Hippocrates used the word prognosis, much as we do today, to mean a foretelling of the course of a disease. In the field of engineering systems health management, prognosis is regarded as science and often called prognostics.

In today's highly competitive marketplace, industries strive to minimize their capital and operational costs by trying to utilize the full life cycle of their equipments or components without sacrificing human, production, or environmental safety. Condition-based maintenance (predictive maintenance) is most useful in predicting equipment failure and avoiding unnecessary maintenance activities. Prognostics is the ability to assess the current health of a part and predict in the future the health of a part for a fixed time horizon or predict the time to failure. Being able to perform reliable prognosis is the key to PHM/CBM. Prognostics is critical to the subsystem or component for improving safety, planning missions, scheduling maintenance costs, and downtime.

We often observe in practice that the life of a piece of production equipment can be divided into two stages. The first stage is referred to as the normal working stage where no significant deviation from the normal operating state is observed. The second stage is called the *failure delay period*, since a defect may be initiated and progressively develop into an actual failure, i.e., the equipment is in a defective stage but still working during this stage. With the help of condition monitoring, hidden defects already present in the equipment may be detected, but for maintenance planning purposes, the prediction of the initiation point of the second stage and, more importantly, the residual life thereafter is important (Wang 2007). This relates to the concept of the delay time originally developed by Christer (1976) as shown in Fig. 8.1.

G. Niu, *Data-Driven Technology for Engineering Systems Health Management*,
DOI 10.1007/978-981-10-2032-2_8

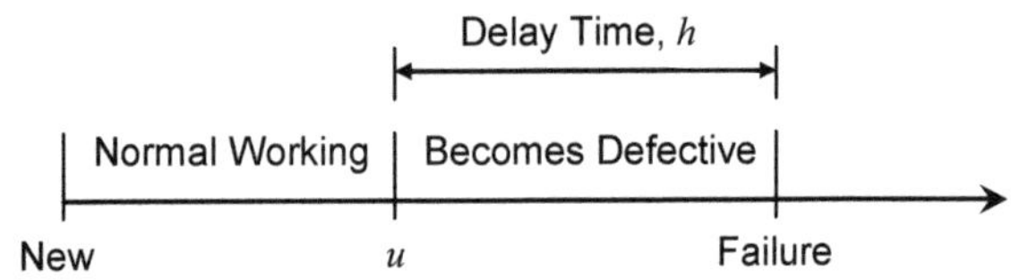

Fig. 8.1 The delay time concept (Christer 1976)

We are interested in how to detect the initial point u and the length of the delay time h. By condition monitoring, it might be easier to detect u, but difficult to assess the severity of the defect and its impact on the residual life. In other words, we are even more interested in assessing the residual life given the available condition information.

A reliable prognostic system is very useful to predict the fault propagation trend in an equipment or component and to provide an alarm before a fault reaches critical levels. An online prognostic system can also be used to improve the reliability of the equipment fault diagnosis by adaptively verifying the diagnostic results and modifying the knowledge (rule) base (Pourahmadi 2001; Wang et al. 2001, 2007).

In recent years, in order to reduce costs and shorten repair time, condition-based predictive maintenance has become an efficient strategy for modern industries which necessitates advanced tools in prognostics. Prognostics is the use of predictive maintenance practices and tools to analyze the trends of equipment or component performance against known engineering limits for the purpose of detecting, analyzing, and correcting problems before failure occurs. More advanced prognostics is focused on performance degradation monitoring and assessment, so that failures can be predicted and prevented. To fulfill the goal of prognostics, three crucial steps are needed. First, the defect or abnormality should be able to be detected at an early stage. Second, the equipment or component performance should be assessed robustly and tracked continuously. Finally, the remaining useful life (or residual useful life) and possible failure mode of the system or component should be effectively predicted. Estimating the remaining useful life is most important in these three steps, because the remaining useful life directly serves decision variables of prognostics.

However, challenges in effectively predicting the remaining useful life of mechanical elements exist. One of the challenges of life prediction is figuring out how to set up an appropriate degradation indicator based on a vibration signal. Usually, the time features, such as average, RMS, kurtosis, or crest factor of the vibration signals, and frequency features such as the average of the amplitudes of the defective frequency and its harmonics over time as the degradation index was often chosen. However, these indexes either have a low sensitivity to incipient defect or do not fit, very well, under highly accelerated tests. In fact, even though a large variety of features can be extracted to describe the characteristics of a vibration signal from other aspects, previous research work has shown that each feature is only effective for a certain defect at a certain stage. A good performance assessment method should take advantage of mutual information from multiple features for system degradation assessment. In addition, a good degradation signal must not only capture the physical transitions that the bearing undergoes during different stages of its life, but must also be easy to operate in an actual situation.

Another challenge is how to establish a life prediction model of the components to effectively estimate failure times. There are many techniques centered on life prediction models. Some of these are based on crack initiation models and crack propagation models, namely *Paris law*. Researchers have also developed physical models for estimating remaining useful life.

In the determination of remaining useful life, one must consider the following fundamental factors:

- Physical deterioration, which is the wear from the use of the asset during its operating life.
- Functional obsolescence which is obsolescence due to technological factors, inadequacy, inefficiency, and/or excessive operating costs.

Depreciation attributable to physical deterioration or functional obsolescence can be divided into two categories: curable and incurable. Curable depreciation is depreciation that can be *reversed* or *cured* by corrective intervention. Incurable depreciation is depreciation due to technological advances based on external factors that cannot be corrected or cured in a cost-effective manner. In the final analysis, if a reasonable remaining lifetime for the current equipment or component or if the equipment or component cannot be converted to production of alternate products, corrective action must be taken. Remaining useful life (or the remaining possibility of use) represents that portion of the normal useful life of an asset that is defined as the period of time expressed in years, running from the date of appraisal to the end of economic use of the asset. This period is conditioned on various factors such as age, conditions of service, physical characteristics of the asset, state of repair, technological advance, and the materials from which the asset is constructed. Normal useful life of an asset is the period of time, expressed in years, running from the date at which the asset is first put into use to the time when the asset is no longer able to be economically productive for its intended use. Proper determination of that portion of the useful life that is remaining is not a matter of simple arithmetic calculation. One must consider the entire maintenance history of the asset, such as major overhauls and upgrades, done over the years that can increase the remaining useful life of the asset. Therefore, the limits of the normal useful life of an asset can be influenced by type and timing of major maintenance that has been performed (Kallberg and Rossi 2007).

Machine learning is a branch of artificial intelligence that employs a variety of statistical, probabilistic, and optimization techniques that allow computers to *learn* from past examples and to detect hard-to-discern patterns from large, noisy, or complex data sets. This capability is particularly well suited to engineering applications. As a result, machine learning is frequently used in fault diagnosis and detection. More recently, machine learning has been applied to fault prognostics and prediction. This latter approach is particularly interesting as it is part of a growing trend toward personalized, predictive system. The ability to predict failures in system before they occur would save time, money, and lives. Mechanical components could be replaced before they caused catastrophic damage.

Most of the earlier research on the subject of damage (failure) detection and identification was concentrated on the development of heuristic methods based on time- or frequency-domain signal processing techniques. More recent research focuses on heuristic methods that can utilize both time- and frequency-domain information. In both cases, these are mainly failure detection methods, i.e., the main emphasis is on the development of a feature vector that will indicate when the system parameters have reached some preset failure values. The main advantage of these methods is that they are easy to implement and sometimes work very well. However, there is no theoretical basis to determine a priori if a given method is going to work well for a particular system without prior experimentation or knowledge.

Model-based methods, in contrast, overcome some of the limitations of heuristic methods at the expense of more complex development and higher implementation costs. They are general in the sense that if some properties of the system or damage physics change, models can be readjusted to accommodate the change. Their main advantage is that knowing a model structure gives the ability to tie the changes in feature vectors to model parameter changes. However, in many cases, one does not know the appropriate model for the system, and since most damage processes are nonlinear, model development costs are high. This problem is addressed by the methods that use neural networks, genetic algorithms, and other data-based modeling techniques to replace physics-based models. However, in this approach, we lose the ability to directly tie damage evolution to changes in a system's physical parameters. It should be mentioned that most of the model-based schemes, just like most heuristic methods, are used primarily as damage *detection* methods.

Equipment or component condition prognosis means the use of available (current and previous) observations to forecast upcoming states of it. Several temporal patterns can be used for equipment or component condition prognosis, such as vibration features and debris properties of the lubrication oil. The vibration-based monitoring, however, is a well-accepted approach due to the ease of measurement and analysis, and thus, it is used in this chapter.

Bearings are among the most precise components in mechanical assemblies and are manufactured to very tight tolerances. They are normally found in most rotational equipment. The condition and health of bearings play an important role in the functionality and performance of these equipments. This work focuses on using vibration analysis to monitor the condition of rolling element bearings and to predict bearing residual life in real time. The significance of using vibration analysis lies in the degradation process of the bearing. Bearing fatigue normally begins with subsurface cracks initiating within the raceway material. During service, cracks propagate and eventually reach the surface of the race dislodging a piece of metal from the surface. This results in what is known as a spall and in many applications is defined as the onset of failure. Rolling elements (balls or rollers) rotate over the race causing repetitive impacts each time they pass over the spall. Consequently, a distinctive frequency known as the defective frequency is excited. The frequency is a function of the number of balls, the rotational speed, geometry of the bearing, and types of defects and their location. Thus, vibration analysis becomes the most suitable condition monitoring technique for investigating the evolution of these defective frequencies over time once a spall has occurred.

Several vibration-based techniques have been proposed in the literature for time-series prediction. The classical approaches mainly use stochastic models such as the autoregressive (AR) model (Tong and Lim 1980), the bilinear model (Rao 1981), and multivariate adaptive regression splines (Friedman 1991). However, these stochastic models are usually difficult to implement in forecasting the dynamic response of complex systems. Recent studies in time-series prediction have focused on the use of flexible models, such as neural networks (NNs) and fuzzy systems. NN predictors are built automatically by training, without the need for the identification of model structures and parameters. NNs have two typical connection architectures: feedforward and recurrent networks, and both have been employed in system behavior forecasting (Husmeier 1999; Connor et al. 1994; Atiya et al. 1999). From the modeling point of view, a feedforward network is a special case of nonlinear AR models, whereas a recurrent network is a nonlinear AR moving average model. From this, we can see that recurrent network predictors have the advantages over feedforward network predictors in much the same way as the AR moving average models have over the AR models. This conclusion was verified by Tse and Atherton (1999) using simulation and practical tests.

Fuzzy system uses linguistic rules for system behavior forecasting. It starts from highly formalized insights about the dynamic behavior of the system and then formulates expert knowledge in fuzzy if-then rules in a way to mimic humans to deal with a forecasting task (Vukovic 2001). However, fuzzy systems lack learning capability. Sometimes, it is difficult to properly determine the fuzzy rules and to optimize the membership functions, especially when more input variables are applied. A solution to overcoming these disadvantages is to use the integrated systems, e.g., using NNs to train the fuzzy structure and parameters. Jang et al. (1997) proposed a neurofuzzy (NF) system for time-series prediction. Through simulation, they found that the NF predictor gave higher forecasting accuracy than both the classical AR models and the feedforward NNs.

8.2 Prognostic Approaches

There are essentially three different approaches for the development of prognostics reasoner techniques. The first are physical models. These are models that have been developed by experts in the component field and validated on large sets of data to show that they are indeed accurate. The second are systems that embody rules of thumb that have been developed and refined by human maintenance experts. Examples of these systems are rule-based expert systems and fuzzy logic systems. The third are statistical models that "learn" from examination of real data that contain nominal and known fault conditions. Examples of these are neural networks and data mining systems.

Physical models and rule-based systems have the good feature of containing information for anticipated fault events that have yet to occur on the component that is being monitoring. This is in contrast to "learning" systems. Learning systems are only as good as the data from which they have been trained.

On the other hand, "learning" systems have the good feature that they can process a wide variety of data types and potentially have performance superior to rule-based systems because they exploit the nuances in the data that are not covered by general rules. This is particularly true for new sources of data for which expert analysis, physical models, and rules have not been developed.

8.2.1 Rule-based Approach

By definition, *artificial intelligence* (AI) programs are intended to give the appearance of human behavior. AI programming languages supplied tools for programs to be built that closely resemble human logic and expertise in their implementation. These programs are called *expert systems.*

Rule-based expert systems are one of the most commonly used approaches for developing expert systems. In this approach, rules are used to represent heuristics, or "rules of thumb," that specify a set of actions to be performed for a given situation. We usually think of the rule-based expert system as a list of "if" and "then" statements. The "if" portion of a rule is a series of patterns which specify the facts (or data) which cause the rule to be activated. The expert system tool provides a mechanism, called the *inference engine,* which automatically matches facts against expert systems that have a major advantage over all the "learning" systems in that real data are not required in order to develop the system. However, a knowledgeable expert is required. Also, the rules tend to follow the structure of the rule development environment that is being used. As such, possible nonlinear and correlated relationships may be lost.

8.2.2 Fuzzy Logic Approach

Fuzzy sets were introduced by Zadeh (1965). Fuzzy sets were first invented as a means of generalizing conventional set theory to model the realities of everyday life (Bezdek and Pal 1992), including applications to industrial problems (Yen et al. 1995). We have all heard the examples; the fuzzy set of operating temperatures of a machine can be (hot, cold), while the actual measurement is a "crisp" number, i.e., $T = 516.7°$. Based on T is the machine "hot" or "cold"? Fuzzy set theory would assign a membership to both "hot" and "cold." Fuzzy set theory has been extended to include fuzzy logic for prognostic reasoner applications. The critical step is the development of ways to aggregate fuzzy measurements to form fuzzy rules. Standard software packages are available for the development of fuzzy systems. The fuzzy membership functions can be developed "by hand" using the developer's intuition/expertise as the shape of the functions. Or the functions can be estimated from training data. There, the developer specifies the function parametric shape (i.e., trapezoid or Gaussian) and the parameters are estimated from the data. The advantage of fuzzy rules over those developed with an expert system is that the degree of membership is carried out through the computation and a "hard" decision is only generated at the very last step.

8.2.3 Model-Based Approach

Model-based approaches utilize an explicit mathematical model for the equipment or component being monitored (Gertler 1998). The models can be physical models or statistical models (Box and Jenkins 1976). Physical models are useful in accounting for all operating conditions. Component life models are included here. As mentioned above, these may be based on the physics of the components being monitored. However, for most applications, these are based on statistics collected over a large set of samples and are used to define an average component.

Model-based approaches are only as good as the models developed. For very simple systems, for example with a single input/output pair that is linearly related, likely the model developed is good. As inputs and complexity of the real system grow, the chances of the model being "exact" diminish. For statistical models, data from all modes of operation and fault conditions must be collected. This is often not possible. Physical models potentially cover these holes. However, to validate a complicated physical model, data from all modes of operation and fault conditions must also be collected.

Several model-based approaches are presented as follows:

(1) *Physics-based model*

A physics-based model is a technically comprehensive modeling approach that has been traditionally used to understand component failure mode progression. Physics-based models provide a means to calculate the damage to critical components as a function of operating conditions and assess the cumulative effects in terms of component life usage. By integrating physical and stochastic modeling techniques, the model can be used to evaluate the distribution of remaining useful component life as a function of uncertainties in component strength/stress properties, loading, or lubrication conditions for a particular fault. Statistical representations of historical operational profiles serve as the basis for calculating future damage accumulation. The results from such a model can then be used for real-time failure prognostic predictions with specified confidence bounds.

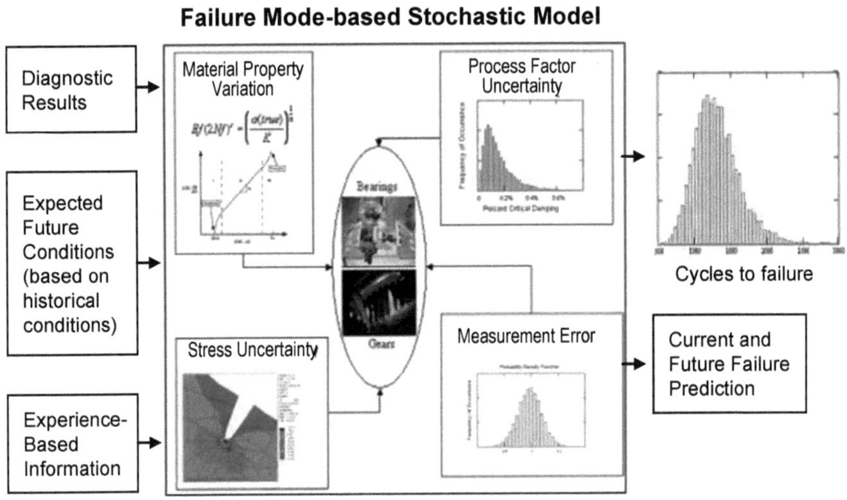

Fig. 8.2 Physics-based modeling approach (Vachtsevanos 2006)

A block diagram of this prognostic modeling approach is given in Fig. 8.2. As illustrated at the core of this figure, the physics-based model utilizes the critical, life-dependent uncertainties so that current health assessment and future remaining useful life (RUL) projections can be examined with respect to a risk level.

Model-based approaches to prognostics differ from feature-based approaches in that they can make RUL estimates in the absence of any measurable events, but when related diagnostic information is present (such as the feature described previously), the model can often be calibrated based on this new information. Therefore, a combination or fusion of the feature-based and model-based approaches provides full prognostic ability over the entire life of the component, thus providing the valuable information for planning which components to inspect during specific overhaul periods. While failure modes may be unique from component to component, this combined model-based and feature-based methodology can remain consistent across different types of critical components or line replaceable units (LRUs).

To perform a prognosis with a physics-based model, an operational profile prediction must be developed using the steady-state and transient loads, temperatures, or other online measurements. With this capability, probabilistic critical component models can then be "run into the future" by creating statistical simulations of future operating profiles from the statistics of past operational profiles or expected future operating profiles.

Based on the expert's past experience correlating operational profile statistics and component or LRU life usage, the nonlinear nature associated with many damage mechanisms is dependent on both the inherent characteristics of the profiles and operational mix types. Significant component damage resulting from the large variability in the operating environment and severity of the missions directly affects

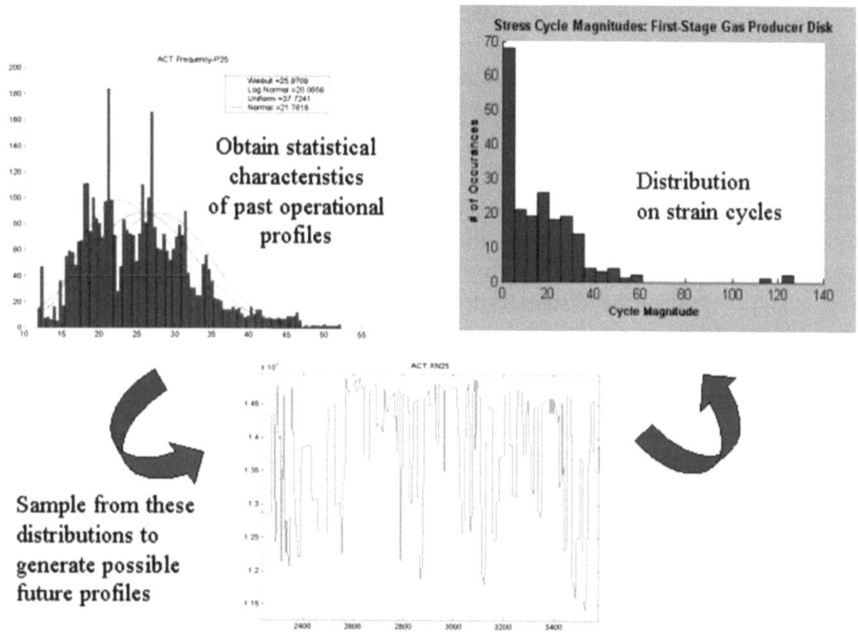

Fig. 8.3 Operation profile and loading model for prognosis (Vachtsevanos 2006)

the vehicle component lifetimes. Very often, component lives driven by fatigue failure modes are dominated by unique operational usage profiles or a few, rare, severe, randomly occurring events, including abnormal operating conditions and random damage occurrences. For this reason, it recommends a statistical characterization of loads, speeds, and conditions for the projected future usage in the prognostic models as shown in Fig. 8.3.

Moreover, the formulation of physical phenomena through mathematical formulation can be used for prognostics of the system.

For example, crack propagation model is one of the interests from the prognosis viewpoint. Fatigue crack growth in such typical system components as bearings, gears, shaft, and aircraft wings is affected by variety factors, including stress state, material properties, temperature, lubrication, and other environmental effects. Variation of available empirical and deterministic fatigue crack propagation models is based on Paris formula:

$$\frac{d\alpha}{dN} = C_0(\Delta K)^n \tag{8.1}$$

where α is the instantaneous length of dominant crack, N is the running cycles, C_0, n are material-dependent constants, and ΔK is the range of stress-intensity factor over loading cycle.

The crack propagation model was applied to determine the cycles required to propagate the crack from the initial α inches to the critical crack length. During a crack growth simulation, at each step of propagation, the direction and extent of crack growth at the each point along the crack front have to be determined.

(2) *System dynamic model*

In many cases involving complex system, it is very difficult or impossible to derive dynamic models based on all the physical process involved. In such case, it is possible to assume a certain form from the dynamic model and then use observed inputs and outputs of the system to determine the model parameters needed so that the model indeed serves as an accurate surrogate for the system. This is known as *model identification*. To introduce this method, we will begin by considering a classic identification/prediction problem: given input-output data of dynamic system, then how do we develop a model to predict the system's future behavior.

The clear example of system dynamic model prediction is presented by Luo et al. (2003). The interested readers are suggested to refer to this reference. They developed an integrated prognostic process based on data collected from model-based dynamic system. Prognostic model is constructed based on different random load conditions (modes). Remaining life predictions are performed by mixing mode-based life predictions via time-averaged mode probabilities.

(3) *Probabilistic model*

Probabilistic methods for prognosis are very effective in such situations. These methods require less detailed information than model-based techniques because the information for prognosis resides in various probability density functions (PDF), not in dynamic differential equations. Advantages are that the required PDF can be found from observed statistical data and that the PDF are sufficient to predict the quantities of interest in prognosis. Moreover, these methods also generally give confidence limits about the result, which are important in giving a feeling for the accuracy and precision of the predictions. The probabilistic methods that can serve prognosis are Bayesian theory (Lewis 1986) and Weibull model (Groer 2000; Schomig and Rose 2003).

The Weibull cumulative distribution is:

$$F(x) = 1 - e^{-(x/\beta)^{\alpha}}$$

where x is the data value (either time or value of an independent variable), α is the shape (or slope) parameter, and β is the location parameter. Parameters can be estimated with maximum-likelihood method, probability plotting, etc. Weibull distribution has been used to provide reasonable model for lifetime of equipments such as ball bearing, mill, and aircraft engine. However, this yields mainly a mean residual life estimate, and thus, it is somewhat unreliable in practice. Practical, dynamic residual life estimates based on monitored condition-related variables such as vibration acceleration are required.

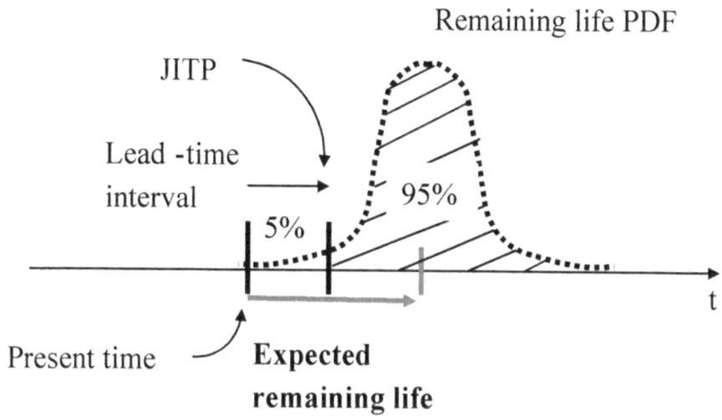

Fig. 8.4 A probability density function (PDF) for prognosis (Engel et al. 2000)

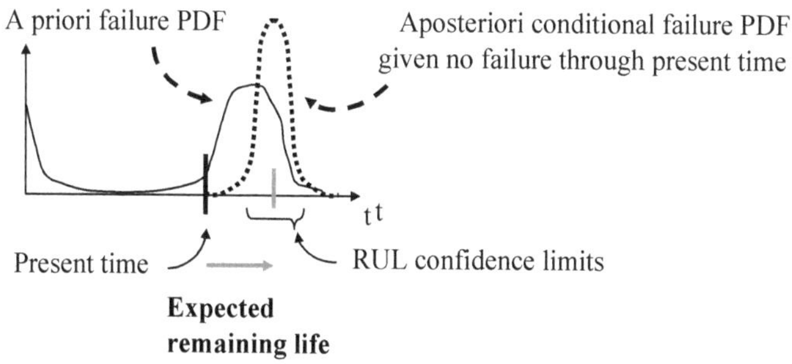

Fig. 8.5 Updated prognostic probability density function (Engel et al. 2000)

A comprehensive description on the probabilistic techniques in prognosis as related to predicting the remaining useful life (RUL) is given by Engel et al. (2000). The seminal notions presented in this paper serve to clarify our thinking about remaining useful life prediction. A key concept in this framework is the remaining useful life failure probability density function (PDF). In this representation, a component or LRU is recommended to be removed from service prior to attaining a high probability of failure, set based on the criticality. This concept is depicted in Fig. 8.4, in terms of the RUL PDF, where a *just-in-time point* is defined for removal from service that corresponds to a 95 % probability that the component has not yet failed.

A key issue, unfortunately, is that the RUL PDF is actually a conditional PDF that changes as time advances. In fact, one must recompute the RUL PDF at each time t based on the new information that the component has not yet filed at that time. This concept is shown in Fig. 8.5. One starts with a priory of PDF similar to hazard function.

Then, as time passes, one must recompute the a posteriori RUL PDF based on the fact that the failure has not yet occurred. This involves renormalizing the PDF at each time so that its area is equal to one. As time passes, the variance of the

RUL PDF decreases; that is, the PDF becomes narrower. This corresponds to the fact that as time passes and one approaches the failure point, one becomes more and more certain about the time of failure and its predicted value becomes more accurate.

8.2.4 Trend-Based Evolutionary Approach

A trend-based evolutionary prognostic approach relies on the ability to track and trend deviations and associated rates of change of these deviations of specific features or measurements from their normal operating condition. Figure 8.6 is an illustration of such a technique.

Evolutionary prognostics may be implemented on systems or subsystems that experience conditional or slow degradation-type faults such as an efficiency loss in a turbo-machinery. Generally, trend-based prognostics works well for system-level degradation because conditional loss is typically the result of interaction of multiple components functioning improperly as a whole. This approach requires that sufficient sensor information is available to assess the current condition of the system or subsystem and relative level of uncertainty in this measurement. Furthermore, the parametric conditions that signify known performance-related fault must be identifiable. While a physical or statistical model that can help classify a specific fault is beneficial, it is not a requirement for this technical approach. An alternative to the physical model is built-in "expert" knowledge of the fault condition and how it manifests itself in the measured and extracted features.

Incipient faults and performance degradations in electrical and mechanical systems exhibit detectable features that provide a means to diagnose and predict the

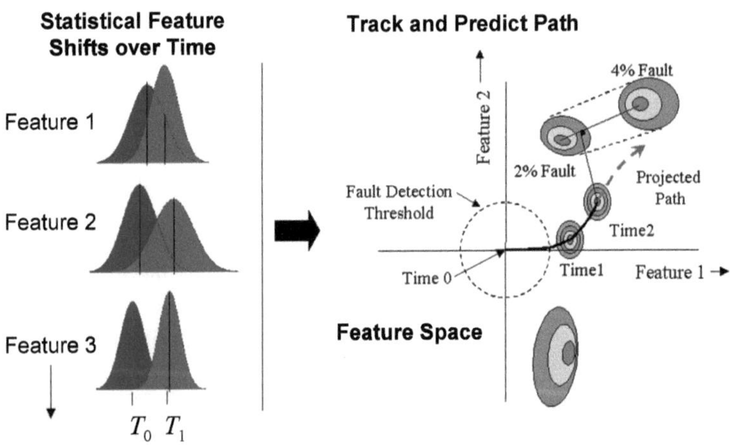

Fig. 8.6 Trend-based or evolutionary approach (Romer et al. 2001)

future progression of that fault under known operating conditions. Feature-based prognostics can be implemented for electronic systems based on changes in a variety of measurable quantities including temperature, current, and voltage at various locations in the system. Features such as heat generation and power consumption that correlate with known faults can be extracted from the sensed data. Once these features are obtained, they can be tracked and trended over the component's life and compared with remaining useful life estimates to provide corroborative evidence of a degrading or failing condition.

8.2.5 Data-Driven Model Based Approach

In many instances, one has historical fault/failure data in terms of time-domain plots of various signals leading up to the failure, or statistical data sets. In many of these cases, it is either difficult or impractical to determine a physics-based model for prediction purposes. In such situations, one may use nonlinear network approximators that can be tuned using well-established formal algorithms to provide desired outputs directly in terms of the data. Nonlinear networks include the neural network, which is based on signal processing techniques in biological nervous systems, and fuzzy logic systems, which are based on the linguistic and reasoning abilities of humans. These are similar in that they provide structured nonlinear function mappings with very desirable properties between the available data and the desired outputs.

Hence, with an understanding of how the fault/failure signature is related to specific measurable or inferred features from the system being monitored, a data-driven modeling approach is a commonly utilized approach. Based on the selected input features that correlate with the failure progression, a desired output prediction of the time to failure is produced based on a training process in which the network will automatically adjust its weights and thresholds based on the relationships it sees between the time to failure and the correlated feature magnitudes. Figure 8.7 shows an example of a network after being trained by some vibration feature data sets for predicting a gear failure. The difference between the network output and the "ground truth" probability of failure curve is due to error that still exists after the network parameters have optimized to minimize this error. Once trained, the network architecture can be used to predict the same feature progressions for a different test under similar operating conditions.

The techniques can be utilized to perform data-driven-based model are as follows:

(1) *Neural networks*

Artificial neural networks or neural nets (NNs) are an attempt to model the brain by the dense interconnection of a large set of simple processing elements. Details of neural net processing and development can be found in several books

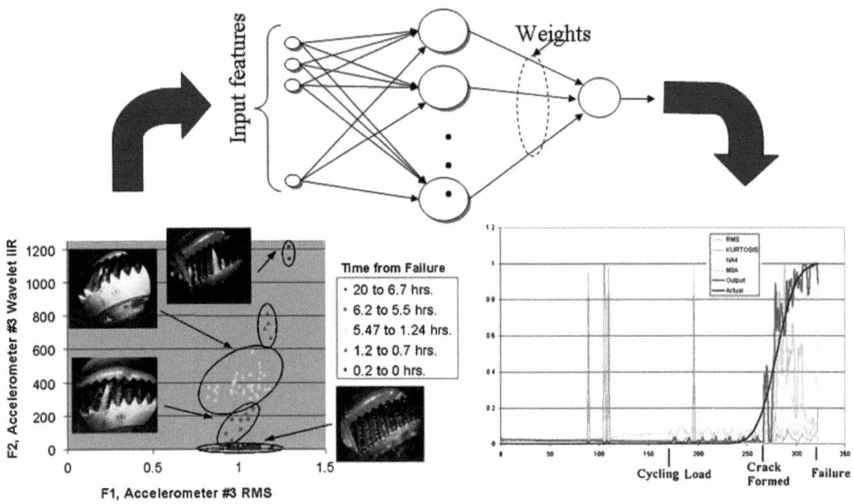

Fig. 8.7 Data-driven modeling approach (Romer et al. 2001)

(Heicht-Nielsen 1990; Haykin 1994). There are two easy-to-read papers that give good introductions to the two most popular neural nets, the multilayer perceptron (MLP), and radial basis function (RBF) neural network (Hush and Horne 1993). Neural nets have proven useful in a variety of areas: detection, classification, multidimensional function approximation, and predictive modeling of data. They are ideal for developing nonlinear transformations to map input data to outputs. Thus, they can be used for classification as well as prediction. NNs are "trained" by presenting examples of input/output pairs of data. For most applications, the output data have been "labeled" as to the correct class or function response. During training, the parameters in the neural net are adjusted until the neural net classification performance reaches an acceptable level.

Two types of NNs that are used extensively are the MLP and RBF NNs. MLP NN has been shown to be excellent for nonlinear function approximation and for solving classification problems. They have been shown to be Bayesian classifiers. However, MLPs are not capable of performing novelty detection: An MLP cannot detect when data for which it has not been trained are present at the inputs. The RBF NN has also been demonstrated to be an excellent classifier and function approximator. For the standard implementation of the RBF, all processing elements in the middle layer apply a multidimensional Gaussian function to the input data. The output layer is the weighted sum of the basis function outputs. The RBF can be thought of as a nearest-neighbor classifier.

As such, it can perform novelty detection. This is important for many applications, as we would like the system to let us know when inputs do not match anything seen before. It also has the ability to let the user know "why" the NN performed as it did. Both the MLP and RBF NNs require real data that represent known fault classes to be trained. As such, when first used for a prognosis problem with limited training data, they likely will not operate very well. However, as more data are collected and examples of fault classes of interest are examined, the NN can be updated using the new data.

(2) *Support vector regression (SVR)*

Support vector machine (SVM), introduced originally by Vapnik (1995), is one of the machine learning methods and AI techniques which have been rapidly developed and applied for classification and regression problems (Gun 1998; Muller et al. 1997). SVM is quite satisfying from a theoretical point of view and can lead to great potential and superior performance in practical applications. This is largely due to the structural risk minimization (SRM) principle in SVM, which has greater generalization ability and is superior to the empirical risk minimization (ERM) principle as adopted in neural networks. Furthermore, SVM is adaptive to complex system and robust in dealing with nonlinear data.

Recently, the application of SVM to time-series prediction, called support vector regression (SVR), has shown many breakthroughs and plausible performance, such as travel-time prediction (Wu et al. 2004), wind speed prediction (Mohandes et al. 2004), electricity load forecasting (Li et al. 2005), and water lake prediction (Khan and Coulibaly 2006). However, paper that concerns with prediction system of machines and mechanical equipments are much fewer. The reported paper in this area is an applied research conducted by Yang and Zhang (2005). They used SVM for condition trend prediction of generator machine sets. The influence of cost functions, kernel functions, and parameters on prediction performance of SVM was studied.

Since there is much evidence from previous research results of time-varying application with SVR prediction, it motivates the research in using SVR for machine prognosis system modeling. In this chapter, SVR is applied to predict time series of failure trending data of machines. The aims are to investigate the feasibility and to evaluate the performance and reliability of SVR in failure trending data prediction and also to develop a reliable prognosis method for equipment or component condition prediction.

(1) *Linear support vector regression*

The basic idea of support vector regression (SVR) is mapping the data $\mathbf{x}$ into a high-dimensional space via nonlinear mapping and performing linear regression in this feature space. The linear equation of SVM can be expressed in the form

$$f(\mathbf{x}_i) = \langle \mathbf{w}, \mathbf{x}_i \rangle + b \tag{8.2}$$

where $\langle , \rangle$ denotes the dot product in R^n.

Flatness in the case of Eq. (7.2) means the one seeks small $\mathbf{w}$. One way to ensure this is to minimize the Euclidean norm, i.e., $\|\mathbf{w}\|^2$. Formally, the problem of Eq. (7.2) can be written as convex optimization problem by requiring

$$\begin{aligned} &\text{minimize} \quad \tfrac{1}{2}\|\mathbf{w}\|^2 \\ &\text{subject to} \quad \begin{cases} \mathbf{y}_i - \langle \mathbf{w}, \mathbf{x}_i \rangle - b \leq \varepsilon \\ \langle \mathbf{w}, \mathbf{x}_i \rangle + b - \mathbf{y}_i \leq \varepsilon \end{cases} \end{aligned} \tag{8.3}$$

The tacit assumption in Eq. (8.3) is that such a function $f(\mathbf{x})$ actually exists that approximately all pairs (x_i, y_i) with ε precision, or in other words, the convex optimization is feasible. However, this may not be the case, or we also may want to allow some errors. Analogously to the soft margin in Vapnik (1995), one can introduce slack variables ξ_i, ξ_i^* to cope with otherwise infeasible constraints to optimization Eq. (8.3). Hence, we present the formulation described by Vapnik (1995).

$$\begin{aligned} &\text{minimize} \quad \tfrac{1}{2}\|\mathbf{w}\|^2 + C\sum_{i=1}^{l}(\xi_i + \xi_i^*) \\ &\text{subject to} \quad \begin{cases} \mathbf{y}_i - \langle \mathbf{w}, \mathbf{x}_i \rangle - b \leq \varepsilon + \xi_i \\ \langle \mathbf{w}, \mathbf{x}_i \rangle + b - \mathbf{y}_i \leq \varepsilon + \xi_i^* \\ \xi_i, \xi_i^* \geq 0 \end{cases} \end{aligned} \tag{8.4}$$

The constant $C > 0$ determines the trade-off between the flatness of $f(\mathbf{x})$ and the amount up to which deviations larger than ε tolerated. The formulation above corresponds to dealing with a so-called ε-insensitive loss function $|\xi|_\varepsilon$ described by

$$|\xi|_\varepsilon = \begin{cases} 0 & \text{if} |\xi| \leq \varepsilon \\ |\xi| - \varepsilon & \text{otherwise} \end{cases} \tag{8.5}$$

Figure 8.8 depicts situation graphically. Only the points outside the shaded region contribute to the cost insofar, as the deviations are penalized in a linear fashion. It turns out that the optimization problem in Eq. (8.4) can be solved more easily in its dual formulation. The dual function provides the key for extending SVM to nonlinear functions.

The calculation can be simplified by converting into the equivalent *Lagrangian* dual problem, which will be

$$\begin{aligned} L(\mathbf{w}, b, \alpha) = &\frac{1}{2}\|\mathbf{w}\|^2 + C\sum_{i=1}^{l}(\xi_i + \xi_i^*) - \sum_{i=1}^{l}\alpha_i(\varepsilon + \xi_i - \mathbf{y}_i + \langle \mathbf{w}, \mathbf{x}_i \rangle + b) \\ &- \sum_{i=1}^{l}\alpha_i^*(\varepsilon + \xi_i^* - \mathbf{y}_i + \langle \mathbf{w}, \mathbf{x}_i \rangle - b) - \sum_{i=1}^{l}(\eta_i \xi_i + \eta_i^* \xi_i^*) \end{aligned} \tag{8.6}$$

Then, the task is minimizing Eq. (8.4) with respect to primal variables ($\mathbf{w}$, b, ξ_i, ξ_i^*) that have to be vanished for optimality.

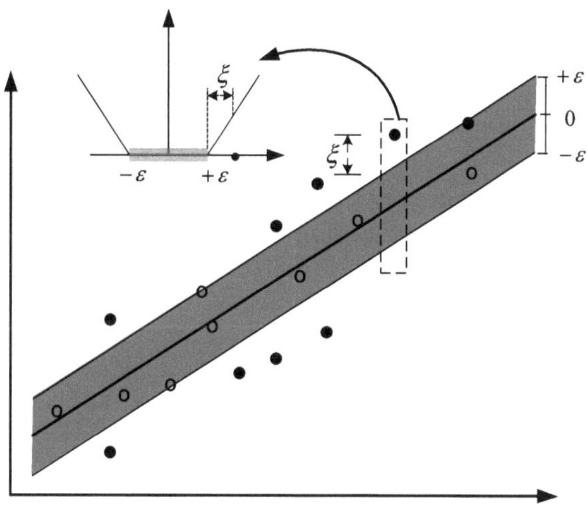

Fig. 8.8 The soft margin loss setting for linear SVR

$$\frac{\partial L}{\partial \mathbf{w}} = \mathbf{w} - \sum_{i=1}^{l} (\alpha_i^* - \alpha_i)\mathbf{x}_i = 0 \tag{8.7}$$

$$\frac{\partial L}{\partial b} = \sum_{i=1}^{l} (\alpha_i^* - \alpha_i) = 0 \tag{8.8}$$

$$\frac{\partial L}{\partial \xi_i^{(*)}} = C - \alpha_i^{(*)} - \eta_i^{(*)} = 0 \tag{8.9}$$

Substituting Eqs. (8.7)–(8.9) into Eq. (8.6) yields the dual optimization problem.

$$\begin{aligned} &\text{minimize} && \left\{ \frac{1}{2}\sum_{i,j=1}^{l} (\alpha_i - \alpha_i^*)(\alpha_j - \alpha_j^*)\langle \mathbf{x}_i, \mathbf{x}_j \rangle - \varepsilon \sum_{i=1}^{l} (\alpha_i + \alpha_i^*) + \sum_{i=1}^{l} \mathbf{y}_i(\alpha_i - \alpha_i^*) \right. \\ &\text{subject to} && \left\{ \begin{array}{l} \sum_{i=1}^{l} (\alpha_i - \alpha_i^*) = 0 \\ \alpha_i, \alpha_i^* \in [0, C] \end{array} \right. \end{aligned} \tag{8.10}$$

In deriving Eq. (8.10), the dual variables η_i, η_i^* have eliminated through condition Eq. (8.9), as the variables did not appear in the dual objective function anymore but only were present in the dual feasibility conditions. Equation (8.7) can be rewritten as follows:

$$\mathbf{w} = \sum_{i=1}^{l} (\alpha_i - \alpha_1^*)\mathbf{x}_i \tag{8.11}$$

And therefore, Eq. (8.12) can be expressed as follows:

$$f(\mathbf{x}) = \sum_{i=1}^{l} (\alpha_i - \alpha_i^*)\langle \mathbf{x}_i, \mathbf{x} \rangle + b \tag{8.12}$$

This is the so-called support vector expansion, i.e., can be completely described as a linear combination of the training patterns $\mathbf{x}_i$. The Lagrange multipliers α_i and α_i^* represent solutions to the above quadratic problem, which act as forces pushing predictions toward target value $\mathbf{y}_i$. Only the nonzero values of the Lagrange multipliers in Eq. (8.10) are useful in forecasting the regression line.

For the variable b in Eq. (8.12), it can be computed by applying the Karush–Kuhn–Tucker condition, and in this case, the interested readers can refer to Muller et al. (1997) for detail explanation.

(2) *Nonlinear support vector regression*

In most real applications, linear function approximation is of limited practical use. The solution is to map the input data in a higher dimensional feature space, in which the training data may exhibit linearity and then to perform linear regression in this feature space. Let $\mathbf{x}_i$ be mapped into a feature space by a nonlinear function $\phi(\mathbf{x}_i)$, and the decision function becomes

$$f(\mathbf{x}_i) = \langle \mathbf{w}, \phi(\mathbf{x}_i) \rangle + b \tag{8.13}$$

Similarly, the nonlinear regression problem can be expressed as optimization problem given in Eq. (8.4) with modification of term $\langle \mathbf{w}, \phi(\mathbf{x}_i) \rangle$.

$$\begin{aligned} &\text{minimize} \quad \tfrac{1}{2}\|\mathbf{w}\|^2 + C\sum_{i=1}^{l}(\xi_i + \xi_i^*) \\ &\text{subject to} \quad \begin{cases} \mathbf{y}_i - \langle \mathbf{w}, \phi(\mathbf{x}_i) \rangle - b \le \varepsilon + \xi_i \\ \langle \mathbf{w}, \phi(\mathbf{x}_i) \rangle + b - \mathbf{y}_i \le \varepsilon + \xi_i^* \\ \xi_i, \xi_i^* \ge 0 \end{cases} \end{aligned} \tag{8.14}$$

Figure 8.9 shows the concept of nonlinear SVR, corresponding to Eq. (8.13) Then, the dual form of the nonlinear SVR can be expressed as follows:

$$\begin{aligned} &\text{minimize} \quad \left\{ \tfrac{1}{2}\sum_{i,j=1}^{l}(\alpha_i - \alpha_i^*)(\alpha_j - \alpha_j^*)\langle \phi(\mathbf{x}_i), \phi(\mathbf{x}_j) \rangle - \varepsilon\sum_{i=1}^{l}(\alpha_i + \alpha_i^*) + \sum_{i=1}^{l}\mathbf{y}_i(\alpha_i - \alpha_i^*) \right. \\ &\text{subject to} \quad \begin{cases} \sum_{i=1}^{l}(\alpha_i - \alpha_i^*) = 0 \\ \alpha_i, \alpha_i^* \in [0, C] \end{cases} \end{aligned} \tag{8.15}$$

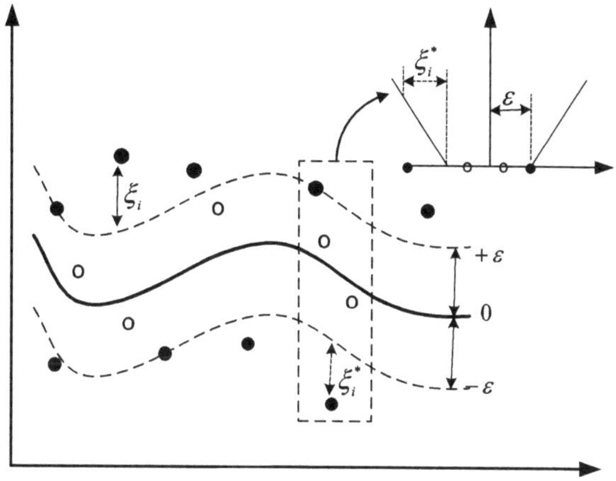

Fig. 8.9 Nonlinear SVR with Vapnik's ε-insensitive loss function

Unfortunately, the computation of $\langle \phi(\mathbf{x}_i), \phi(\mathbf{x}_j) \rangle$ in the feature space may be too complex to perform. Therefore, an advantage of SVM is useful to overcome this difficulty that the nonlinear function $\phi(\mathbf{x}_i)$ need not be used. The computation in input space can be performed using kernel function.

In Eq. (8.15), the dot product of $\langle \phi(\mathbf{x}_i), \phi(\mathbf{x}_j) \rangle$ can be replaced with kernel function $K(\mathbf{x}_i, \mathbf{x})$. Kernel functions enable the dot product to be performed in high-dimensional feature space using low-dimensional space data input without knowing the transformation. All kernel function must satisfy the Mercer's condition (Cristianini et al. 2000) that corresponds to the inner product of some feature space. In this chapter, the RBF kernel function is used for performing nonlinear regression

$$K(\mathbf{x}_i, \mathbf{x}) = \exp\left\{-\gamma|\mathbf{x}_i - \mathbf{x}|^2\right\} \tag{8.16}$$

The verification performance statistic, such as the root-mean-square error (RMSE) and correlation statistic coefficient (R), are used to examine the system. RMSE provides a general illustration of the overall accuracy of the predictions they show the global goodness of fit, given as follows:

$$\text{RMSE} = \sqrt{\frac{\sum_{i=1}^{N}(y_i - \hat{y}_i)^2}{N}} \tag{8.17}$$

where N represents the total number of data points in the test set, y represents the observed value, and $\hat{y}$ represents the predicted value. The R measure, the linear correlation between the actual and predicted values, can be calculated as follows:

$$R_{y,\hat{y}} = \frac{\text{Cov}(y, \hat{y})}{\sigma_y \sigma_{\hat{y}}} \tag{8.18}$$

where R is the correlation coefficient and Cov(y, $\hat{y}$) is the covariance between observed and predicted values, which can be calculated as follows:

$$\mathrm{Cov}(y,\hat{y}) = \frac{1}{N}\sum_{i=1}^{N}(y_i - \bar{y})(\hat{y}_i - \bar{\hat{y}}) \tag{8.19}$$

where $\bar{y}$ is the mean of the observed value and $\bar{\hat{y}}$ is the mean of the predicted value. The standard deviation of the observed and predicted values, σ_y and $\sigma_{\hat{y}}$, respectively, can be calculated as follows:

$$\sigma_y = \left(\frac{1}{N-1}\sum_{i=1}^{N}(y_i - \bar{y})^2\right)^{1/2} \tag{8.20}$$

$$\sigma_{\hat{y}} = \left(\frac{1}{N-1}\sum_{i=1}^{N}(\hat{y}_i - \bar{\hat{y}})^2\right)^{1/2} \tag{8.21}$$

8.2.6 State Estimator-Based Approach

State estimation techniques such as Kalman filters or various other tracking filters can also be implemented as a prognosis technique. The Kalman filter (Lewis 1986; Drexel 2001) is a dynamical systems tool for estimating unknown states by combining current measurements with the most recent state estimate. It can be considered as a virtual sensor in that it takes current available sensor measurements and provides optimal estimates (or predictions) of quantities of interest that may in themselves not be directly measurable. Knowledge of noise processes is used to minimize the estimation error covariance, via the optimal determination of the so-called Kalman gain.

It is typically implemented with the use of a linear system model, but can also be extended to nonlinear systems through the use of the extended Kalman filter algorithm that linearizes the system about an operating point. The discrete-time system with internal state x_k and sensor measurements z_k may be described in terms of the recursive difference equation

$$\begin{aligned} x_{k+1} &= A_k x_k + B_k u_k + G_k w_k \\ z_k &= H_k x_k + v_k \end{aligned} \tag{8.22}$$

where u_k is a control input, w_k is a *process noise* that captures uncertainties in the process dynamics, such as modeling errors and unknown disturbances (e.g., wind gusts in aircraft), and v_k is a *measurement noise*. This is depicted in Fig. 8.10.

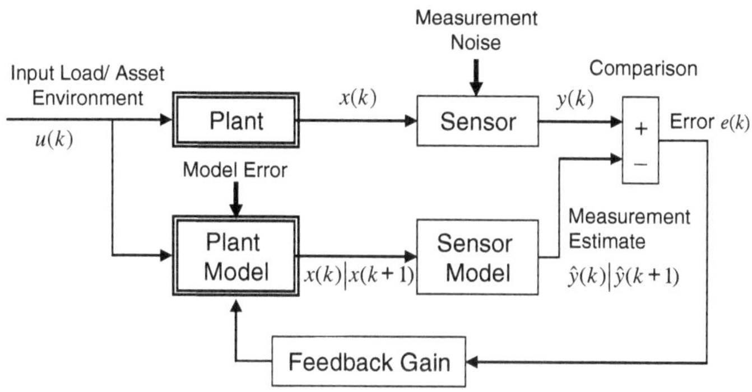

Fig. 8.10 State estimation Kalman filter implementation approach

In this type of application, the minimization of error between a model and measurement can be used to predict future feature states and hence the behavior of the modeled system. Either fixed or adaptable filter gains can be utilized (Kalman is typically adapted, while alpha-beta-gamma is fixed) within an nth-order state variable vector.

In a slightly different application of the Kalman filter, measured or extracted features f can be used to develop a state vector as shown below.

$$x = \left[f \dot{f} \ddot{f}\right]^T \tag{8.23}$$

Then, the state transition equation can be used to update these states based on a model. A simple Newtonian model of the relationship between the feature position, velocity, and acceleration can be used if constant acceleration is assumed. This simple kinematic equation can be expressed as follows:

$$f(n+1) = f(n) + \dot{f}(n)t + \frac{1}{2}\ddot{f}(n)t^2 \tag{8.24}$$

where f is again the feature and t is the time period between updates. There is an assumed noise level on the measurements and model related to typical signal-to-noise problems and unmodeled physics. The error covariance associated with the measurement noise vectors is typically developed based on actual noise variances, while the process noise is assumed based on the kinematic model. In the end, the tracking filter approach is used to track and smooth the features related to predicting a failure.

8.2.7 Statistical Reliability and Usage-Based Approach

In situations where sophisticated prognostic models are not warranted due to the lower level of criticality or low failure occurrence rates and/or there is an insufficient sensor network to assess condition, a statistical reliability or usage-based prognostic approach may be the only alternative. This form of prognostic algorithm is the least complex and requires the component of line replaceable unit (LRU) failure history data and/or operational usage profile data. Typically, failure and/or inspection data are compiled from legacy systems and a Weibull distribution or other statistical failure distribution can be fitted to the data (Groer 2000; Schomig 2003). An example of these types of distributions is given in Fig. 8.11. Although simplistic, a statistical reliability-based prognostic distribution can be used to drive interval-based maintenance practices that can then be updated on regular intervals. An example may be the maintenance scheduling for an electrical component or airframe component that has few or no sensed parameters and is not critical enough to warrant a physical model. In this case, the prognostics of when the component will fail or degrade to an unacceptable condition must be based solely on the analysis of past experience or reliability. Depending on the maintenance complexity and criticality associated with the component, the prognostics system may be set up for a maintenance interval (i.e., replace every 1000+/−20 engine flight hours) and then updated as more data become available. The benefit of having a regularly updated maintenance database as happens in autonomic logistics applications is significant for this application.

The next logical extension to this type of reliability-based statistical model is to correlate the failure rate data with specific operational usage profiles that are more directly related to the way a specific vehicle is used. In this manner, statistical

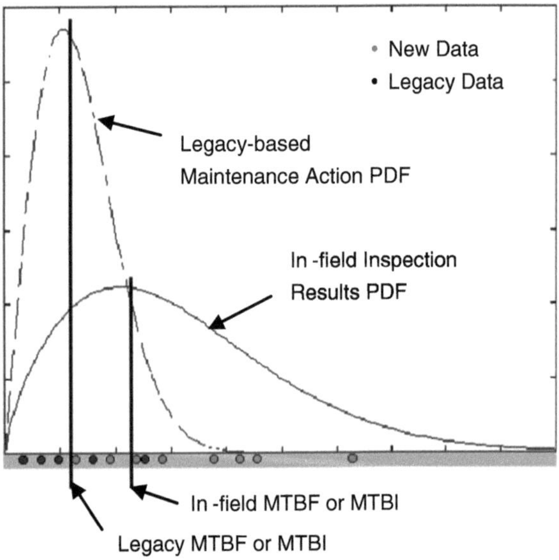

Fig. 8.11 Statistical reliability-based approach

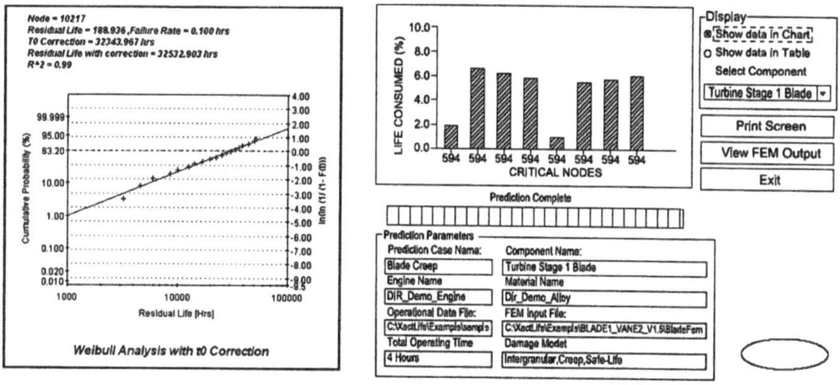

Fig. 8.12 Usage-based damage accumulation approach

damage accumulation models or usage models for specific components/LRUs can be directly tied to the loading profiles inferred from the high-level operations data sets, for example, fatigue cycles that are a function operating conditions such as speed or maneuvering conditions. An example of this is shown in Fig. 8.12, where a usage model (in this case damage accumulation model) was developed based on the operating speed of an engine. This type of usage models is often referred to as regime recognition in the helicopter community.

It is important to recognize that this is not another form of reliability-centered maintenance, in which we replace components based on a conservative safe-life operational time. It is a method to include the operational profile information and up-to-date reliability/inspection data in an automated algorithm that will augment existing fault detection conclusions or provide a prediction when more accurate means are not justified. More accurate prognostic methods are described further.

8.2.8 Adaptive Prognostics

As a direct extension to the concept presented above, the idea of updating the prognosis PDF based on additional state awareness (fault detection and diagnosis) information that can become available over time is also desirable. The adaptive prognostics concept entails that information available at the current time (which may or may not be diagnostic in nature) be used to modify future predictions, hence updating the prognosis PDF. This idea is illustrated in Fig. 8.13 (Engel et al. 2000) and briefly described next.

Consider point d_0 to be the mean initial damage condition for a prognostic model. A prognosis of life, from time k to predetermined damage level, is found to be represented by RUL_0 or remaining useful life. Suppose that some imperfect measurement $z(k)$ regarding the damage state becomes available at time $k = k + p\Delta t$. The challenge is to find optimal current damage state to reinitialize the model and/or

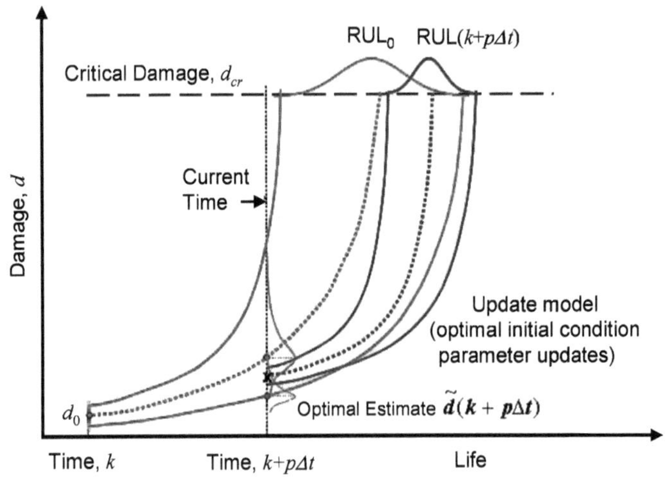

Fig. 8.13 Adaptive prognosis concept (Engel et al. 2000)

adjust model parameters so that a calibrated and more accurate prognosis can be established.

Though the utilization of a new initial condition, $\tilde{d}(k)$ at time $k = k + p\Delta t$ as shown in Fig. 8.13, it is apparent that the prediction mean has shifted and the confidence bounds on the resulting *RUL* have less variance than the original (blue prediction). The prediction accuracy improvement would generally mean that a decision to take action based on failure probability will likely reduce the lost operational availability over a run-to-failure maintenance plan.

8.2.9 Data Mining and Automated Rule Extraction

Data mining and automated rule extraction are synonymous. Data mining and rule extraction have been in development in the financial community for a number of years. In that field, managers are interested in processing data to target customers for marketing. There are a variety of commercially available software packages to perform data mining.

Those tools can be applied to the development of the prognostic reasoner as well. In most rule extraction processing, rules are extracted from input data by brute force examination of the data. Input data must consist of fault class-labeled samples. Rule extraction has several advantages over neural nets:

- Comprehensibility: Something that humans can understand.
- Explanation: Let the user know "why" the system did what it did.
- Validation: The user can explore all possible set of inputs to ensure the system operates as expected under all conditions.

- Discovery: Find something "new" in the features of data not known before.
- As mentioned, rule extraction is performed by brute force examination of the input data. Typical rules are found as a binary tree.

Data mining is a powerful new method with which to develop classifiers. The advantages are as described above. However, as with the other "trained" classifiers, real data that are representative of all the conditions expected to be encountered need to be collected.

8.2.10 Distributed Prognostic System Architecture

The cornerstone of an effective prognostics and health management (PHM) system is the information/data architecture and the ability for understanding and managing the anomaly, diagnostic, and prognostic (A/D/P) information from the LRU level all the way up through to the subsystem- and vehicle-level reasoners. This concept is briefly illustrated in Fig. 8.14, where faults detected and predicted at the LRU level are assessed through the hierarchy of reasoners in order to determine the root causes of vehicle malfunctions and contingency option for impending failures.

In general, the A/D/P technologies implemented at the lower levels (LRUs) are used to detect and predict off-nominal conditions or damage accumulating at an accelerated rate. In the distributed PHM architecture, this information is analyzed through the hierarchy of reasoners to make informed decisions on the health of the

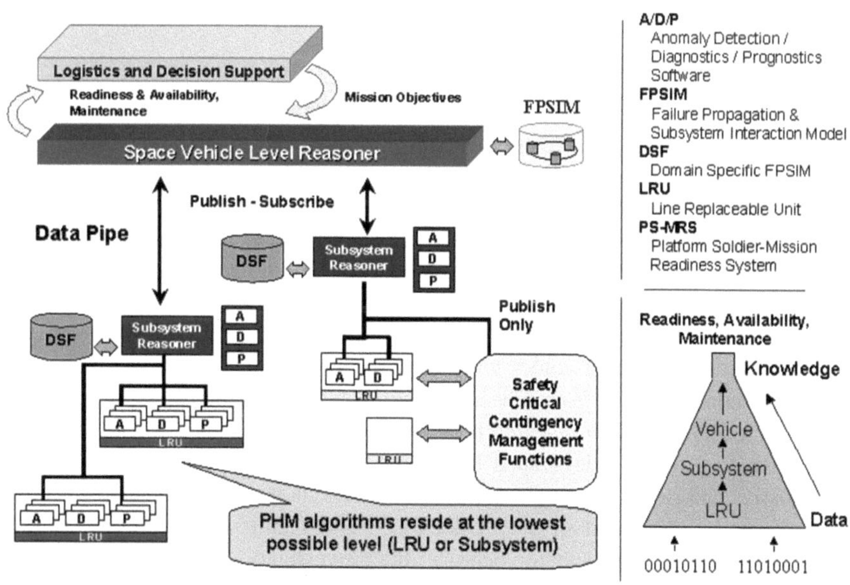

Fig. 8.14 Distributed prognostic system architecture

vehicle subsystems/systems and how they affect total vehicle capability. This integration across LRUs, subsystems, and systems is vital to correctly isolate the root cause of failures and understanding the propagation of up/downstream effects of the faults. Integration of the individual subsystem PHM results is eventually accomplished with the vehicle-level reasoner, which will assess the intrasystem A/D/P results in order to prioritize the recommended maintenance action(s) to perform in order to correct the problem.

A distributed PHM architecture, such as that shown below, has many benefits including:

- Optimal computational resource management (i.e., placing high bandwidth processing at the lowest level and only passing up critical features);
- Supports the concept of "smart LRU/subsystem," where the most detailed "intelligence" about the system exists (i.e., supplier/designer responsibility);
- Provides the ability to isolate and assess the extent of multiple faults and battle damage, hence improving survivability of the vehicle;
- Hierarchical reasoners have a "built-in" data management capability for containing erroneous information and utilizing multiple data and information sources;
- Ability to capture and localize system degradations (as opposed to only hard failures), based on increased health awareness of the lowest level LRUs, hence providing a more accurate vehicle availability assessment.

8.3 Applications

8.3.1 Bearing Prognostics

There has been a clear motivation to develop adequate systems monitoring and process prediction techniques due to high cost associated with equipment downtime in the manufacturing industry. Since rotating machinery exists in the almost every mechanical equipment, it is important to develop reliable techniques for the prediction of rotating element failure early enough to facilitate preventive maintenance. The failure of rolling element bearing is one of the primary causes of breakdown in rotating machinery. Bearing failure can be catastrophic in certain situations, such as in the cases of helicopter rotors and automatic processing machines. To prevent disastrous consequences from a bearing failure, fatigue life prediction and bearing operating endurance limit research have been in strong demand.

Romer et al. (2001) developed the method presented in Fig. 8.15 for bearing prognostics and health management module. Various sources of diagnostic information can be combined in the model-based and feature-based prognostic integration algorithm for the specific bearing under investigation, utilizing probabilistic update process.

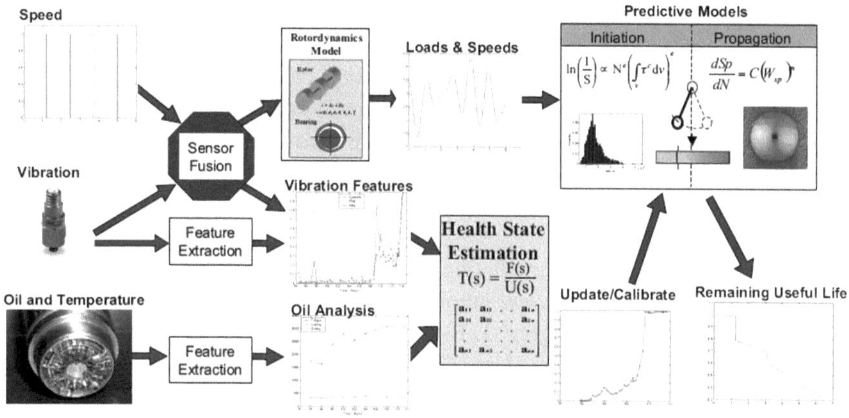

Fig. 8.15 Prognostic architecture through data fusion (Romer et al. 2001)

Knowledge of how the specific bearing is being loaded, historical failure mode information, and inspection data feedback can also be accommodated within the generic prognostic architecture. Moreover, in the developed system, data fusion also can be accommodated for determining RUL of bearing. Fusion can play a role in advanced prognostics process in terms of producing useful features, combining features, and incorporating model-based information.

As there are many different architectures for fusion, there are also many different algorithms themselves for performing the fusion. Most sophisticated techniques, specially Bayesian and Dempster–Shafer combination, are briefly described. Bayesian inference can be used to determine the probability that a diagnostic is correct, given a piece of a priori information. Analytically, Bayes theorem is expressed as follows:

$$P(f_1|O_n) = \frac{P(O_n|f_1) \cdot P(f_1)}{\sum_{j=1}^{n} P(O_n|f_1) \cdot P(f_1)} \tag{8.25}$$

where $P(f|O)$ is the probability of fault (f) given a diagnostic output (O), $P(O|f)$ is the probability that a diagnostic output (O) is associated with fault (f), and $P(f)$ is the probability of fault (f) occurring.

In the Dempster–Shafer approach, uncertainty in the conditional probability is considered. The Dempster–Shafer methodology hinges on the construction of a set, called the frame of discernment, which contains every possible hypothesis. Every hypothesis has a belief denoted by a mass probability (m). Beliefs are combined with the following equation:

$$\text{Belief}(H_n) = \frac{\sum_{A \cap B = H_n} m_i(A) \cdot m_j(B)}{1 - \sum_{A \cap B = 0} m_i(A) \cdot m_j(B)} \tag{8.26}$$

Model-based approaches to prognostics differ from feature-based approaches in that they can make RUL estimating in the absence of any measurable events, but when related diagnostic information is present (such as feature described previously), the model can often be calibrated based on the new information. For model-based prognostics of bearing, there are two well-known approaches as follows:

(1) *Spall initiation model*

A variety of these exist for predicting spall initiation from bearing dimensions, loads, lubricant quality, and view empirical constants. Many modern theories are based on the Lundberg–Palmgren (L-P) model (Lundberg and Palmgren 1947) that was developed in 1940. Yu and Harris (Y-H) model (Yu and Harris 2001) proposed a stress-based theory in which relatively simple equations are used to determine the fatigue life purely from the induced stress. The fundamental equation of Y-H model is stated as follows:

$$\ln\left(\frac{1}{S}\right) \propto N^{e}\left(\int_{V} \tau^{c} \mathrm{d}V\right) \tag{8.27}$$

This equation relates the survival rate (S) of the bearing to a stress-weighted volume integral as shown below. The model utilizes a new material property for the stress exponent (c) to represent the material fatigue strength and the conventional Weibull slope parameter to account for dispersion in the number of cycles (N). The fatigue initiating stress (τ) may be expressed using sines multi-axial fatigue criterion for combined alternating and mean stresses, or as a simple Hertz stress.

(2) *Spall progression model*

Once initiated, a spall usually grows relatively quickly, producing increased amounts of oil debris, high vibration levels, and elevated temperatures that eventually lead to bearing failure. While spall progression typically occurs more quickly than spall initiation, a study by Kotzalas and Harris (2001) showed that 3–20 % of the useful life of a particular bearing remains after spall initiation. The study identified two spall progression regions. Stable spall progression is characterized by gradual spall growth and exhibits low broadband vibration amplitudes.

Kotzalas and Harris also presented a spall progression model. The model relates the spall progression rate ($\mathrm{d}S_p/\mathrm{d}N$) to the spall amplitude (W_{sp}) using two constants (C and m). The spall similitude is defined in terms of the maximum stress ($\sigma_{\max}$), average shearing stress (τ_{avg}), and the spall length (S_p).

$$\frac{\mathrm{d}S_p}{\mathrm{d}N} = C(W_{sp})^{m} \tag{8.28}$$

$$W_{sp} = (\sigma_{\max} + \tau_{avg})\sqrt{\pi S_p} \tag{8.29}$$

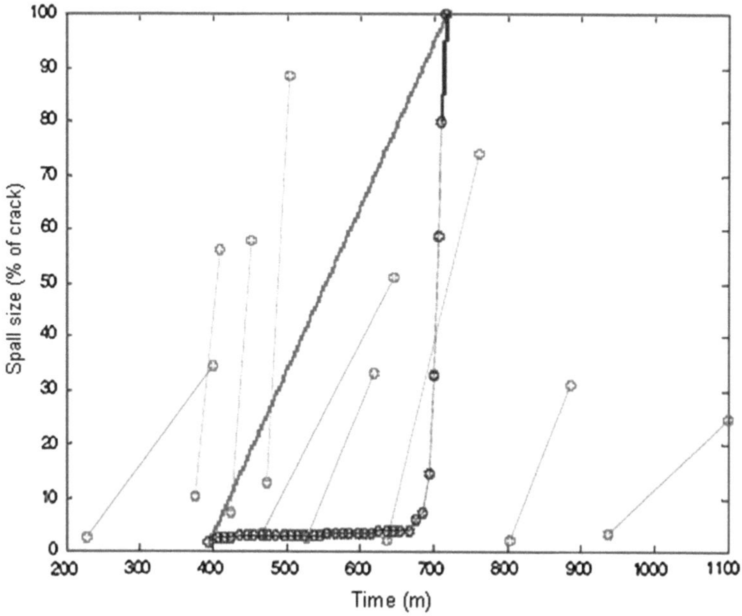

Fig. 8.16 Spall progression data (Kotzalas and Harris 2001)

Spall progression data were collected during Kotzalas–Harris study using ball/V-ring test device. The data were collected at the Pennsylvania State University. Each of the lines shown in Fig. 8.16 represents spall progression of a ball bearing. For one ball, indicated by multiple markers, the test was periodically suspended to measure the spall length. Only two spall length measurements, initial and final values, were taken for the other balls. The figure shows a large dispersion of the spall initiation life, which is indicated by the lowest point of each line. The figure also shows a large dispersion of the spall progression life, which can be inferred from the slope of the lines.

8.3.2 Gear Prognostics

In the case of gear prognostics, a high fidelity stress, fatigue and crack propagation model with fits within generic prognostics module architecture is shown in Fig. 8.17, configured for components that fail via fatigue mechanisms.

This comprehensive approach allows for the integration of material-level models, systems-level data/information fusion algorithms, and adaptive modeling techniques for tuning key failure mode variables at a local material/damage site. The output failure rate prediction is developed within a probabilistic framework to identify directly confidence bounds associated with the specific component failure

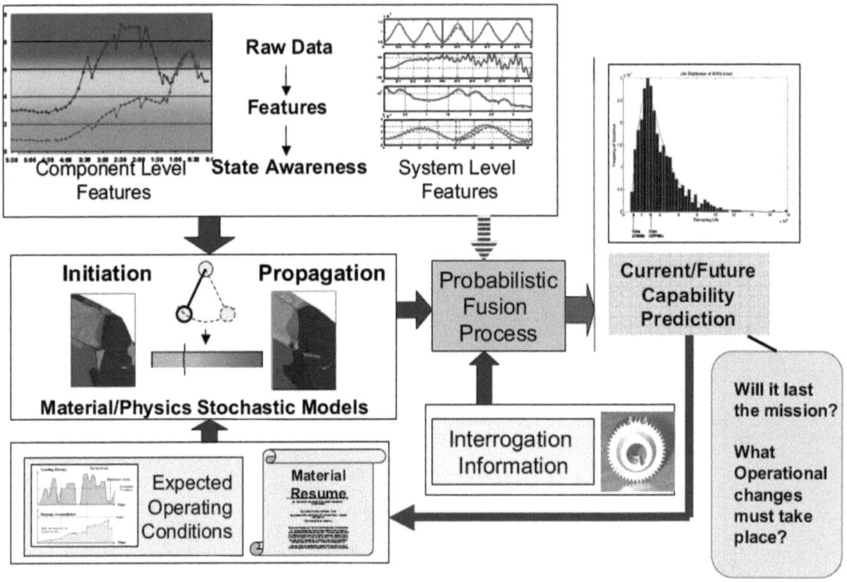

Fig. 8.17 Gear prognostics approach (Roemer et al. 2001)

mode progression. By providing continuous updates to critical parameters used by the probabilistic fatigue/damage models, based on observing systems-level measurements, more accurate failure rate prediction is possible throughout the life of the component.

Before performing prognostics, it is important to know the failure mode and characterization of gear. The principal failure mode considered in this gear prognostics module development was a single fatigue crack propagating from the filet region for a single tooth. This is an important failure mode to consider because the separation of a single tooth can lead to complete loss of power transmission under the certain conditions. Common failure modes for gears involve the following:

- Tooth separation due to a fatigue crack initiating in the fillet region. Crack initiation in this area is driven by high tensile bending stresses along fillet. A surface defect may not be required for a failure to occur.
- Partial tooth separation due to fatigue crack initiating from surface damage sites such as one or more pits, and scoring and scuffing marks.
- Excessive contact surface damage due to pitting, scoring, or scuffing that can lead to high levels of vibration and noise.
- Propagating cracks can initiate from accidental nicks during assembly or surface damage from oil debris.

Model-based gear prognostics can be developed through FEM model that generates the model of gear properties and behavior including material, load, and failure model scenario.

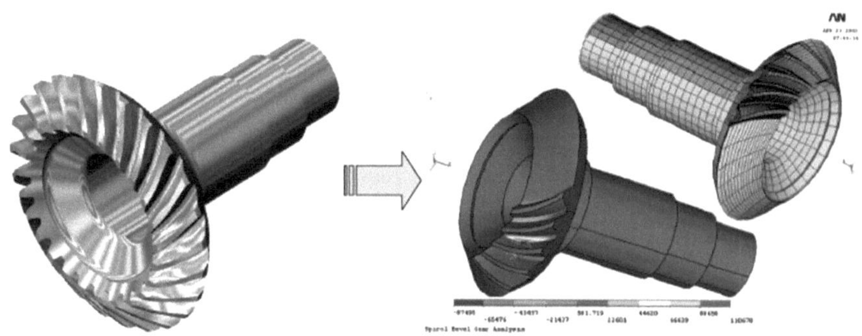

Fig. 8.18 Gear using FEM model (Lewicki 2001)

Impact technologies generated model of gear pinion geometry and associated load definition, which represented the pinion's load history through a complete engagement with the mating gear shown in Fig. 8.18. This model is developed using ANSYS that three pinion teeth defined about the global origin of the pinion and a load history, which encompassed one tooth having load applied from initially being at zero to a maximum and then unloaded back to zero.

A typical failure scenario would involve formation and propagation of a crack in the tooth fillet region. The primary stress field in this area is nominally uniaxial and is oriented along the fillet. Therefore, the initial crack formation and propagation would be perpendicular to the base fillet. However, computational fracture mechanics models show that once crack grows past the field of influence of the fillet, the crack trajectory is influenced by the rim design for the gears.

In order to simulate the proposed prognostics approach, an EDM notch was placed at the root of one of the gear teeth to accelerate the crack initiation process. For predicting the total fatigue life of a component without notches, strain-life data obtained from tests on uniaxial specimen should not be directly used because they are subject to uniform strain across the entire test cross section. For notched component, there is a severe stress gradient within the notch region, and therefore, special attention must be paid to this region by dividing it into subzones and calculating fatigue life for each zone independently. This approach facilitates the definition of crack initiation and thereby provides an unambiguous way to transition from the crack initiation phase to the crack propagation phase and circumventing an historical dilemma. The dilemma is that if crack initiation is defined as creation of a defect of an arbitrary length, then the associated stress intensively range (ΔK) from the Paris law can be made arbitrarily small or large. If the crack tip will be smaller than the $\Delta K_{\text{threshold}}$ for the material (depending on the R ratio) and therefore based on the fracture mechanics principles, the crack will not propagate. On the contrary, if crack initiation is defined as a large crack, then the predicted cycles for crack propagation could be rather small. Concisely, the prediction of the total fatigue life is greatly influenced by the definition of what constitutes the crack.

References

Atiya A, El-Shoura S, Shaheen S, El-Sherif M (1999) A comparison between neural-network forecasting techniques—case study: river flow forecasting. IEEE Trans Neural Networks 10 (2):402–409

Bezdek IC, Pal SK (eds) (1992) Fuzzy models for pattern recognition: methods that search for structure in data. IEEE Press, New York

Box GEP, Jenkins GM (1976) Time series analysis: forecasting and control. Holden-Day, San Francisco

Christer AH (1976) Innovative decision making. In: Brown KC, White DJ (eds) Proceedings of the NATO conference on role and effectiveness of theory of decision practice. Hodder and Stoughton, UK, pp 368–377

Cristianini N, Shawe-Taylor NJ (2000) An introduction to support vector machines. Cambridge University Press, Cambridge

Connor J, Martin R, Atlas L (1994) Recurrent neural networks and robust time-series prediction. IEEE Trans Neural Networks 5(2):240–254

Drexel M, Ginsberg JH (2001) Mode isolation: a new algorithm for modal parameter identification. J Acoust Soc Am 110(3):1371–1378

Engel SJ, Gilmartin BJ, Bongort K, Hess A (2000) Prognostics, the real issues involved with predicting life remaining. Aerosp Conf Proc 6:457–469

Friedman J (1991) Multivariate adaptive regression splines. Ann Stat 19(1):1–67

Gertler JJ (1998) Fault detection and diagnosis in engineering systems. Marcel Dekker, Inc, NewYork

Groer PG (2000) Analysis of time-to-failure with a Weibull model. In: Proceeding of the Maintenance and Reliability Conference, Knoxville

Gun SR (1998) Support vector machine for classification and regression. Technical Report, University of Southampton, UK

Haykin S (1994) Neural networks—a comprehensive foundation. Macmillan College Pub. Co, New York

Heicht-Nielsen R (1990) Neurocomputing. Addison-Wesley Publishing Co., Redwood City

Hush DR, Horne BG (1993) Progress in supervised neural networks. IEEE signal processing magazine. 10(1):8–39

Husmeier D (1999) Neural networks for conditional probability estimation: forecasting beyond point prediction. Springer, London

Jang J, Sun C, Mizutani E (1997) Neuro-fuzzy and soft computing. Prentice-Hall Inc., Englewood Cliffs

Kallberg KT, Rossi A (2007) Remaining useful life. www.italamericon.com/usefullife.htm

Khan MS, Coulibaly P (2006) Application of support vector machine in lake water level prediction. J Hydrol Eng 11(3):199–205

Kotzalas M, Harris T (2001) Fatigue failure progression in ball bearings. Trans ASME J Tribol 123:238–242

Lewicki (2001) Gear crack propagation pat studies—guideline for ultrasave design, NASA/TM-2001-211073

Lewis L (1986) Optimal estimation: with an introduction to stochastic control theory. Wiley, New York

Li X, Sun C, Gong D (2005) Application of support vector machine and similar day method for load forecasting. LNCS 3611:602–609

Lundberg G, Palmagen A (1947) Dynamic capacity of rolling bearings. Acta Polytechnica Mechanical Engineering Series 1, Royal Swedish Academy of Engineering Sciences

Luo J, Namburu M, Pattipati K, Qiao L, Kawamoto M, Chigusa S (2003) Model-based techniques. In: Proceedings of IEEE AUTOTESTCON, pp 330–340

Mohandes MA, Halawani TO, Rehman S, Hussain AA (2004) Support vector for wind speed prediction. Renewable Energy 29:939–947

Muller KR, Smola AJ, Scholkopf B, Kohlmorgen J, Vapnik V (1997) Predicting time series with support vector machine. In: Proceedings of the international conference artificial neural network

Pourahmadi M (2001) Foundations of time-series analysis and prediction theory. Wiley, New York

Rao T (1981) On the theory of bilinear models. J Roy Stat Soc B 43:244–255

Roemer M, Kacprzynsky G (2001) Development of diagnostic and prognostic technologies for aerospace health management application. In: IEEE aerospace conference, Montana

Schomig A, Rose O (2003) On the suitability of the Weibull distribution for the approximation of machine failure. In: Proceeding of the industrial engineering research conference, Portland

Tong H, Lim S (1980) Threshold autoregression, limited cycles and cyclical data. J Roy Stat Soc B 42:245–292

Tse P, Atherton D (1999) Prediction of machine deterioration using vibration based fault trends and recurrent neural networks. J Vib Acoust 121:355–362

Vachtsevanos G, Lewis F, Roemer M, Hess A, Wu B (2006) Intelligent fault diagnosis and prognosis for engineering system. Wiley, New Jersey

Vapnik VN (1995) The nature of statistical learning theory. Springer, New York

Vukovic P (2001) One-step ahead predictive fuzzy controller. Fuzzy Sets Syst 122:107–115

Wang W (2007) A two-stage prognosis model in condition based maintenance. Eur J Oper Res 182:1177–1187

Wang W, Ismail F, Golnaraghi F (2001) Assessment of gear damage monitoring techniques using vibration measurements. Mech Syst Signal Process 15:905–922

Wang W, Ismail F, Golnaraghi F (2007) A neuro-fuzzy approach to gear system monitoring. IEEE Trans Fuzzy Syst (in press)

Wu CH, Ho JM, Lee DT (2004) Travel-time prediction with support vector regression. IEEE Trans. Intell Transp Syst 5(4):276–281

Yang J, Zhang Y (2005) Application research of support vector machines in condition trend prediction of mechanical equipment. Lect Notes Comput Sci 3498:857–864

Yen J, Langari R, Zadeh LA (1995) Industrial applications of fuzzy logic and intelligent systems. IEEE Press, New York

Yu WK, Harris T (2001) A new stress-based fatigue life model for ball bearings. Tribol Trans 44 (1):11–18

Zadeh L (1965) Fuzzy set. Inf Control 8:338–353

Chapter 9
Data Fusion Strategy

9.1 Introduction

With industry system becoming more and more complex, maintenance personnel tend to utilize the multi-type sensors to collect signals from the machine being monitored. By comparing these acquired signals with healthy signals (benchmarking signals), the maintenance personnel are able to assess the running condition of the monitored machines. Moreover, they may also be able to identify the special faults based on their knowledge of fault signatures. However, this is a hard work even for experts. One reason is that the signal qualities acquired from different types of sensors may be different. Given various types of sensors, a requirement is to know the performance of different types of sensors in condition monitoring and fault diagnosis. Another reason is that the signals collected from industrial environments are often masked by random noise so do not appear to be in accordance with the standard signatures.

In recent years, numerous intelligent systems have been employed to assist system fault diagnosis and prognosis tasks by correctly interpreting the fault and degradation data such as expert systems, artificial neural networks, support vector machines, and fuzzy logic systems, and the results of these techniques are promising. However, many researches have shown that an individual decision system with a single data source can only acquire a limited classification or prediction capability, which may be not enough for a particular application. Therefore, the application of data fusion methodology has received a lot of interests in recent years, and researchers have achieved considerable successes from this approach to solve complex pattern recognition and prediction tasks.

The *fusion* is a relatively new term to the engineering system in health management community. A "*fused*" definition, which fits many examples in engineering, is identified as the process of combining data and knowledge from different sources with the aim of maximizing the useful information content, for improved

G. Niu, *Data-Driven Technology for Engineering Systems Health Management*,
DOI 10.1007/978-981-10-2032-2_9

reliability or discriminant capability, while minimizing the quantity of data ultimately retained.

Most application of data fusion can be found at decision level. *Decision fusion* is also called multiple classifiers fusion, combination of classifiers, multiple experts, and hybrid method. Due to the integration of different decisions from multiple classifiers, the technique can boost the accuracy of recognition. The multiple classifiers fusion can be categorized into two classes: the static and the dynamic methods. The static combination strategies are simple but only concentrate on the output of the classifiers, such as majority voting (Ruta and Gabrys 2000), minimum and maximum (Kuncheva 2002), and average (Taniguchi and Tresp 1997). By comparison, the dynamic methods are more elaborate which take into account the information from the training phase on the behavior of the classifiers which include the Bayesian method, behavior knowledge space, and Dempster–Shafer theory (Xu et al. 1992; Hang and Suen et al. 1993; Yang and Kim et al. 2006)

This chapter describes the data fusion strategies for condition monitoring, fault diagnosis, and prognosis, which integrates data sources of different features or sensors and decisions of multiple classifiers or prediction tools.

9.2 Fusion Application Areas

In the sensor and signal processing area, fusion is used to synthesize the results of two or more sensors. In the decision and recognition area, fusion is used to deduce the most reliable conclusion from different decisions.

For data-driven PHM/CBM, a machine is repaired or serviced only when an intelligent monitoring system indicates that it cannot fulfill its mission requirements. Condition monitoring system should be processed large amount of data if proper assessment of the machine's health is to be ensured. The data may contain vibration, temperature, pressure, oil analysis, and other measurements that encapsulate the parametric properties of the system and can aid in its condition assessment. The examination of data can be tedious and sensitive to errors.

An important aspect of condition monitoring is the fidelity of information received by the sensor units. The data acquired must be consistent and as much noise-free as possible. After the data have been acquired at the source level, it passes through to the preprocessing for digital conversion and proper manipulation. The processed data are then routed according to the level of fusion sought, i.e., raw data, feature, or decision-level fusion. The selection of fusion level must be made depending upon the application.

Data-driven PHM/CBM has been an integral part of maintenance strategy. In order to monitor the deteriorating state, the analysis of vibration, acoustic emission, and wear debris was conducted. Data fusion encompasses the theory, techniques, and tools conceived and employed for exploiting the synergy in the information acquired from multiple sources (sensor, databases, information gathered by human, etc). The resulting decision or action is better than the information sensed by a single sensor without exploiting the synergy of many sensors. Data fusion techniques combine data from multiple sensors and retrieved related information from the associated databases to achieve improved accuracies and with more specific inferences than that could be achieved by the use of a single sensor. Moreover, decisions from several systems can be fused and deduced to acquire the most reliable conclusion.

Depending on the nature of the available information, various models can be used for its representation. These models range from fully quantitative frameworks, such as probability theory, belief function-based evidence theory, and upper and lower probabilities, to more qualitative ones, such as possibility theory and non-monotonic or para-consistent logics.

9.3 Data Fusion Architectures

Identifying an optimal fusion architecture and approach at each level is a vital factor, assuring that the realized system truly enhances health monitoring capabilities. A brief explanation of fusion architectures will be provided here. In terms of different fusion phases of measured information, fusion architectures can be categorized into three levels: data level, feature level, and decision level (Hall and Llinas 1997).

9.3.1 Data-Level Fusion

All sensor raw data from a measured object are combined directly, and a feature vector is then extracted from the fused data. At this stage, a pattern recognition process is performed as shown in Fig. 9.1. Methods for this feature-based identity declaration include neural networks, template methods, and pattern recognition methods such as cluster algorithms. Fusion of data at this level contains the most

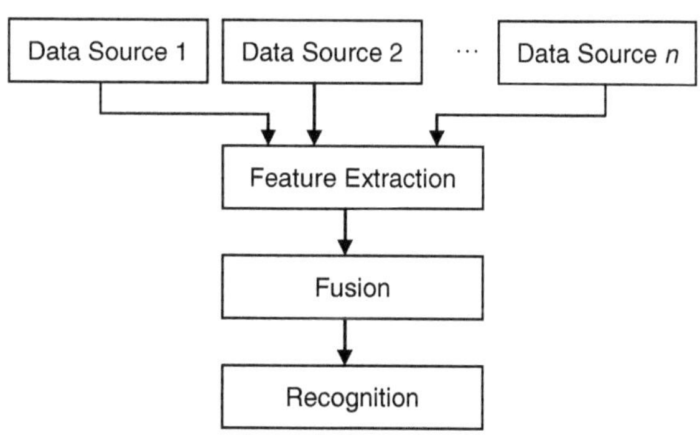

Fig. 9.1 Data-level fusion

information and can give good results. However, the sensors used in this level must be commensurate. That means the measurement has to be the same or has similar physical quantities or phenomena such as a vibration signal. As a consequence, the data-level application is limited in the real environment where there are many physical quantities having to be measured for synthesis analysis.

9.3.2 Feature-Level Fusion

In this level, features are extracted from each sensor according to the type of raw data. Then, the information of these non-commensurate sensors is combined at the phase of the feature level. All the features vectors are combined in turn into a bigger single feature vector, which is then used in a special classification model such as neural network or cluster algorithm for decision making as shown in Fig. 9.2. The function of feature vector normalization should still be performed prior to linking the feature vectors from individual sensor into a single larger feature vector in order to scale them into a same range.

9.3.3 Decision-Level Fusion

In this structure, the processes of feature extraction and pattern recognition are applied for single-source data obtained from each sensor. Then, the generated decision vectors are fused using decision-level fusion techniques such as voting strategy, Bayesian method, behavior knowledge space, and Dempster–Shafer theory, as shown in Fig. 9.3.

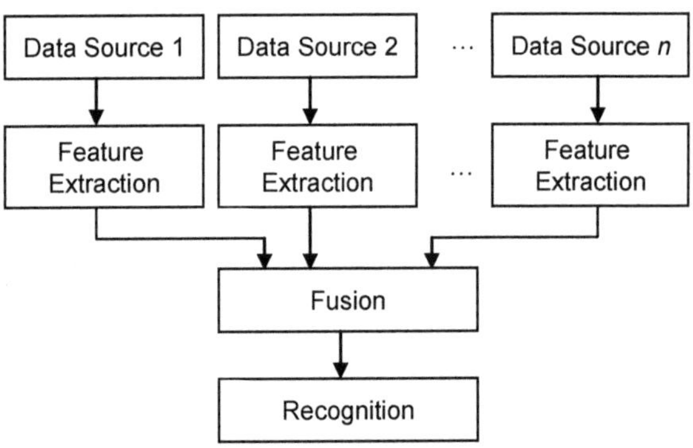

Fig. 9.2 Feature-level fusion

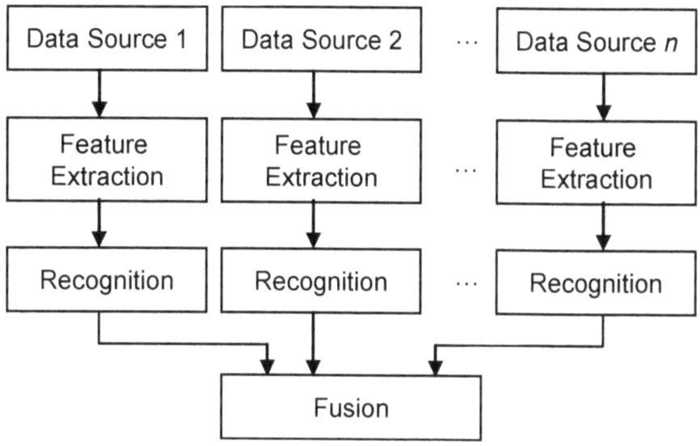

Fig. 9.3 Decision-level fusion

9.4 Data Fusion Techniques at Decision Level

There is a large of techniques for performing data, feature, or decision fusion. Because of this fact, sorting through which technique is the best can be a daunting and involved task. In addition, there are no hard and fast rules about what fusion techniques or architectures work best for any particular application. This section will describe some common fusion approaches at decision level.

According to the characteristics of output information of the classifiers, decision fusion methods can be divided into three styles (Xu et al. 1992):

- The abstract style: A classifier *C* only generates a single class output with an input *x;*
- The rank style: A classifier *C* ranks all classes in a queue and chooses the top one; and
- The measurement style: A classifier *C* evaluates each class using a probability value that the *x* subjects to the class.

Among the styles mentioned above, the required information for a classification increases in sequence, and the abstract style contains the least information, while the measurement style contains the most information. Accordingly, the classifier fusion algorithms of the measurement style can produce the best results. However, the classifiers being able to output each class's probability are seldom available. As a result, the classifier fusion algorithms belonging to an abstract style are commonly used. The methods used in abstract style mainly consist of voting, Bayesian, BKS, and Dempster–Shafer theory.

9.4.1 Voting Method

There are various voting strategies such as *first past the post, threshold, majority,* and *Borda count. First past the post* or *winner-takes-all* is a method in which a single winner is chosen in a given constituency by having the most votes, regardless of whether or not he or she has a majority of votes. By comparison, *threshold voting* method considers a specified threshold; the term will be selected, provided it is equal to or greater than that threshold.

Borda count (BC) (Verma et al. 2001) is defined as a mapping from a set of individual rankings to a combined ranking leading to the most relevant decision. For a particular class c_k, Borda count $B(c_k)$ is defined as a sum of the number of classes ranked below class c_k by each classifier. The magnitude of the BC reflects the level of agreement that the input pattern belongs to the considered class. To a certain degree, the BC can be treated as a generalization of the majority voting rule. This method is based on the assumption of additive independence among the contributing classifiers. It is easy to implement and does not require any training. Weak point of this technique is that it treats all classifiers equally and does not take into account the confidence values produced by various classifiers.

Totally, in various voting methods, *majority voting* may be the most popular method. In this method, the class voted by most of classifiers will be regarded as the result of fusion decision. If no class won more than half of the votes, the input is rejected. The method is simple and easy to realize. Nevertheless, it does not consider the characteristics of each classifier which are related with the performance of classifier fusion.

As an improvement of traditional majority algorithm, *weighted majority algorithm* (WMA) (Littlestone and Warmuth 1989) is a meta-learning algorithm used to construct a compound algorithm from a pool of prediction algorithms, which could be any type of learning algorithms, classifiers, or even real human experts. The algorithm assumes that we have no prior knowledge about the accuracy of the algorithms in the pool, but there are sufficient reasons to believe that one or more will perform well.

Assume that the problem is a binary decision problem. To construct the compound algorithm, a positive weight is given to each of the algorithms in the pool. The compound algorithm then collects weighted votes from all the algorithms in the pool and gives the prediction that has a higher vote. If the compound algorithm makes a mistake, the algorithms in the pool that contributed to the wrong prediction will be discounted by a certain ratio β where $0 < \beta < 1$.

9.4.2 Bayesian Belief Fusion

Bayesian belief algorithm (Xu et al. 1992) offers a soft fusion strategy, which is based on the assumption of mutual independency of classifiers and considers the error of each classifier. For a multiple class recognition problem with classes 1 through M, the error for kth classifier can be represented by a two-dimensional confusion matrix.

$$\mathbf{PT}_k = \begin{bmatrix} n_{11} & n_{12} & \cdots & n_{1M} \\ n_{21} & n_{22} & \cdots & n_{2M} \\ \cdots & \cdots & & \cdots \\ n_{M1} & n_{M2} & \cdots & n_{MM} \end{bmatrix}, \quad (k = 1, \ldots, K) \tag{9.1}$$

where the rows stand for classes c_1, …, c_M which consist of input sample $\mathbf{x}$, and the columns indicate the classes which consist of the input sample assigned by the classifier e_k. The element n_{ij} illustrates the input samples from class c_i while assigned to class c_j by the classifier e_k. On the basis of the confusion matrix, a belief measure of recognition can be calculated for each classifier by the belief function:

$$Bel(x \in c_i/e_k(x)) = P(x \in c_i/e_k(x) = j_k) \tag{9.2}$$

where $i, j = 1, \ldots, M$ and

$$P(x \in c_i/e_k(x) = j_k) = n_{ij}^{(k)} / \sum\nolimits_{i=1}^{M} n_{ij}^{(k)} \tag{9.3}$$

Combining the belief measures of all fusion classifiers can result in the final belief measure of the multiple classifier system and is shown as follows:

$$Bel(i) = P(x \in c_i) \frac{\prod_{k=1}^{K} P(x \in c_i/e_k(x) = j_k)}{\prod_{k=1}^{K} P(x \in c_i)} \tag{9.4}$$

For practical implementation, an approximation of Eq. (9.4) is often used as follows:

$$Bel(i) = \eta \prod_{k=1}^{K} P(x \in c_i/e_k(x) = j_k) \tag{9.5}$$

where η is a constant and has

$$\frac{1}{\eta} = \sum_{i=1}^{M} \prod_{k=1}^{K} P(x \in c_i/e_k(x) = j_k) \tag{9.6}$$

The highest combined belief measure *Bel*(*i*) is chosen as the final classification decision. However, one of the significant limitations of Bayesian method is that it requires mutual independencies among multiple classifiers, which does not usually hold in real application.

9.4.3 Behavior Knowledge Space (BKS)

When compared with the Bayesian method, this method does not emphasize independence of the decisions made by classifiers. BKS (Hang and Suen 1993) is a k-dimensional space where each dimension corresponds to a single classifier. Each classifier could produce $N + 1$ crisp decisions, N class labels, and one rejection decision. The intersection of the decisions of every single classifier occupies one unit of the BKS, and each unit contains three elements: the total number of incoming samples, the best representative class, and the number of incoming samples of each class. The unit, which is the intersection of the classifiers' decisions of the current input, is called the focal point. For an unknown test sample, the decision of the individual classifiers indexes a unit of BKS and the unknown sample is assigned to the class with the most training samples in the BKS unit. In the BKS method, large numbers of training data are required to build the BKS so that the lack of enough training data often is a problem.

Constantinidis et al. (2000) suggested an improved BKS method called *augmented behavior knowledge space* (ABKS), which is an extended and enhanced implementation of the established BKS technique. ABKS contains two rules:

Rule 1: The ABKS classification rule using information for the best representative class:

$$E(x) = R_{e(1),\ldots,e(K)} \quad \text{when} \quad T_{e(1),\ldots,e(K)} > 0 \tag{9.7}$$

and

$$\frac{n_{e(1),\ldots,e(K)}\left(R_{e(1),\ldots,e(K)}\right)}{T_{e(1),\ldots,e(K)}} \gg \lambda \tag{9.8}$$

where λ is the accuracy level threshold.

If the focal unit addressed by the unknown sample has been initialized during the ABKS training process and the ratio of the total number of samples belonging to the best representative class divided by the total number of samples that have entered the focal unit is greater than the desired accuracy level λ, then the sample is classified according the most frequent class $\left(R_{e(1),\ldots,e(K)}\right)$. Therefore, different performance profiles are generated according to the desired accuracy threshold level λ $(0 \leq \lambda \leq 1)$.

Rule 2: The ABKS classification rule using the individual expert confidence levels:

$$E(x) = CC_{e(1),\ldots,e(K)} \quad \text{when} \quad T_{e(1),\ldots,e(K)} = 0 \tag{9.9}$$

If the focal unit addressed by the unknown sample has not been initialized during the ABKS training process, the sample is classified according to the decision of the classifier that demonstrates the highest level of confidence.

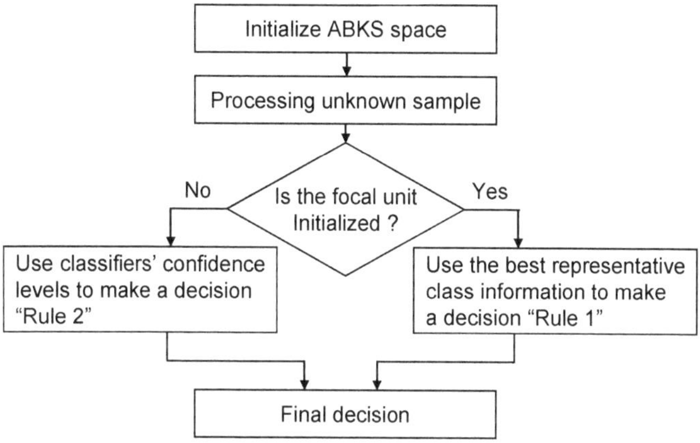

Fig. 9.4 Schematic of ABKS decision procedure

Figure 9.4 illustrates the overall decision model. Firstly, the ABKS space is initialized with the training data set. Then, in the sample classification stage, all the unknown samples are processed. If the experts point to a focal unit that has been initialized during the training phase of the ABKS, then the sample is classified according to the Rule 1. Otherwise, if the focal unit was not initialized, the unknown sample is classified according to the Rule 2.

9.4.4 Dempster–Shafer Theory

Evidence theory is initially based on Dempster's work (Dempster 1967) concerning lower and upper probability distribution. From these mathematical foundations, Shafer (1976) has shown the ability of the belief functions to model uncertain knowledge. The basic concepts and mechanisms of the Dempster–Shafer theory are introduced in this section. For detailed tutorials on the subject, the reader can refer to Basir and Yuan (2007), Smets and Kennes (1994), Yager (2001), and references cited therein.

Let Ω be a finite non-empty set of mutually N exhaustive and exclusive hypotheses about some problem domain. Evidence theory first assumes the definition of a set of hypotheses $\Omega = \{A_1, A_2, \ldots, A_N\}$ called the frame of discernment. Let us denote 2^Ω, the power set composed with the 2^N proposition A of Ω:

$$2^\Omega = \{\phi, \{A_1\}, \{A_2\}, \ldots, \{A_N\}, \{A_1 \cup A_2\}, \{A_1 \cup A_3\}, \ldots, \Omega\} \tag{9.10}$$

A key point of evidence theory is the basic belief assignment (BBA). The mass of belief in an element of Ω is quite similar to a probability distribution, but differs by the fact that the unit mass is distributed among the elements of 2^Ω, that is to say not only on the singletons A_n in Ω but also on composite hypotheses too. The belief

m_j assigned to an information source S_j is thus defined by m_j: $2^\Omega \rightarrow [0, 1]$. This function verifies the following properties:

$$m_j(\phi) = 0, \quad \sum_{A \subseteq \Omega} m_j(A) = 1 \tag{9.11}$$

The mass m_j (A) represents how strongly the evidence supports A which, in the case of a disjunction of hypotheses, has not been assigned to a subset of A because of insufficient information. This mass can be reassigned more precisely to the subsets of A if additional information is available. Each subset $A \subseteq \Omega$ such as $m_j(A) > 0$ is called the focal element of m. The belief function $Bel_j(A)$ and the plausibility function $Pl_j(A)$ are defined, respectively, for all $A \subseteq \Omega$ as

$$Bel_j(A) = \sum_{B \subseteq A} m_j(B), \quad Pl_j(A) = 1 - Bel_j(\overline{A}) = \sum_{B \cap A \neq \phi} m_j(B) \tag{9.12}$$

where $\bar{A}$ is the negation of a hypothesis A.

$Bel_j(A)$ represents the minimal (necessary) support for A and can be interpreted as a global measure of one's belief that hypothesis A is true. Belief of A is the sum of all the belief masses allocated to hypothesis B. Pl_j (A) provides a maximal (potential) support for A and is the sum of the parts of belief that are allocated to hypothesis B and are compatible with A.

From the basic belief assignment denoted by m_j obtained for each information sources S_j, it is possible to use a combination rule in order to provide combined masses synthesizing the knowledge of the different sources. These belief masses can then be used by a decision process with the benefit of the whole knowledge contained in the belief functions given by each source. Evidence can be combined by calculating the orthogonal sum using Dempster's rule of combination where evidence A and evidence B are used for calculating a new belief function for a focal element C as follows:

$$m(C) = m_1 \oplus m_2(C) = (1 - K)^{-1} \times \sum_{A \cap B = C} m_1(A) m_2(B) \tag{9.13}$$

where

$$K = \sum_{A \cap B = \phi} m_1(A)\, m_2(B)$$

The combination rule states that representing the combination of m_1 and m_2 apportions the total amount of belief among the subsets of Ω by assigning $m_1(A)$ $m_2(B)$ to the set resulting from interaction of sets A and B. K is often interpreted as a measure of conflict between the two sources and is a normalization factor. The larger the K is, the more the sources are conflicting and the less sense their combination.

A belief function has to be transformed into a probability function for decision making. The only transformation satisfying elementary rationality requirements was shown by Smets and Kennes (1994) to be the pignistic transformation, in which each mass of belief $m(A)$ is distributed equally among the elements of A for all $A \subseteq \Omega$. This leads to the pignistic probability distribution of class ω defined as (Denoeux 2000)

$$BetP(\omega) = \sum_{\omega \in A} \frac{m(A)}{|A|}, \quad \forall \omega \in \Omega \tag{9.14}$$

where $|A|$ denotes the cardinality of $A \subseteq \Omega$.

Dempster–Shafer theory is considered uncertainty using belief function and has some characteristics as follows:

- Evidential reasoning method which is an extension of Bayesian inference that overcomes some of the drawbacks;
- Can be used without prior probability distributions;
- Ability to deal with ignorance and missing information;
- Can deal with any union of classes (in the classification problem);
- Frame of discernment (hypothesis space) Ω contains every single hypothesis considered and the null set; and
- Dempster–Shafer reasoning considers every element of the power set 2^{Ω} and results complex and computation time necessary.

9.4.5 Multi-Agent Fusion

Multi-agent fusion algorithm (Kou and Zhang 2003) is an extension of Bayesian belief theory. This method absorbs the properties of multi-agent system into the algorithm of classifiers fusion. It integrates Bayesian belief at the starting phase and majority voting at the final phase. A codecision matrix is set up for information exchange between the classifier agents, so that Bayesian belief matrix can be modified dynamically until a predetermined criterion is satisfied. Finally, a combination decision is made. The flowchart of multi-agent fusion algorithm is shown in Fig. 9.5.

Confusion matrix is firstly created as a training parameter, which accumulates the errors of each classifier. Then, an initial belief matrix can be calculated easily for each test sample based on the training parameter. In the initial belief matrix, the rows indicate kth classifier, where $k = 1, \ldots, K$, and columns stand for class $c_1, \ldots, c_M$. The elements in kth row show the probabilities of an input sample x belonging to different classes estimated by kth classifier using Eq. (9.15). The processes to calculate the confusion matrix and initial belief matrix are based on the Bayesian belief method.

Fig. 9.5 Flowchart of multi-agent fusion algorithm

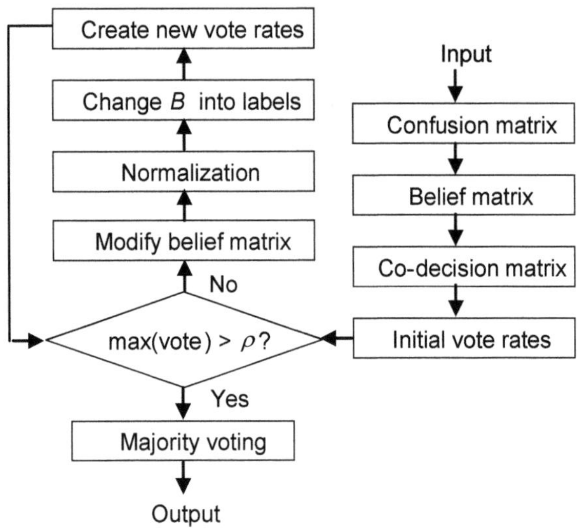

After calculating the two matrixes, a five-dimensional codecision matrix is required as the last training parameter. Each cell in the codecision matrix stands for decision correlation between two classifiers, which is calculated through Eq. (9.15) where $E = i$ is the expectation of input sample x, that is, the real class of x ranges from c_1 to c_M, j_1, and j_2, respectively, and stands for the decision of classifiers k_1 and k_2 where $k_1 \neq k_2$. U_2 is the training sample set of the fusion model. Each element in the matrix shows the probability of classifier k_1 classifying x as j_1 class and classifier k_2 assigning x as j_2 class.

$$\begin{aligned} d_{j_1,j_2,i,k_1,k_2} &= P(E = i|e_{k_1} = j_1, e_{k_2} = j_2) \\ &= \frac{|\{x|E(x) = i, e_{k_1}(x) = j_1, e_{k_2}(x) = j_2, \forall x \in U_2\}|}{\sqrt{|\{x|E(x) = i, e_{k_1}(x) = j_1, \forall x \in U_2\}|} \cdot \sqrt{|\{x|E(x) = i, e_{k_2}(x) = j_2, \forall x \in U_2\}|}} \end{aligned} \quad (9.15)$$

After obtaining the necessary statistical parameters, the confusion matrix and codecision matrix, the initial vote rates for input sample x can be calculated. The column class corresponding to the maximum of kth row of belief matrix is regarded as kth classifier's decision. By doing this, the belief matrix can be transformed into a decision label vector. Then, the voting strategy can be employed. The original vote rate of each class is calculated for input x.

Next, an accordance criterion is set to compare with the maximum vote rate. Higher accordance criterion is set to allow for less different decisions. If the maximum vote rate is less than the threshold, a repeating modification scheme is fired and the original belief degrees have to be modified dynamically. The exchange of information of the two classifiers based on the codecision matrix is added to the vote rates using the following equation:

$$b_{ki} = b_{ki} + \frac{1}{K} \sum_{k_n=1, k_n \neq k}^{K} d_{j,j_n,i,k,k_n} \cdot \sqrt{b_{ki} \cdot b_{k_n,i}} \tag{9.16}$$

where the original belief matrix b is acquired by the confusion matrix, K is the number of total fusion classifiers, b_{ki} represents the belief probability of classifier k to class i and d_{j,j_n,i,k,k_n} which is the exchange of information between kth classifier and k_nth classifier.

After the modified belief degree, a normalization process is required to bring the summation of each row probabilities of new belief matrix equals to one. Then, the new belief matrix b can be transformed into a decision vector, so the new vote rates can be acquired. If the maximum vote rate is still less than the predetermined criterion, the repeating modification process will continue until the maximum vote rate reaches the threshold. Finally, an improved majority voting method is utilized for the output of fusion decision, which only chooses the class gaining the most votes as the fusion decision and needs not beyond half of votes as original voting strategy.

9.4.6 Decision Templates (DTs)

Decision templates (Kuncheva et al. 2001) are a robust classifier fusion scheme that combines classifier outputs by comparing them to a characteristic template for each class.

Let $Z = \{z_1, \ldots, z_N\}, z_j \in R^n$ be the crisply labeled training data set.

Definition 1 Let $\{D_1(x), \ldots, D_L(x)\}$ be the set of L classifiers, which denote the output of the ith classifier as $D_i(x) = [d_{i,1}(x), \ldots, d_{i,c}(x)]^T$, where $d_{i,j}(x)$ is the *degree of support* given by classifier D_i to the hypothesis that x comes from class j.

We construct $\hat{D}$, the fused output of the L first-level classifiers as $\hat{D}(x) = \varphi(D_1(x), \ldots, D_L(x))$, where φ is called *aggregation rule*. The classifier outputs can be organized in a *decision profile* (DP) as shown in Fig. 9.6.

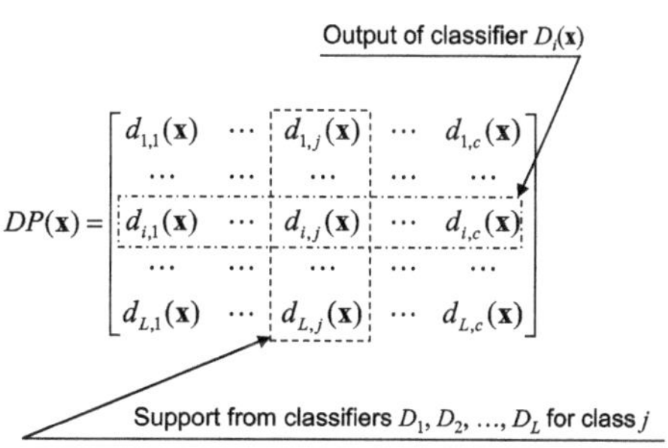

Fig. 9.6 Organization of classifier outputs

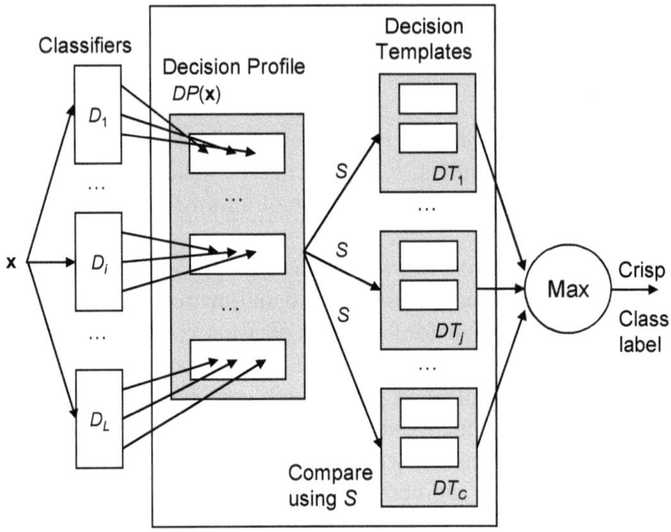

Fig. 9.7 Architecture of the decision templates classifier fusion scheme

Definition 2 The decision template $DT_i(Z)$ of class i is the $L \times c$ matrix $DT_i(Z) = [dt_i(k,s)(Z)]$ whose (k,s) the element is computed by

$$dt_i(k,s)(Z) = \frac{\sum_{j=1}^{N} Ind(z_j,i)d_{k,s}(z_j)}{\sum_{j=1}^{N} Ind(z_j,i)}, \; k = 1,\ldots,L, \; s = 1,\ldots,c, \tag{9.17}$$

where $Ind(z_j,i)$ is an indicator function ith value 1 if z_j has crisp label i, and 0, otherwise.

To simplify the notation $DT_i(Z)$ will be denoted by DT_i. Figure 9.7 illustrates how the DT scheme operates. The decision templates are calculated in advance using Z in Eq. (9.17). The decision template DT_i for class i is the average of the decision profiles of the elements of the training set Z labeled in class i. When $x \in R^n$ is submitted for classification, the DT scheme matches $DP(x)$ to DT_i, $i = 1,\ldots,c$, and produces the soft class labels.

$$\mu_D^i(x) = \varphi(DT_i, DP(x)), \quad i = 1,\ldots,c, \tag{9.18}$$

where φ is interpreted as a *similarity measure.*

The higher the similarity between the decision profile of the current $x(DP(x))$ and the decision template for class $i(DT_i)$, the higher the support for that class $(\mu^i_{\widehat{D}}(x))$. A method to determine similarity is based on *Euclidean distance* between matrices DP and DT_i,

$$N(DP, DT_i) = \mu^i_{\widehat{D}}(x) = 1 - \frac{1}{Lc}\sum_{k=1}^{L}\sum_{s=1}^{c}(dt_i(k,s) - d_{k,s}(x))^2 \tag{9.19}$$

The label with maxim similarity is assigned to the test sample.

9.5 Data Fusion for Condition Monitoring

As far as real-time condition monitoring is concerned, timely indicating potential failures relies on reliable degradation indicator and appropriate alarm setting. Failures can often be attributed to many correlated degradation processes, which could be reflected by multiple degradation indicators extracted from sensor signals (Lee et al. 2006). These features are the major source of information regarding the health of the component under monitoring; however, the failure boundary is hard to define using these features. In reality, the same feature vector could be attributed to totally different combinations of the underlying degradation processes and their severity levels. In other words, the failure boundary is gray by monitoring the degradation features. Each degradation indicator has its own merits and shortcomings and is only effective for certain failure at certain stage (Boutros and Ling 2007). Another important aspect is the difficulty in threshold setting when different indices are used, each with its own threshold setting scheme.

Therefore, fusing multiple indicators can provide a simple and reliable solution to degradation monitoring. Also, alarm setting determines a division between normal and degradation operating condition of system, which is important, so that a slight change in a threshold can mean a drastic change in false alarms, missed detections, and prognosis prediction time horizons.

9.5.1 A Proposed Fusion System for Condition Monitoring

In this section, an intelligent condition monitoring system is introduced based on the data fusion strategy. The flowchart of the proposed system is shown in Fig. 9.8. The proposed system is based on the OSA-CBM architecture and focused on the condition monitoring module and prognosis module. Data fusion strategy is emphasized and employed in the whole system to enhance the reliability and universality of the system health assessment. This section will focuses on the description of fusion condition monitoring, and the introduction of fusion prognostics will be addressed in Sect. 9.7. The procedures of the system can be summarized as follows.

Condition monitoring determines where a system or component is on the indication curve. Is it "nominal"? Does some "anomaly" condition exist? Or, is it somewhere between those two extremes? Determining where we are on the health curve is the first step in prognosis.

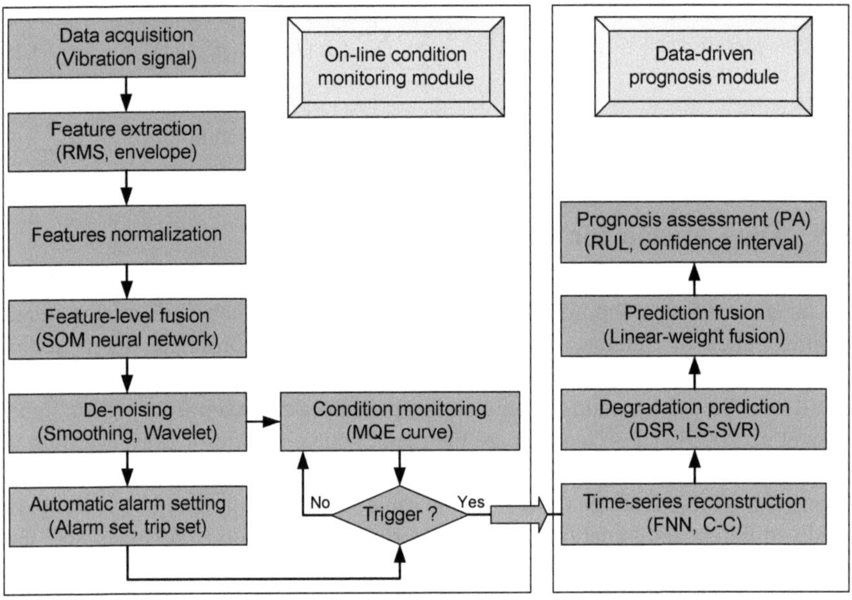

Fig. 9.8 The flowchart of proposed system

In this module, at first, signals of multi-sensors attached on the operating machine are collected and features embossing the operating state are extracted. Then, those calculated features are normalized and grouped as input set for a feature-level fusion algorithm. The enormous mutual information reflecting equipment health is expected to generate a robust health degradation indicator for machine monitoring, diagnosis, and prognosis. Self-organizing map (SOM) neural network is chosen as a feature-level fusion algorithm. The fused output, minimum quantization error (MQE), has the merits of unsupervised learning, short training time, easy operation, and consistent tracking performance.

Next, the process of denoising is considered for filtering process noise of features extraction and fusion. The methods of smoothing and wavelet decomposition are suggested. Smoothing is an easy but efficient way, which can capture important patterns in the data, while leaving out noise. Furthermore, wavelet decomposition is performed to delete the detail noise and increasing smoothness. After the denoised process, a clear tracking trend of operating state is picked out. Next, an automatic alarm setting strategy is suggested based on the largest time constant of machine and statistical properties of the candidate baseline, which can be employed to solve the problem of non-RMS indicator alarm setting. Finally, condition monitoring can be carried out and a comparison is exerted continuously between the alarm threshold and each coming MQE value. If the monitored indicator reaches to the threshold, a data-driven prognosis module would be triggered.

9.5.2 Degradation Indicator Using SOM Neural Network Fusion

9.5.2.1 Theoretical Background of SOM

The SOM is a neural network concept developed by Kohonen in 2001. It forms a one- or two-dimensional presentation from multi-dimensional data. The topology of the data is kept in the presentation such that data vectors, which closely resemble one another, are located next to each other on the map. In contrast to traditional methods, such as principal component analysis, the SOM can also be created from highly deviating, nonlinear data (Jounela et al. 2003).

The map units of SOM, or neurons, form usually a two-dimensional regular lattice. Each neuron i of the SOM is represented by a n-dimensional weight, or model vector, $m_i = [m_{i1}, m_{i2}, \ldots, m_{im}]^T$ (n is the dimension of the input vector). The neurons of the map are connected to adjacent neurons by a neighborhood relation, which indicates the topology, or the structure, of the map. Usually, rectangular or hexagonal topology is used.

After the input data are normalized, the SOM is trained iteratively. In each training step, one sample vector X from the input data set is chosen randomly, and the distance between it and all the weight vectors of the SOM, which is originally initialized randomly, is calculated using some distance measure, such as Euclidian distance. The Best Matching Unit (MBU) is the map unit whose weight vector is closest to X. After the BMU is identified, the weight vectors of the BMU as well as its topological neighbors are updated so that they are moved closer to the input vector in the input space. The vectors are updated following the learning rule: $m_i(t+1) = m_i(t) + \alpha(t) \cdot h(n_{BMU}, n_i, t)(X - m_i(t))$ where $h(n_{BMU}, n_i, t)$ is the neighborhood function, which is monotonically deceasing with respect to the distance between the *BMU* n_{BMU} and n_i in the grid, and the training time; $\alpha(t)$ is the learning rate, a decreasing function with $0 < \alpha(t) < 1$.

At the end of the learning process, the weight vectors are grouped in clusters depending on their distance in the input space. The SOM can be interpreted by labeling the units according to input vectors, whose type or operation state is known. Unlike networks based on supervised learning which require that target values corresponding to input vectors are known, the SOM can be used for clustering data without knowing the class membership of the input data. It can be used to detect feature inherent to the problem. Therefore, the SOM is an effective tool for condition monitoring and system degradation detection.

9.5.2.2 SOM Degradation Detection and Fusion

If data from multiple operation regions as well as faulty operation states are available, the trained SOM can be treated as a state space, where different clusters represent different operation states. The condition of the machine can be described by its matching region in SOM. And the operation state changes can be described by the trajectory of its BMUs in SOM. In normal operation, the BMUs should follow well-defined paths or trajectories in normal regions. When an incipient fault appears, its BMU would deviate from the normal region. The deviation will depend on the type and severity of the abnormality. By plotting the trajectory of current data on a labeled map, the machine condition can be followed over time. Furthermore, if a probability of the next BMUs is available, a prediction of the next possible machine state can be assessed.

Given the fact that most of the time, it is hard to acquire a data set representative of the whole failure space, whereas the normal operation space can be characterized very accurately, fault detection can be based on the quantization error away from the normal feature space. At first, the SOM is trained with normal operation data, which reduces, even avoids, the influence of the local minima problem. Then, the feature vector corresponding to the unidentified measurement is compared with the weight vectors of all map units, and if the smallest difference exceeds a predetermined threshold, the process is probably in a fault situation. This conclusion is based on the assumption that a large quantization error corresponding to the operation point belonging to the space not covered by the training data. Therefore, the situation is new and something is possibly going abnormal. Depending on how far away the current process is deviating from the normal operation state, a quantitative degradation index can be calculated (Qiu et al. 2003).

Practically, the quantitative degradation assessment can be determined by the calculation of minimum quantization error (MQE) of the mew measurement data to an SOM trained using normal operation data sets. From a degradation monitoring point of view, the distance between the BMU and the input data actually indicates how far the input data deviate from the region of normal operation. Thus, the MQE can be defined as follows:

$$\mathrm{MQE} = \|D - m_{\mathrm{BMU}}\| \tag{9.20}$$

where D is the input data vector and m_{BMU} stands for the weight vector of the BMU. An extremely high MQE value may occur for two reasons: Either the testing feature vector is an outlier or it belongs to a fault class (Kang 2003). Therefore, the condition degradation can be quantized and visualized by following the trends of MQE.

The advantages of MQE index are summarized by Huang et al. (2007) as follows:

- It takes advantage of mutual information from multiple features for the system and uses it for degradation assessment;
- Its learning process is unsupervised;
- Training the SOM does not require very much time, and the MQE index is easy to operate in an actual situation; and
- The MQE index is more consistent than any other.

Usually, it would be very difficult to establish a deterministic model that can accurately describe the variable process of defect propagation. Therefore, a practical approach to describe the component degradation should base on the combination of trend analysis and robust performance assessment that derives from the understanding of the historical behaviors and in-process condition symptoms.

9.5.3 Automatic Alarm Setting Strategy

The alarm setting is based on applying an alarm coefficient that is multiplied by the established baseline value. Alarm is used to provide a warning that a defined value of indicator has been reached or a significant change has occurred, at which remedial action may be necessary. In general, if an alarm situation occurs, operation can continue for a period, while investigations are carried out to identify the reason for the change in indicator value and to define any remedial action.

The alarm values may vary considerably, up or down, for individual machines. The values chosen will normally be set relative to a baseline value determined from the experience for the measurement position or direction for that particular machine. Where there is no established baseline (e.g., with a new machine), the initial alarm setting should be based either on experience with other similar machines or relative to agreed acceptance values. After a period of time, the steady-state baseline value will be established and the alarm setting should be adjusted accordingly. Where the baseline signal is non-steady and non-repetitive, some method of time averaging of the signal is required. This could be achieved with the aid of a computer.

By comparison, trip value is considered to specify the magnitude of indictor beyond which further operation of the equipment may cause damage. If the trip value is exceeded, immediate action should be taken to reduce the indicator value or the equipment should be shut down. Basically, trip value is a fixed value corresponding to the maximum value of operating indicator to which the equipment should be subjected.

Recently, Ginart et al. (2006) put forward a new alarm setting mechanism based on the largest time constant of system and statistical properties of the candidate baseline, which can be employed to solve the problem of fused indicator alarm setting. Firstly, the baseline candidate is decided by calculating the mean and deviation of fused indicator data set in good operational and health condition. Then, the indicator is checked and an alarm coefficient is determined according to a proposed table, which is established according to the extensive references from heuristics, norms, basic mathematical structure of acceleration model, ration SD/mean, and slope of baseline. Finally, the alarm setting is finished by multiplying an alarm coefficient with the established baseline. Two criteria are considered in checking the acceptance of monitoring indicator and determining alarm coefficient as follows:

9.5.3.1 Criteria Based on Largest Time Constant

The largest time constant of the system is related with the size of the system and the processes themselves. For large industrial equipment such as huge pumps, blowers, and centrifuges, 20 min is usually a very conservative estimation. In general terms, it is reasonable to establish a time window t that is at least three times of the largest time constant of the system. This allows the system hearing up fully and reaching at least a quasi-steady state, which can allow only changes coming from the failure not from the process itself.

The response for a step functions in the first-order system is:

$$c(t) = 1 - \mathrm{e}^{-t/\tau} \tag{9.21}$$

where τ is the time that taken for any first-order system to reach 67 % of the final steady state (Ogata 2009). Its slope of Eq. (9.21) can be computed as follows:

$$\frac{\partial c(t)}{\partial t} = \frac{1}{\tau}\mathrm{e}^{-t/\tau} = m \tag{9.22}$$

The first criterion for selecting a baseline in mathematical terms is:

$$|m| < 0.017 \tag{9.23}$$

where m is the slope of the linear regression of the candidate baseline.

9.5.3.2 Criteria Based on Statistics (6σ)

This criterion is based on the very low probability, assuming Gaussian distribution, of obtaining a consistence value greater than six times the standard deviation (6σ) normalized by the mean (μ) of the indicator value (Fig. 9.9). This restriction minimizes the possibility down to very low random probabilities of having a false alarm when the machine is in good operational and health condition (Byington et al. 2003). Based on this phenomenon, the acceptable baseline usually complies with the following relation:

$$k = \frac{\sigma}{\mu} < \frac{1}{6} \tag{9.24}$$

Based on the criteria above, Fig. 9.9 shows the examples of acceptable and unacceptable baselines.

Totally, using parameters to establish acceptable baselines, the bounded limit derived from acceleration model, norms, and heuristic knowledge is proposed as the following alarm coefficients as shown in Table 9.1.

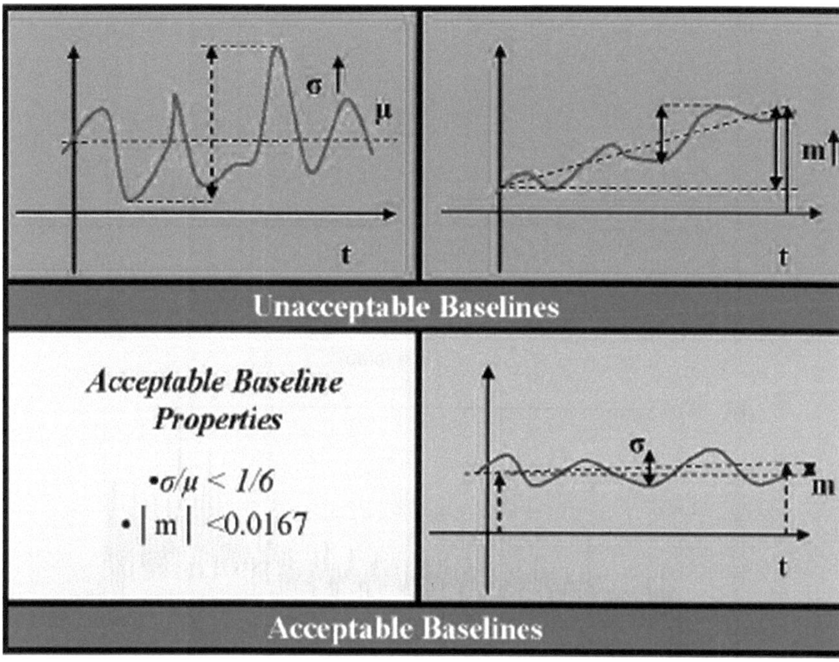

Fig. 9.9 Unacceptable and acceptable baselines examples

Table 9.1 Proposed alarm coefficients

	$\alpha/\mu < 0.04$	$\alpha/\mu < 0.082$	$\alpha/\mu < 0.167$
$m < 0.004$	2	3	4
$m < 0.0082$	3	4	5
$m < 0.0167$	4	5	5

Finally, the proper alarm setting can be calculated as follows:

$$\text{Alarm}_{\text{mean(deviation)}} = \text{Alarm coefficient} \times \text{Baseline}_{\text{mean(deviation)}} \tag{9.25}$$

9.5.4 Condition Monitoring of Compression Using Fusion Techniques

In this section, an experiment based on the proposed method in Fig. 9.8 is described. The experiment object is a low methane compressor driven by an induction motor 440 kW, 6600 V, and 2 poles with operating speed of 3565 rpm. The built online monitoring system consists of acceleration sensors located in only the horizontal direction of four locations, namely drive-end motor, non-drive-end

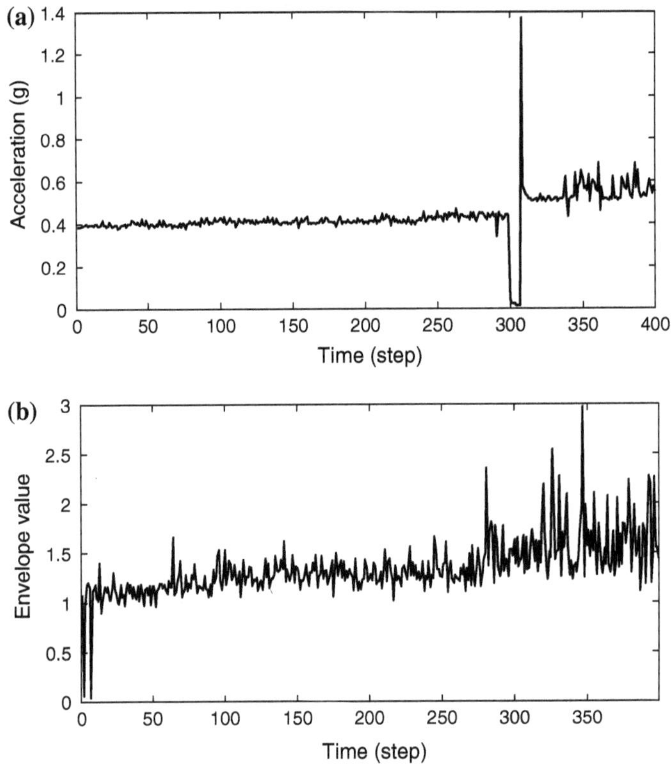

Fig. 9.10 Original time-series RMS and envelope plot. **a** RMS plot. **b** Envelope plot

motor, male rotor compressor, and suction part of compressor. In addition, a ground-coupling cable was utilized to filter the influence of electromagnetism noise to the collected signal. A record of the vibration signals of the methane compressor from August 15, 2005, to November 22, 2005, was analyzed, predicted, and validated. The sampling period of the originally measured signals was 6 h and four times measurements per day. Thus, a total of 400 time-series samples were obtained which involved a process of performance degradation from normal to abnormal running condition.

Two monitoring indicators, RMS and envelope features, are extracted from collected raw signals as shown in Fig. 9.10. It is shown that the collected time-series data set involves a process of performance degradation from healthy to abnormal operating sate.

After extracting features of vibration signals, a process of normalization is conducted to transform values of features into a common scale and group them as input set for feature-level fusion. Next, SOM-based neural network, explained in Sect. 9.5.2, is employed to combine the input set into a single out indicator, MQE, as shown in Fig. 9.11. The correlated training parameters of SOM, that is, the

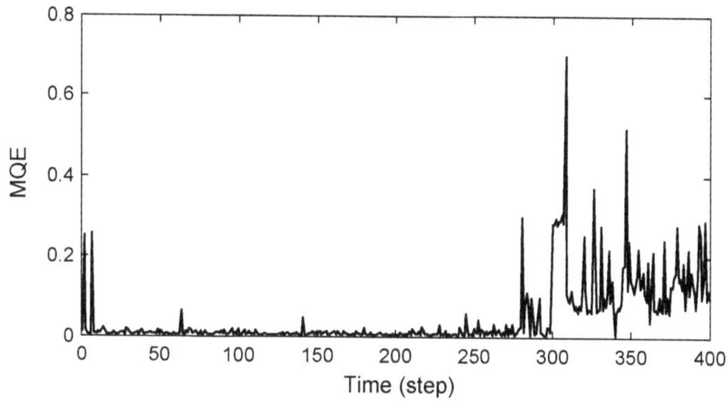

Fig. 9.11 SOM fusion indicator plot

Fig. 9.12 SOM training topology

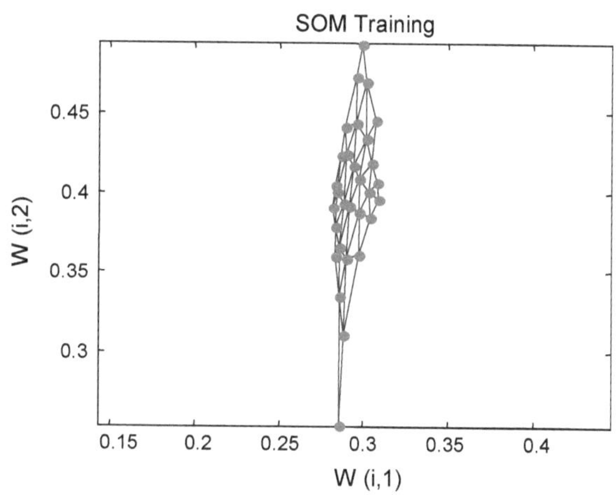

number of neurons and epochs, are set up as 30 and 10, respectively. The SOM training map is shown in Fig. 9.12. If more train samples are given, better map topology will be obtained and SOM will be trained well. Comparing with the plots of RMS and envelope features, it can be seen that MQE indicator maintains a more steady state than the envelop curve, while it enhances a degradation trend than the RMS curve, which is particularly appropriate for initial fault detection and health degradation prediction. Therefore, MQE is considered as a good health monitoring indicator for followed analysis.

Furthermore, the step of denoising is performed to reduce the influence of random noise generated in the processes of feature extraction and fusion, which is contributable in following trend analysis.

Based on the denoised MQE indicator, alarm value is set up using the proposed methods in Sect. 9.5.3. A data set in healthy and steady operating stage, ranging from point 50 to point 200, was picked out for baseline statistic. The calculated

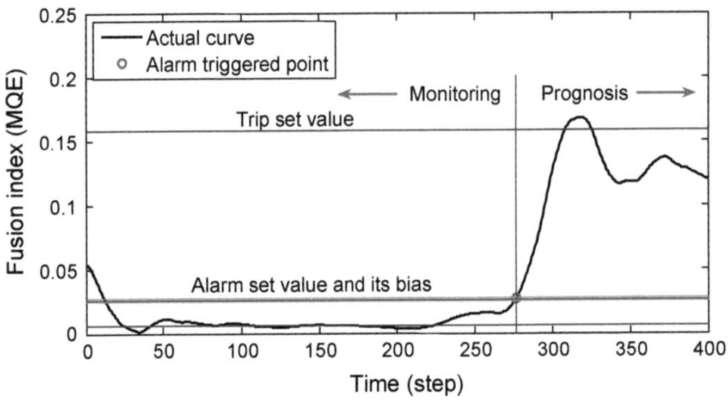

Fig. 9.13 Condition monitoring and alarming

baseline mean value is 0.00696 with the deviation interval [0.00673, 0.00719]. According to the operating history of this machine, the largest constant time was chosen as 20 min. And the span of time window is set as three times of largest constant time. Referencing the calculated values m and k and Table 9.1, the alarm coefficient was selected as 4. As a result, alarm mean value was calculated as 0.02784 with the deviation interval [0.02692, 0.02876]. In addition, the trip mean value was set as six times of alarm mean value in this experiment, which was based on the estimated maximum vibration to which the machine may be subjected.

With the continuous condition monitoring to this machine, the indicator MQE shows an ascending trend as shown in Fig. 9.13. At the time point of 277, the state curve crosses through the given threshold of alarm. This action triggers a prognosis module and declares a start of performance degradation. Initial fault should be detected; meanwhile, prediction and evaluation of RUL together with its uncertainty interval of this machine should be carried out.

9.5.5 Detection Matrix

Faults are malfunctions that are observed in the monitoring system. Two types of errors can occur during automated fault detection: (1) missed detections or (2) false alarms. The practical consequence of either type of error is that a failed component may not be replaced when necessary, or alternatively, may be unnecessarily serviced due to a false alarm (Bock 2006).

The false alarm problem is particularly troublesome for condition-based maintenance and prognostics problems where fault signatures necessarily need to be detected at lower levels. It is well known that two types of errors are possible during automated fault detection: (1) An incipient fault is present, but remains unrecognized by the condition indicators (CI). This is a missed detection or false negative.

(a)

Detection Metrics			
	Cases with a Fault (F1)	Cases with No Fault (F0)	Total
Fault Indicated (D1)	*A:* Number of detected faults	*B:* Number of false alarms	*A+B:* Total number of alarms
Fault Not Indicated (D0)	*C:* Number of missed faults	*D:* Number of correct rejection	*C+D:* Total number of no alarm
	A+C: Total number of faults	*B+D:* Number of fault free cases	*A+B+C+D:* Total number of cases
	Detection rate: A/(A+C)	False alarm rate: B/(B+D)	Accuracy: (A+D)/(A+B+C+D)

(b)

Metric	Equation
Probability of Detection	P(D1/F1) = A / (A+C)
Probability of False Alarm	P(D1/F0) = B / (B+D)
Accuracy	P(D1/F1&D0/F0) = (A+D) / (A+B+C+D)
Receiver Operating Characteristic (ROC)	P(D1/F1) *vs* P(D1/F0)

Fig. 9.14 Detection metrics. **a** Cells of detection metrics. **b** Statistical equations

(2) A fault alarm is raised by the CI, when in truth no fault is present. Here, a false alarm is said to have occurred.

An estimate of the fault detection performance can be computed directly using a decision matrix approach (Fig. 9.14a), which is based on a hypothesis testing methodology and represents the possible fault detection combinations that may occur with an automated fault detection system. As shown in the figure, Column 2 of the matrix lists all the cases in which a fault is present in a monitored component. Column 3 consists of all cases in which no fault occurs in the system. Row 2 represents an indicated fault by the detection logic, and Row 3 represents when a fault is not indicated by the monitoring logic. Cells A, B, C, and D are populated by the results using the fault detection logic, and the performance metrics can by computed from these data.

Figure 9.14b lists the detailed metrics that are used for statistical detection performance analysis. Many well-developed metrics for analysis of detection exist; however, a consolidated tool that incorporates raw data, algorithm interfaces, and metrics-based analysis would prove valuable for development and long-term support for specific applications.

9.6 Data Fusion for Fault Diagnosis

Individual diagnosis system with a single data source can only acquire a limited classification capability that may not be enough for a particular application. Meanwhile, with the rapid development of computer in hardware techniques, stronger calculation capacity of chips and larger storage of memory make it possible for the application of data fusion to fault diagnosis, and many related theory and methods are built up to assist decision as partly introduced in Sect. 9.4.

9.6.1 Classifier Selection

It is essential for data fusion at decision level to have a proper method of classifier selection because the combination of different classifiers can affect fusion accuracy. Given various classifiers and sensor data sets, how to select them is often a problem before a final fusion strategy is employed. A proper classifier team should be robust and can generate the best fusion performance. It also should be optimal so that it can reduce the time for calculation and saving the data in the memory.

The process of classifier selection can be seen the analogy with the feature selection problem, where the techniques for choosing those features that are most effective for preserving class reparability have been developed. Accordingly, techniques for evaluating the degree of error diversity of classifiers that form an ensemble have been used for classifier selection purposes. The goal of the choice phase is to select the subset C^* of classifiers that can be combined to achieve optimal accuracy. There are various approaches to choose the classifiers for successful fusion.

Classifiers selection technique is an ongoing active research area in recent years. Most of the selection methods are based on the statistic theory such as Q statistic (Kuncheva et al. 2003), generalized diversity (Partridge and Krzanowski 1997), and agreement (Petrakos and Benediktsson et al. 2001), among which agreement theory received much attentions as its availability. Agreement measurement mainly contains two methods: *correlation* measure and *kappa* measure. In this section, correlation measure (Goebel et al. 2002) will be introduced.

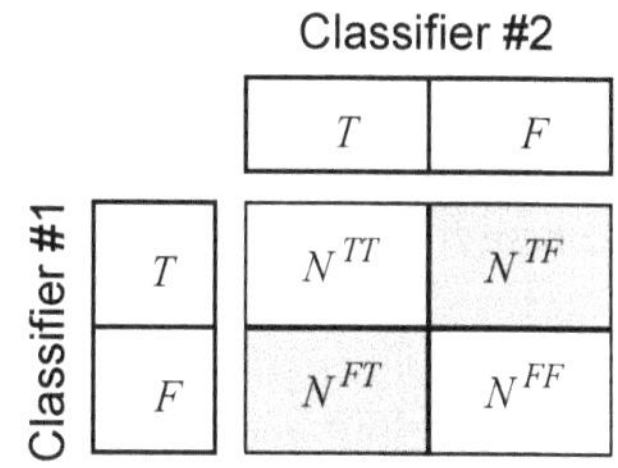

Fig. 9.15 Correlation analysis matrix

Samples correctly classified by both classifiers.	Samples correctly classified by the first and incorrectly classified by the second
Samples incorrectly classified by the first and correctly classifier by the second	Samples incorrectly classified by both classifiers

Fig. 9.16 Explanation of Fig. 9.15

9.6.1.1 2-Classifier Correlation Analysis

Based on the classifier outputs on the labeled training data, a 2×2 matrix N shown in Fig. 9.15 can be created for each classifier pair. And each element in the matrix is explained in Fig. 9.16.

The cell at the lower right part of the table gives the number of samples that will inevitably be misclassified; if divided by the total number of the classified samples, it also gives an upper limit of the classification accuracy. On the contrary, the cell at the upper left side gives the number of samples we cannot possibly misclassify and if divided by the total number of samples gives a lower limit of the classification accuracy of any combined classifier. Then, the off-diagonal cells provide the relative goodness of the two classifiers. Classifiers with fewer samples in the off-diagonal cells have little to gain from their combination no matter how successfully it is performed, because they are correlated. What one classifier decides tell us a lot about what the other's decision will be. The degree of correlation between two classifiers is defined as follows:

$$\rho_2 = \frac{2 \times N^{FF}}{N^{TF} + N^{FT} + 2 \times N^{FF}} \tag{9.26}$$

9.6.1.2 Multi-classifier Correlation Analysis

Based on the formula, Goebel recommended an effective method for classifier selection using the correlation degree of n different classifiers and is shown in Eq. (9.27).

$$\rho_n = \frac{nN^f}{N - N^f - N^r + nN^f} \tag{9.27}$$

where N^f means the number of samples which are misclassified by all classifiers, N^r means those data which are classified correctly by all classifiers, and N is the total number of experiments samples. Generally, smaller correlation degree ρ can lead to better performance of classifier fusion because the independent classifiers can give more effective information.

According to the correlation measurement principle, a team of classifiers can be selected and the steps of classifier selection can be summarized as follows:

Step 1: Select an appropriate performance measure as the initial evaluation criterion, such as accuracy rate which is the ratio of number of samples classified correctly to the total samples;
Step 2: Find the best performing classifier as the first classifier of the team;
Step 3: Calculate the correlation degree between the first classifier and the other classifiers, respectively, using Eq. (9.27);
Step 4: Select the classifier having the "low correlation" for fusion. A practical improvement is that when a similar low correlation degree appears for more than one classifier, the classifier that has the highest recognition rate is chosen;
Step 5: Repeat step 3 to step 4 between selected classifiers and the classifiers unselected yet until all the classifiers are determined;

Finally, the optimal sequence of classifiers can be found.

9.6.2 Decision Fusion System

In this section, a whole decision fusion system (Niu 2007a, b) is introduced as shown in Fig. 9.17, which considers vibration and current signal; the similarity can be extended to other input source. Here, v_i and c_i indicate label vectors of vibration and current data of the electric motors assigned by classifier i, $i = 1, \ldots, k$.

9.6.2.1 Level 1 to Level 3: Preparation for Decision Fusion

The levels 1–3 are regarded as a process of preparation for decision fusion. First, vibration and stator current signals are collected from accelerometers and current probes, respectively. Then, a step of features calculation was exerted. Although the time-series data contain abundant feature information, the important part cannot show intuitively and that much unnecessary information also is contained. Therefore, the feature extraction is essential for effectual estimation conditions of machine. Statistical parameters, calculated in the time domain, frequency domain,

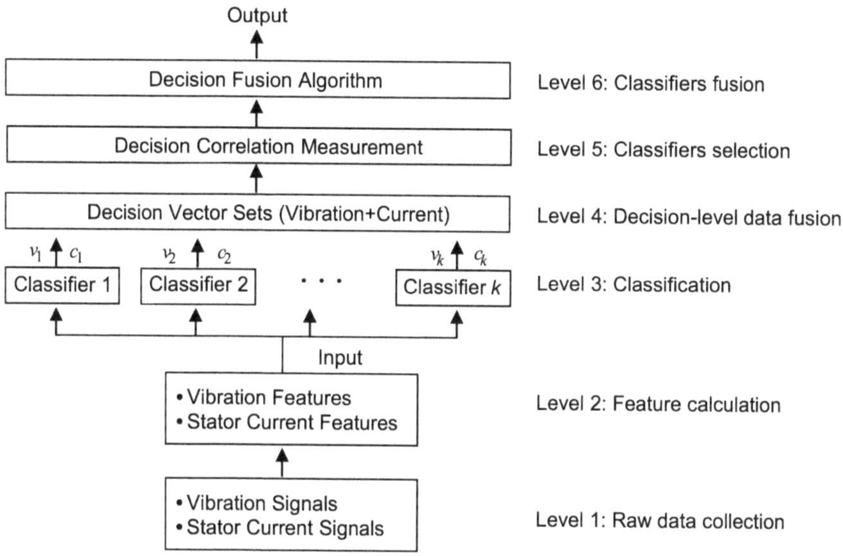

Fig. 9.17 Flowchart of the fusion decision system

and autoregression, are generally used to define average properties of acquired data. Next, several classifiers are utilized to classify the calculated features of vibration and current and generate the decision vectors which are regarded as input source of decision fusion parts from level 4 to level 6.

9.6.2.2 Level 4: Decision-Level Data Fusion

For fault diagnosis of rotating machinery, current signal analysis is equally important as vibration analysis. Therefore, data fusion of the two types of sensors is expected to provide more accurate information to multi-classifiers system.

The labels of a vector are usually diverse for different classifiers with the same data set. From character of relativity, one can notice that the outputs could also be changed for different data sets classified by a same classifier so that two data sets classified by a same classifier separately can be seen as one data set assigned by two different classifiers. From this view, the structure of multi-sensors fusion at decision level (Fig. 9.3) can be improved and shown in Fig. 9.18. As a result, k classifiers with i data sets can be regarded as $i \times k$ classifiers owning one data set.

9.6.2.3 Level 5: Classifier Selection

After feature extraction and data fusion, a classifier selection process is often welcomed for the $i \times k$ classifiers (a collection of labels vectors) by correlation

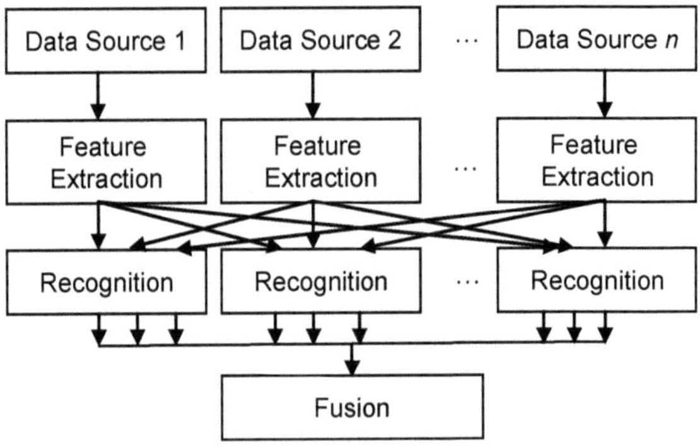

Fig. 9.18 An improved model of multi-sensors fusion at decision level

measure method. As a result, an optimal team of classifiers containing classes information both vibration and current signals is formed to improve classification accuracy.

9.6.2.4 Level 6: Classifiers Fusion

The last step of this system is a multi-classifier of decisions fusion, and the selected labels vectors generated by multi-classifier would be sent into a decision fusion model to deduce the most reliable conclusion of fault diagnosis.

9.6.3 Faults Diagnosis of Test Rig Motors Using Fusion Techniques

To demonstrate the effectiveness of the proposed decision fusion strategy, an experiment was carried out using a self-designed test rig motors.

9.6.3.1 Data Acquisition

The test specimens consist of six 0.5 kW, 60 Hz, and 4-pole induction motors to create the data needed. This motor was set to operate at full-load conditions. One of the motors is normal (healthy), which is used as a benchmark in comparison with faulty motors. The others are faulty motors with broken rotor bar, bowed rotor, bearing outer race fault, rotor unbalance, rotor eccentricity, and phase unbalance. Three AC current probes and three accelerometers were used to measure the stator current of three-phase power supply and vibration signals. The maximum frequency of the signals was 3 kHz, the number of sampled data was 16,384, and the measured time was 2.1333 s. For each condition, 40 samples were measured, 20 of

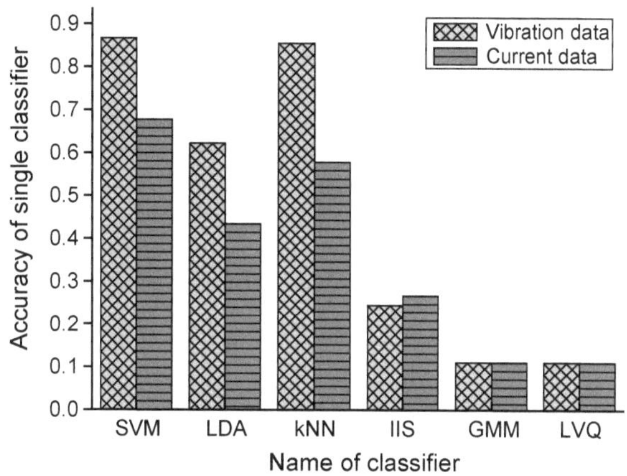

Fig. 9.19 Comparison of single classification accuracy

them were used for training parameters of the classifiers, 10 for training parameters of the multi-agent fusion model, and the other 10 for testing.

9.6.3.2 Feature Calculation and Classification

After data acquisition, a process of features calculation was exerted. A total of 63 features (21 parameters × 3 signals) are calculated for vibration and current signals. Next, six classifiers were utilized to classify the calculated features of vibration and current, which are *support vector machine (SVM), linear discriminant analysis (LDA), k-nearest neighbors (k-NN), improved iterative scaling (IIS), Gaussian mixture model (GMM),* and *learning vector quantization (LVQ).*

Figure 9.19 shows the training accuracy of the six classifiers. It is shown that the classification accuracy of vibration signal is far better than the ones obtained from stator current signal. The best classification accuracy is using vibration data by SVM and *k*-NN classifiers with a value of 0.867. When using current signal by SVM, the accuracy is 0.678. As far as performance of the six classifiers is concerned, SVM and *k*-NN produced superior results.

9.6.3.3 Classifiers Selection and Fusion

The six classifiers were used to classify the features data of the vibration samples, and the generated decision vectors were named as vectors 1–6 in sequence. The process was then repeated for the current data with the result vectors named from 7–12. Next, the 12 decision vectors were sent for classifiers selection in order to find the best sequence of classifier fusion. Finally, multi-agent algorithm was utilized for decision fusion. The comparison of classification accuracy between the fusion data and single data using the vibration or current signal is shown in Fig. 9.20. It is

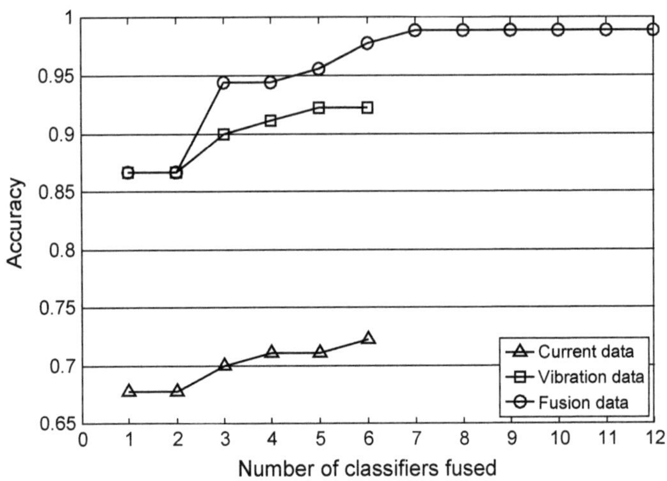

Fig. 9.20 Comparison of classification fusion accuracy

Table 9.2 Fusion performances of vibration and current signals

Classifier	Signal	SVM	LDA	*k*-NN	IIS	GMM	LVQ
Serial numbers of classifiers	Vibration	1	2	3	4	5	6
	Current	7	8	9	10	11	12
Fusion sequence of classifiers	Vibration	No. 1	No. 7	No. 5	No. 10	No. 11	No. 4
	Current	No. 6	No. 8	No. 3	No. 9	No. 12	No. 2
Fusion accuracy	Vibration	0.867	0.989	0.956	0.989	0.989	0.944
	Current	0.967	0.989	0.944	0.989	0.989	0.867

shown that the highest accuracy of single-source data, after classifiers selection and combination, is 0.922 for vibration data and 0.722 for current data. Table 9.2 shows the fusion performance of vibration and current data. The fusion sequence of classifiers is acquired by the selection step for the 12 decision vectors using classifier selection methods (Goebel et al. 2002).

9.6.3.4 Classifiers Fusion Comparison

In this case, three classifiers fusion algorithms were used for comparison, i.e., majority voting, Bayesian belief, and multi-agent algorithms. The results are shown in Fig. 9.21. The maximum fusion accuracy for multi-agent method was 0.989, while the accuracy using Bayesian belief and majority voting methods were 0.989 and 0.911, respectively. Figure 9.21 shows that the multi-agent and Bayesian belief methods are far superior to the majority voting strategy.

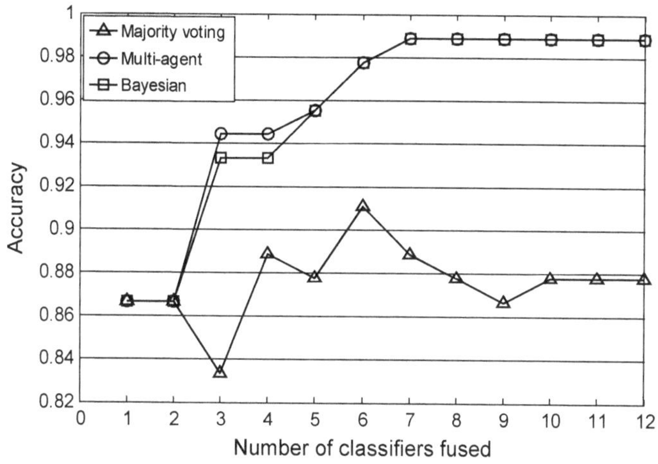

Fig. 9.21 Comparison of multi-agent, Bayesian, and majority voting method

Fig. 9.22 Experimental apparatus

9.6.4 Faults Diagnosis of Elevator Motor Using Fusion Techniques

The next example is to demonstrate the effectiveness of decision fusion in real-world operating conditions; a case of experiment is investigated to an elevator motor system. The elevator is driven by a motor connected to the sheave at the top of the elevator system as shown in Fig. 9.22.

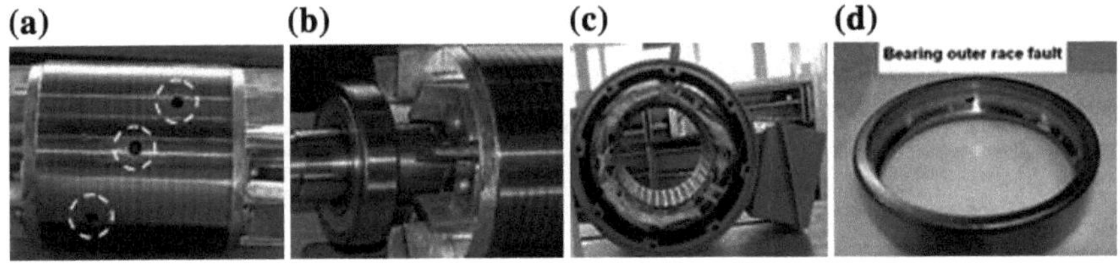

Fig. 9.23 Some faults on the induction motors. **a** Broken rotor bar, **b** rotor unbalance, **c** stator eccentricity and **d** outer race fault

Table 9.3 Description of fault types of the motor tested

Fault types	Fault details
Rotor unbalance	In-phase, 60 g mm/kg
Stator eccentricity	30 % (+0.23 mm)
Rotor eccentricity	Out-of-phase, 80 g mm/kg
Broken rotor bar	3 spots
Bearing housing looseness	Between outer race and housing
Inner race looseness	Between shaft and inner race
Ball fault	Diameter 2 mm, depth 1.5 mm
Outer race fault	Diameter 2 mm, depth 2 mm
Inner race fault	Diameter 2 mm, depth 2 mm

9.6.4.1 Data Acquisition

The test objects are ten 15 kW, 50 Hz, and 4-pole induction motors for elevators. These motors are set to operate at full-load conditions. One of the motors is normal (healthy), which is used as a benchmark in comparison with faulty motors. The others are faulty motors with rotor unbalance, stator eccentricity, rotor eccentricity, broken rotor bar, bearing housing looseness, bearing inner race looseness, ball fault, bearing outer race fault, and inner race fault, as shown in Fig. 9.23. The conditions of faulty induction motors are described in Table 9.3.

Three accelerometers and one AC current probe were used to measure the vibration signals of horizontal, vertical, axial directions, and stator current signal to evaluate the fault diagnosis system. The maximum frequency of sampling signals is 3 kHz, and the number of sampled data is 16,384. Sampling time is 2.133 s, and Hanning window was chosen for filtering. The permitted measuring time for each fault is 15 s containing three running conditions: speedup, steady, and slowdown. A real limitation is that many times of measurement for each fault is nearly impossible, or else the elevator will break down severely. As a result, each fault was measured for two times in our experiment, and then, steady signals were picked out for analysis. Considering the few raw data that are not enough for fusion analysis and the periodicity of steady signal, we employed an "overlap method" to solve the problem. This method picks out each sample from collected steady signal in terms

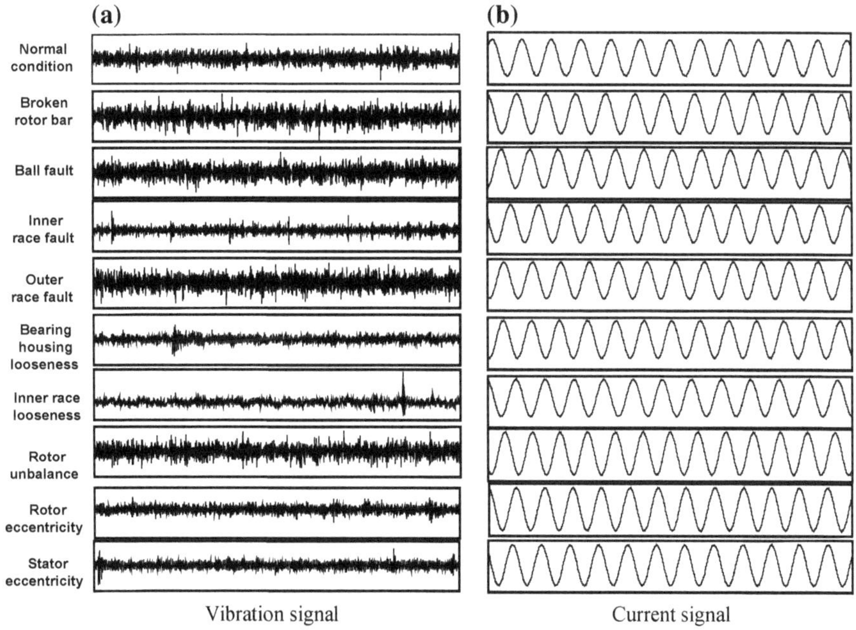

Fig. 9.24 Waveform of time-domain signal of vibration and current. **a** Vibration signal and **b** current signal

of an overlap rate predetermined. The overlap rate was set as 0.75 in this experiment. Using this method, we extended the steady signal of one time measurement into 10 times. So finally, we acquire 20 samples per fault and total samples are 200. Among them, 100 samples were divided for training classifiers, 50 samples for training fusion algorithms, and the left 50 samples for test.

The waveforms of time-domain signal collected from accelerometer and current probe are shown in Fig. 9.24. It is shown that there are some differences in waveform of each condition when observing vibration signal. However, there is no obvious difference in the waveforms of current signal.

9.6.4.2 Feature Calculation and Classification

After the signals are collected from the tested motors, a process of features calculation is exerted. And then, a process of classification is exerted for five classifiers (SVM, ART-KNN, LDA, RF, and *k*-NN).

The comparison of accuracy rates for test samples is shown in Table 9.4. It is shown that classification performance using vibration data is better than the current data obviously. Among them, three classifications can achieve 100 % accuracy rates for vibration data. However, the highest accuracy rate for current data is less than 75 % from SVM. Due to the high classification accuracy rates using vibration

Table 9.4 Accuracy rates of individual classifier

Classifier		SVM	ART-KNN	LDA	RFA	*k*-NN
Accuracy	Vibration signal	100	90	100	100	98
Rate (%)	Current signal	74	60	72	72	66

Table 9.5 Accuracy rates of classification of each fault for current data

Fault types		Classifier				
		SVM	ART-KNN	LDA	RFA	*k*-NN
F1	Broken rotor bar	40	60	60	80	40
F2	Ball fault	80	20	40	40	60
F3	Inner race fault	100	20	100	100	100
F4	Outer race fault	20	0	20	0	0
F5	Bearing housing looseness	0	80	60	40	40
F6	Inner race looseness	100	80	80	100	100
F7	Rotor unbalance	100	100	60	100	80
F8	Rotor eccentricity	100	100	100	80	100
F9	Stator eccentricity	100	40	100	80	40
F10	Normal condition	100	100	100	100	100

feature data, the fusion experiment is intended to focus on the current data for classifiers fusion.

Moreover, a detailed comparison of the accuracy rates for each fault using current feature data is shown in Table 9.5, which tells some valuable information: Types of faults, such as broken rotor bar, ball fault, outer race fault, and housing looseness are hard to detect while constituting major faults in induction motor. As far as classification performance is concerned, SVM is the best and, nevertheless, still gives bad diagnosis results to housing looseness and outer race fault. ART-KNN is the worst, while gives the better diagnosis accuracy to housing looseness fault. The above analysis indicates necessity and potential values of multi-classifier fusion diagnosis.

9.6.4.3 Classifier Fusion

After acquired diagnosis results from each classifier, Bayesian belief algorithm is employed to enhance diagnosis performance by fusing decisions of multi-classifier. Fusion accuracy of each fault for different numbers of classifiers is shown in Table 9.6. The results show that the diagnosis accuracy of types of faults (bearing inner race looseness, rotor unbalance, rotor eccentricity, and stator eccentricity) increased to 100 % remarkably. The fusion performance improves with the increasing in number of classifiers.

Table 9.6 Fusion accuracy of each fault for different numbers of classifiers using Bayesian belief theory

Fault types	Accuracy rates of classifier fused (%)				
	No. 1	No. 1–2	No. 1–3	No. 1–4	No. 1–5
Broken rotor bar	80	60	60	100	100
Ball fault	80	80	100	80	100
Inner race fault	100	40	100	100	100
Outer race fault	0	60	40	100	100
Housing looseness	0	80	60	100	100
Inner race looseness	100	100	100	100	100
Rotor unbalance	100	100	100	100	100
Rotor eccentricity	100	100	100	100	100
Stator eccentricity	100	100	100	100	100
Normal condition	100	100	100	100	100

The experiment results show excellent performance was acquired when compared with single classification method. Multi-classifier fusion strategy can improve the faults diagnosis accuracy remarkably. Fusion accuracy utilizing Bayesian method arrived at ideal results, 100 %, while only 74 % from the best individual classifier, SVM, for current data in this experiment.

9.7 Data Fusion for Failure Prognostics

9.7.1 A Proposed Fusion Strategy for Failure Prognostics

This section discusses data fusion application in terms of the failure prognostics. Integrating fusion technology for condition monitoring introduced in Sect. 9.5, a data fusion-enabled prognostics strategy (Niu and Yang 2010) is proposed, as shown in Fig. 9.25.

Usually, performance degradation trend is reflected as nonlinear or chaotic character. Therefore, state space reconstruction becomes the first step in nonlinear chaotic time-series prediction. To get appropriate reconstruction input vectors, two basic parameters, delay time and minimum embedding dimension, need to be determined at first. In this system, delay time is selected by C–C method and minimum embedding dimension is chosen by false nearest neighbor (FNN) method, respectively, due to their merits, which will be introduced in the following section.

Then, the degradation prediction is conducted. In this process, two data-driven models, Dempster–Shafer regression (DSR) and least square support vector machine (LS-SVM), are particularly appropriate for nonlinear time-series prediction and therefore proposed as tools for data-driven prognosis. In addition, the two models can give predictions of not only degradation trend (or point estimate) but

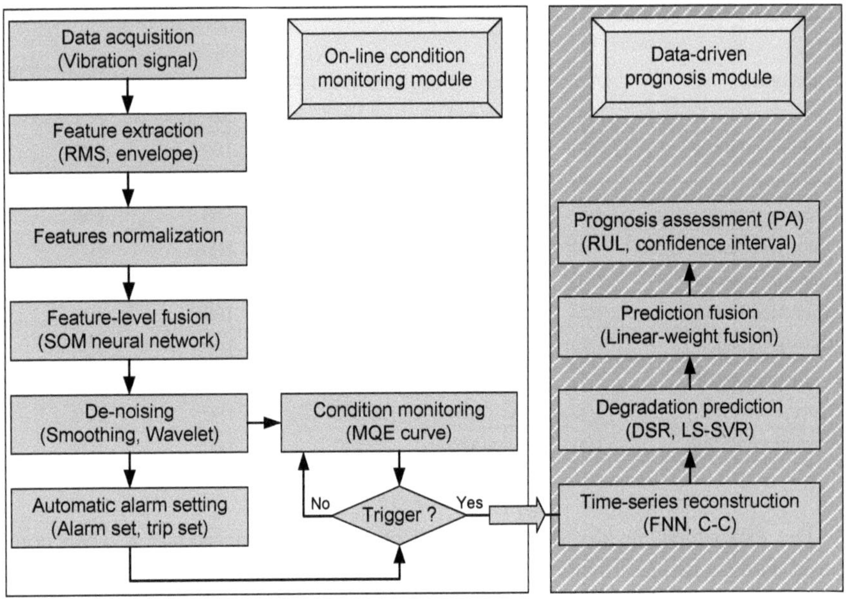

Fig. 9.25 A proposed data fusion strategy for failure prognostics

also uncertainty bounds (or interval estimate). Next, combining the prediction results of DSR and LS-SVM is conducted by a linear weight fusion method. Finally, prognosis assessment (PA) is carried out. The cores of PA is estimating the remaining useful lifetime of a failing component or system and assigning uncertainty bounds to the degradation trend that will provide maintainers with the earliest and the latest (with increasing risk) time to perform maintenance and the associated risk factor when maintenance action is delayed.

9.7.2 *Time-Series Prediction*

The first step of time-series prediction is a process of reconstruction in high dimensions. The state space parameters, delay time, τ, and embedding dimension, m, should be selected appropriately. In addition, we need to determine prediction way that is one-step-ahead (OS) or multi-step-ahead (MS) prediction, and the difference of them and analysis are introduced. At last, two time-series models, DSR and LS-SVM, are explained, which are partially suitable for data-driven prognostics due to its ability of nonlinear time-series prediction and uncertainty bias assessment.

9.7.2.1 State Space Reconstruction

State space reconstruction is the first step in nonlinear chaotic time-series prediction. The fundamental theorem of reconstruction is pioneered by Takens (1981) and subsequently expanded by Sauer et al. (1991). The reconstruction consists of viewing a time series, $x_k = x(k\tau_s)$, where $k = 1, \ldots, N$ in a Euclidean space $\mathbb{R}^m$, where m is the *minimum embedding dimension* and τ_s is the *sampling time*. A standard way to reconstruct the state space is the method of delays (MOD). Using MOD, each m-dimensional embedding vector is formed as follows:

$$\mathbf{x}_k = \left[x_k, x_{k+p}, \ldots, x_{k+(m-1)p}\right]^T \tag{9.28}$$

where p is the multiple integer of τ_s so that the *delay time* τ equals $p\tau_s$. The m coordinates of each point $\mathbf{x}_k$ are samples from the time series covering a *time window* of length $\tau_w = (m - 1)\tau$. An example of embedding of Lorenz System with the effect of different values of m and τ on reconstruction of the original system is shown in Fig. 9.26 where (a) is original Lorenz attractor with x/y/z variables, (b) is a single time series of variable x, (c)–(f) are reconstructed attractors from variable x with various m and τ. From the example, it is clear that for correct state space reconstruction (or reconstruction of the attraction), a fine estimation of the parameters (m and τ) is needed. In practice, with a limited number of possibly noisy observations, the selection of τ and m is rather difficult to ensure the quality of the reconstruction.

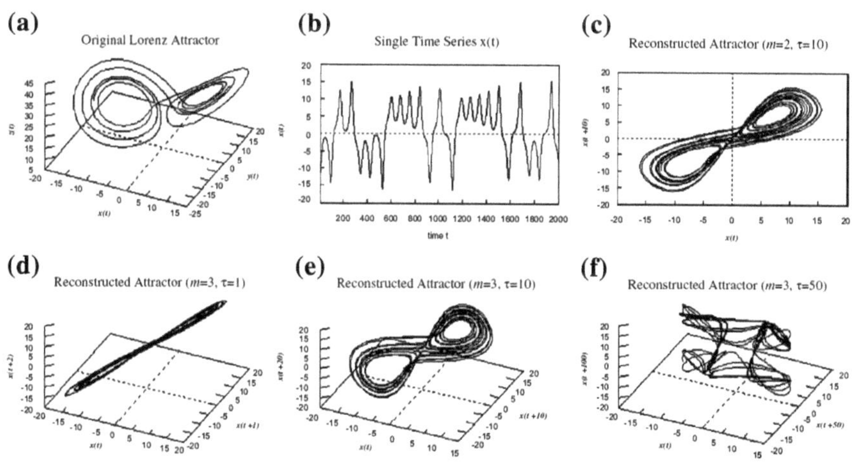

Fig. 9.26 An example of reconstruction of the attraction

9.7.2.2 One-Step-Ahead/Multi-step-Ahead Prediction

The time-series prediction problem is the prediction of future values based on the previous values and the current value of the time series. They are used as inputs for the prediction model. One-step-ahead prediction is needed in general and is referred to as short-term prediction. However, when multi-step ahead predictions are needed, it is called a long-term prediction problem (Sorjamaa et al. 2007).

Unlike the *one-step-ahead* time-series prediction, the *multi-step-ahead* prediction is typically faced with growing uncertainties arising from various sources. For instance, the accumulation of errors and the lack of information make the prediction more difficult.

In some cases, such as slow-changing signal, the predicted time intersection is long enough that the real value at the first step prediction has already been obtained before the second step prediction, and then we can take place the predicted value in real value, which becomes one-step-ahead prediction.

In machinery prognostics, collected data change rapidly with time such as the vibration and acoustic emission signals. Hence, engineers need to acquire information several steps ahead in the future horizon, that is, multi-step ahead prediction (MS model), to assess the remaining useful life (RUL) of the failing component/system, which is regarded as one of the main outputs of prognosis.

There are two typical alternatives to build MS models: One is *iterated* prediction and the other is *direct* prediction. The choice between iterated and direct prediction involves a trade-off between bias and estimation variance.

The iterative multi-step prediction can be divided into three processes as shown in Fig. 9.27. The first steps of prediction stage can be estimated by the actual values of previous sliding window. And then, the next steps can be predicted by the proceeding sliding window with mixed embedding elements of actual values and predicted values. Finally, the left steps can be predicted by only previous predicted values. It is obvious that using the previous predicted values as inputs, it can gradually deteriorate the accuracy of the prediction.

9.7.2.3 Time-Series Model

After selecting the appropriate reconstruction parameters and prediction way in multi-step-ahead according to the real requirement from machinery prognostics, suitable time-series models should be considered to predict future trajectory of performance degradation based on the previous and the current monitored descriptors. Usually, the collected monitored signal of machine is nonlinear and the time-series prediction of it contains uncertainty more or less; hence, the time-series model with the characters of nonlinear and uncertain estimation will be welcome. Regression analysis can be taken as an effective approach for time-series prediction such as Dempster–Shafer regression (DSR) and least square support vector machine (LS-SVM) models.

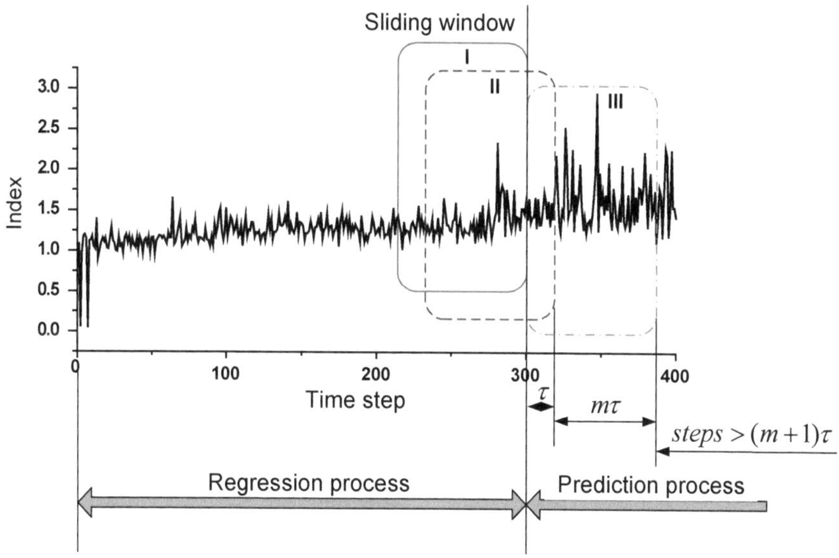

Fig. 9.27 Sketch map of iterative multi-step-ahead time-series prediction (Niu and Yang 2009)

9.7.3 Failure Prognostics of Compression Using Fusion Techniques

After accumulating techniques above, a whole data-driven prognostics experiment about performance degradation of a methane compressor can be carried out following the case of monitoring alarm in Sect. 9.5.4. Particularly, remain useful life (RUL) and its deviation of the operating machine could be predicted, which provide the most valuable maintenance decisions. Due to the difficulty of wide uncertainty, data fusion techniques are proposed to improve the prognosis accuracy and precision.

In this measurement, 400 time-series samples were obtained which involved a process of performance degradation from normal to abnormal running condition. The first 300 samples, involving the main part of a normal condition and a small part of initial abnormal condition indicating a potential fault, are employed to train the DSR and LS-SVM models. Then, the system was employed to predict the degradation trend of the left 100 steps. The predicted results will then be compared with the actual 100 test samples.

Before carrying out time-series prediction, the time-series should be reconstructed in high dimensions. The state space parameters, delay time, τ, and embedding dimension, m, were selected using C–C method and FNN method, respectively, in this case. The delay time is chosen as 61, and embedding dimension is chosen as 4.

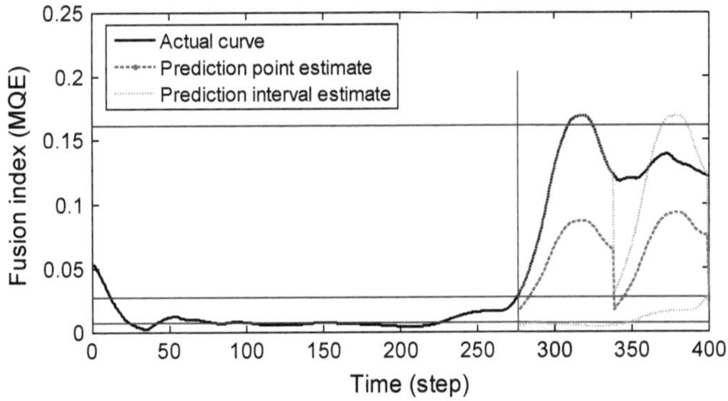

Fig. 9.28 Degradation prediction using DSR model

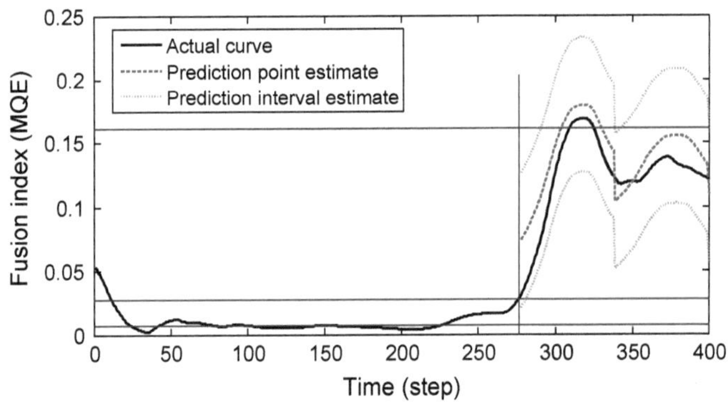

Fig. 9.29 Degradation prediction using LS-SVM model

Next, time-series prediction is performed using DSR and LS-SVM models, respectively, by strategy of iterated multi-step-ahead (MA). According to the time point of alarm triggered shown in Sect. 9.5.4, the previous 277 samples are used for training models and the left 123 samples for validating the degradation prediction. The predicted degradation curve and its prediction interval are shown in Fig. 9.28 for DSR model and in Fig. 9.29 for LS-SVM model. The predicted results are compared to the two models in terms of prediction point estimate and prediction interval estimate.

(1) *Accuracy*: prediction point estimate (PP)
Accuracy is a measure of how close a point estimate of failure time is to the actual failure time. For degradation prediction, if the prediction is earlier than the actual curve, a correct premaintenance can be carried out. On the contrary, if the prediction is made after actual failure occurred, prediction becomes meaningless. Therefore, given the same error size, it is in most situations

preferable to have a positive bias (early prediction), rather than a negative one (late prediction). Comparing PP estimates in Figs. 9.28 and 9.29, DSR does not trigger the trip setting, even though the actual failure occurs, while LS-SVM generates an appropriate positive prediction. Quantitatively, the root-mean-squared error (RMSE) of PP estimate for LS-SVM is 0.019, less than 0.063 for DSR. Therefore, LS-SVM shows higher accuracy than DSR.

(2) *Precision*: prediction interval estimate (PI)
Precision is a measure of the narrowness of an interval in which the remaining life falls. The interval is enclosed by upper and lower bounds. A narrow PI estimate indicates high precision. In particular, for degradation prediction, a positive upper bound is welcome.

$$Narrowness = \frac{1}{n}\sum_{t=1}^{n}\left|\hat{y}_{\mathrm{sup}}(t) - \hat{y}_{\mathrm{inf}}(t)\right| \tag{9.29}$$

where $\hat{y}_{\mathrm{sup}}(t)$ and $\hat{y}_{\mathrm{inf}}(t)$ stand for the upper and lower bound of predicted degradation curve, respectively.

Comparing PI estimates in Figs. 9.28 and 9.29, LS-SVM shows a positive upper bound, while the upper bound in DSR is a little reluctant. Quantitatively, the narrowness of PI estimate for LS-SVM is 0.105, less than 0.112 for DSR. Therefore, LS-SVM also shows higher precision than DSR.

After obtaining the prediction results from DSR and LS-SVM, linear weight fusion is employed to combine the predicted results of the two models. The weights of the DSR and LS-SVM are given as 0.15 and 0.85, respectively, considering the previous compared results. The better one deserves a higher weight. The final fused results are shown in Fig. 9.30. It is shown that the PP estimate is closer to actual trend curve with RMSE of 0.013. The fused PP is improved comparing with single best LS-SVM, with a proper positive upper bound.

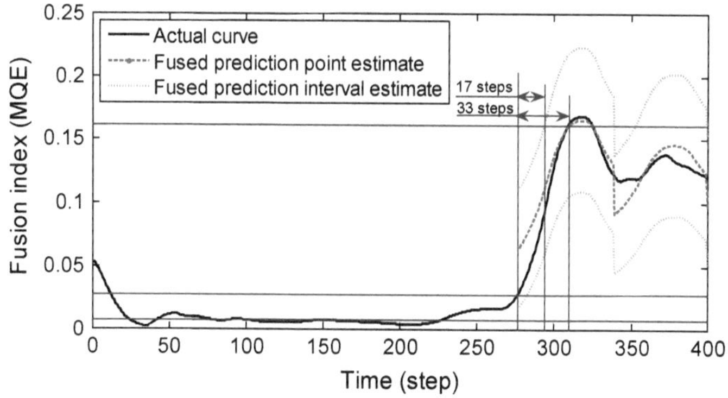

Fig. 9.30 Degradation prediction fusing DSR and LS-SVM (linear weight combination)

According to the fused prediction results, the PP estimate crosses through the predetermined threshold of trip at point 310, and the upper bound crosses through the threshold at point 294. Finally, the predicted RUL is 33 steps (198 h), with maximum uncertain upper bound of 17 steps (102 h). Comparing with the actual RUL with 33 steps (198 h), the results give a positive prediction, with appropriate time in indicting repair to the machine. Hence, the aims of cost-effective maintenance can be acquired.

9.8 A Framework of Cost-Effective and Accurate PHM/CBM System

Ideal data-driven PHM/CBM will allow the maintenance personnel to do only the right things, minimizing spare parts cost, system downtime, and time spent on maintenance. In reality, however, exerting a reliable and effective data-driven PHM/CBM faces two challenges: Above all, stating to use data-driven PHM/CBM is costly. Often, the cost of sufficient instrumentation can be quite large, especially on equipment that is already installed. It is therefore important to decide whether checked equipment is sufficiently important to justify the investment. Secondly, data-driven PHM/CBM stands for accurate maintenance, which is not easy to reach due to the complex outer environment or inner structure of equipment, and abstruse failure mechanisms.

For the purpose of *cost-effective* maintenance, data-driven PHM/CBM tool is suggested to be carried out through the advanced maintenance management mechanism. Furthermore, when the functions of the component and its importance are considered at the same time, reliability-centered maintenance (RCM) also seems an appropriate choice. Usually, the aim of RCM is to maximize the result with regard to system reliability or outage cost reduction (Lehtonen 2006).

For the purpose of *accurate* maintenance, data fusion techniques are suggested containing data-level fusion, feature-level fusion, and decision-level fusion (Hall and Llinas 1997). Applying fusion techniques into engineering practice has been receiving increasing attentions in recent years. In particular, with the rapid progress of advanced sensor and signal processing technologies, fusing large of mutual information becomes possible, which is expected to bring about accurate data-driven PHM/CBM.

This section develops an optimal data-driven PHM/CBM system (Niu, et al. 2010) as shown in Fig. 9.31, which integrates data fusion strategy with traditional OSA-CBM in the architecture of RCM management. RCM is employed for the process management of cost-effective maintenance, while data fusion technology is considered during the CBM processes of monitoring, diagnostics, and prognostics for improving the maintenance accuracy. The procedure of the system can be described as follows.

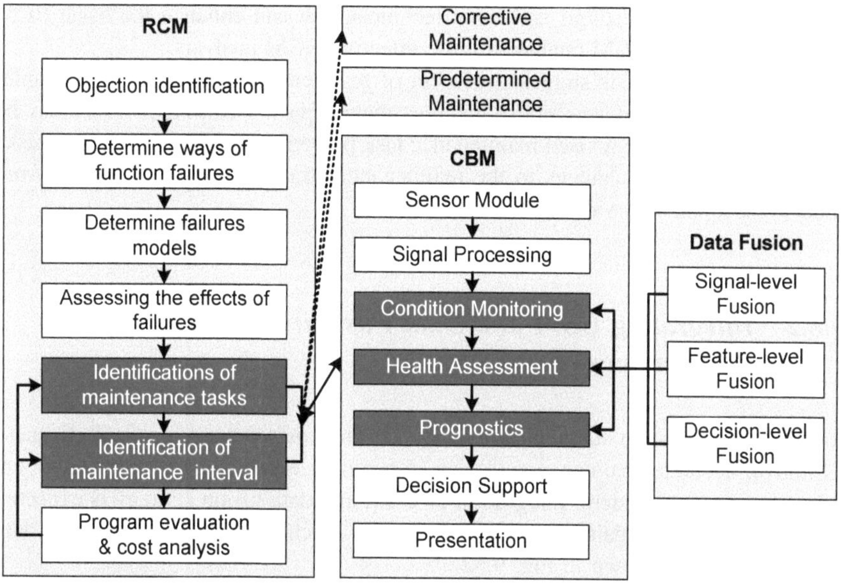

Fig. 9.31 Flowchart of purposed CBM system

9.8.1 *Integrating CBM and RCM: Cost-Effective Maintenance*

The general steps to work through when apply the RCM method can be summarized as seven steps.

(1) Defining system functions, performance standards, and system boundary definition.
(2) Determining the ways in which the system functions may fail.
(3) Determining the significant failure modes.
(4) Assessing the effects and consequences of the failures.
(5) Identifications of maintenance tasks by means of a decision logic scheme.
(6) Identification of maintenance tasks interval.
(7) Auditing, implementation, and feedback.

The RCM analytical approach assists the maintenance manager in identifying potential failures and supporting the selection of viable courses of action. RCM tools help define the optimal failure management strategies. Meanwhile, CBM is built on the foundation of the RCM methodology. CBM is not a process in itself. It is a comprehensive strategy to select, integrate, and focus a number of process improvement capabilities, thereby enabling maintenance managers and their customers to attain the desired levels of system and equipment readiness in the most cost-effective manner. CBM strategy includes a number of capabilities and initiatives, some procedural and some technical, that can enhance the basic RCM tasks. In this way, CBM enables a more effective RCM analysis.

If the RCM analysis suggests revision of maintenance tasks, then the maintenance manager should accomplish an assessment of how CBM capabilities may be applied to support the revised maintenance task procedures. Often, the revised tasks require fundamental changes to the maintenance strategy such as transition from time cycle repair intervals to CBM.

9.8.2 *Integrating CBM and Data Fusion: Accurate Maintenance*

In engineering system maintenance, except for the requirement of cost-effectiveness as above, accurate maintenance is also imperative, especially for core functional components or subsystem. Integration of CBM and data fusion strategy is expected to obtain improved maintenance decisions. A flowchart of CBM integrating data fusion strategy is shown in Fig. 9.32.

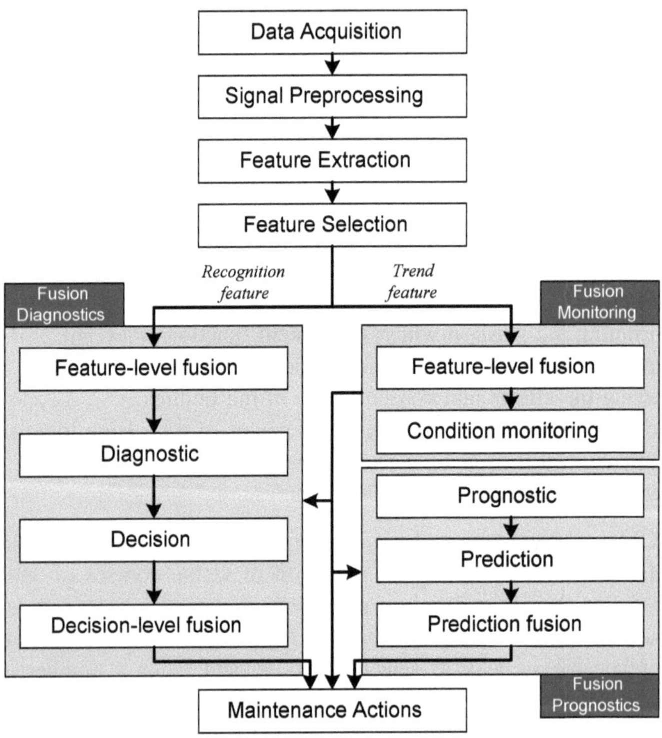

Fig. 9.32 Flowchart of CBM integrating data fusion

First, raw signals are collected, signal preprocessing is exerted, and then, the appropriate features are extracted which indicates the state information of running machine. Next, a step of feature selection is designed, and the features that can recognize different faults clearly are selected for diagnostics analysis, while those features indicating degradation trend of equipment health are chosen for monitoring and prognostics task to predict equipment remain useful life (RUL). Then, fusion strategy is considered and employed in the followed two subsystems: fault diagnostics, condition monitoring, and prognostics.

As far as health degradation is concerned, each degradation indicator has its own merits and shortcomings and is only effective for certain failure at certain stage. Therefore, fusing multiple degradation indicators would provide an accurate and reliable way for degradation monitoring. When the monitored index exceeds the predetermined alarm set, the processes of diagnosis and prognosis are trigged.

In order to get accurate diagnostics result, feature-level or decision-level fusion strategy can be used. Moreover, in order to get a near-ideal prediction of RUL with its uncertainty interval, data fusion at feature-level also can be employed. Finally, optimal maintenance actions can be established based on the results of monitoring, diagnostics, and prognostics.

References

Basir O, Yuan XH (2007) Engine fault diagnosis based on multi-sensor information fusion using Dempster-Shafer evidence theory. Inf Fusion 8(4):379–386

Bock JR, Brotherton T, Grabill P, Gass D, Keller JA (2006) On false alarm mitigation. In: Proceedings of the IEEE aerospace conference, Big Sky, MT, March 4–11

Boutros T, Ling M (2007) Mechanical fault detection using fuzzy index fusion. Int J Mach Tools Manuf 47:1702–1714

Byington CS, Watson M, Kalgren P, Safa-Bakhsh R (2003) Metrics evaluation and tool development for health and usage monitoring system technology. In: AHS Forum 59, Phoenix, Arizona

Constantinidis AS, Fairhurst MC, Rahman AFR (2000) Detection of circumscribed masses in digital mammograms using behaviour-knowledge space method. IEE Electron Lett 36(4):302–303

Dempster AP (1967) Upper and lower probabilities induced by multivalued mappings. Ann Math Stat 38:325–339

Denoeux T (2000) A neural network classifier based on Dempster-Shafer theory. IEEE Trans Syst Man Cybernet A: Syst Hum 30(2):131–150

Ginart A, Barlas I, Goldin J, Dorrity JL (2006) Automated feature selection for embeddable prognostic and health monitoring (PHM) architectures. In: IEEE System Readiness Technology Conference, USA

Goebel K, Yan WZ, Cheetham W (2002) A method to calculate classifier correlation for decision fusion. In: Proceedings of decision and control, pp 135–140

Hall DL, Llinas J (1997) An introduction to multisensor data fusion. IEEE Digit Object Identif 85 (1):6–23

Hang YS, Suen CY (1993) The behavior-knowledge space method for combination of multiple classifiers. In: IEEE, pp 347–352

Huang R, Xi L, Li X, Liu CR, Qiu H, Lee J (2007) Residual life predictions for bass bearings based on self-organizing map and back propagation neural network methods. Mech Syst Signal Process 21:193–207

Jounela SJ, Vermasvuorim M, Enden P, Haavisto S (2003) A process monitoring systems based on the Kohonen self-organizing maps. Control Eng Pract 18(1):78–84

Kang P, Birtwhistle D (2003) Condition assessment of power tansformer on load tap changers using wavelet analysis and self-organizing map: field evaluation, IEEE Trans on Power Deliver 18:78–84

Kohonen T (2001) Self-organizing maps, 3rd edn. Springer, ISBN 3-540-67921-9

Kou JB, Zhang CS (2003) Multi-agent based classifier combination. Chin J Comput 1–5 (in Chinese)

Kuncheva L (2002) A theoretical study on six classifier fusion strategies. IEEE Trans Syst 24 (2):281–286

Kuncheva LI, Bezdek JC, Duin RPW (2001) Decision templates for multiple classifier fusion: an experimental comparison. Pattern Recogn 34:299–314

Kuncheva LI, Whitaker CJ, Shipp CA (2003) Limits on the majority vote accuracy in classifier fusion. Pattern Anal Appl 6:22–31

Lee J, Ni J, Djurdjanovic D, Qiu H, Liao H (2006) Intelligent prognostic tools and e-maintenance. Comput Ind 57:476–489

Lehtonen M (2006) On the optimal strategies of condition monitoring and maintenance allocation in distribution systems. In: 9th International Conference on Probabilistic Methods Applied to Power Systems, June 11–15, 2006, Stockholm, Sweden

Littlestone N, Warmuth M (1989) Weighted majority algorithm. In: IEEE symposium on foundations of computer science, pp 256–261

Niu G, Yang BD (2009) Dempster-Shafer regression for multi-step-ahead time-series prediction towards data-driven machinery prognosis. Mech Syst Signal Process 23:740–751

Niu G, Yang BS (2010) Intelligent condition monitoring and prognostics system based on data-fusion strategy. Expert Sys with Appl 37:8831–8840

Niu G, Yang BS, Pecht M (2010) Development of an optimized condition-based maintenance system by data fusion and reliability-centered maintenance. Reliab Eng Syst Saf 95:786–796

Niu G, Han T, Yang BS, Tan ACC (2007a) Multi-agent decision fusion for motor fault diagnosis. Mech Syst Signal Process 21(3):1285–1299

Niu G, Son JD, Achmad W, Yang BS (2007b) A comparison of classifier performance for fault diagnosis of induction motor using multi-types signals. Struct Health Monit 6(3):0215-15

Ogata K (2009) Modern control engineering, 5th edn. Prentice Hall. ISBN: 9780137133376

Partridge D, Krzanowski W (1997) Software diversity: practical statistics for its measurement arid exploitation. Inf Softw Technol 39(10):707–717

Petrakos M, Benediktsson JA (2001) The effect of classifier agreement on the accuracy of the combined classifier in decision level fusion. Trans Geosci Remote Sens 39(11):2546–2569

Qiu H, Lee J, Lin J, Yu G (2003) Robust performance degradation assessment methods for enhanced rolling element bearing prognostics. Adv Eng Inform 17:127–140

Ruta D, Gabrys B (2000) An overview of classifier fusion methods. IEEE Comput Inf Syst 1–10

Sauer T, Yorke JA, Casdagli M (1991) Embedology. J Stat Phys 65:579–616

Shafer G (1976) A mathematical theory of evidence. Princeton University Press, Princeton

Smets P, Kennes R (1994) The transferable belief model. Artif Intell 66:191–243

Sorjamaa A, Hao J, Reyhani N, Ji Y, Lendasse A (2007) Methodology for long-term prediction of time series. Neurocomputing

Takens F (1981) Detecting strange attractors in turbulence. In: Rand DA, Young LS (eds) Dynamical systems and turbulence. Springer, New York, pp 366–381

Taniguchi M, Tresp V (1997) Averaging regularized estimators. Neural Computation 9(5): 1163–1178

Verma B, Gader P, Chen W (2001) Fusion of multiple handwritten word recognition techniques. Pattern Recogn Lett 22:991–998

Xu L, Krzyzak A, Suen C (1992) Methods of combining multiple classifiers and their applications to handwriting recognition. IEEE Trans Syst Man Cybernet 22(3):418–435

Yager RR (2001) Dempster-Shafer belief structures with interval valued focal weights. Int J Intell Syst 16:497–512

Yang BS, Kim KJ (2006) Application of Dempster-Shafer theory in fault diagnosis of induction motors using vibration and current signals. Mech Syst Signal Process 20(2):403–420

Chapter 10
System Support and Logistics

10.1 Introduction

System support and logistics are the terms used to describe all the processes, personnel, equipment, and facilities necessary to support a system during its intended life. An effective PHM/CBM system is expected to provide early detection and isolation of the precursor and/or incipient fault of components or subelements; to have the means to monitor and predict the progression of the fault; and to aid in making, or autonomously trigger maintenance schedule and asset management decisions or actions. The detected incipient fault condition should be monitored, trended from a small fault as it progresses to a larger fault, until it warrants some maintenance action and/or replacement. By employing such a system, the health of a system, component, or subsystem can be known at any point of time, and the eventual occurrence of a failure can be predicted and prevented, enabling the achievement of near-zero downtime performance. Unnecessary and costly preventive maintenance can be eliminated, maintenance scheduling can be optimized, and lead time for spare parts and resources can be reduced—all of which can result in significant cost savings.

To do so, the operator must have a continuous and accurate assessment of system performance and awareness as early as possible of the presence of faults that will degrade system performance. System support and logistics is a system of systems that delivers the necessary information, material, and skilled labor to sustain the operations of a system and its constituent components.

Developing the ability to identify and characterize faults (diagnosis) and predict the system or component's useful life (prognosis) is only useful if those facts are turned into action. The first objective of any diagnostic or prognostic system is to provide the operator or maintenance technician with the facts: There is a fault in a specific component, and within a given period of time, the system will degrade beyond useful limits. But there is much more to consider because those facts are

G. Niu, *Data-Driven Technology for Engineering Systems Health Management*,
DOI 10.1007/978-981-10-2032-2_10

important to the full spectrum of activities in logistics or product support: getting the right information, material, equipment, and skilled personnel to the right operation or autonomous control at the right time to sustain the system. This chapter will introduce two application approaches for Data-driven PHM/CBM: *Intelligent maintenance* for logistics support and *autonomous control* for system safety operation.

10.2 Intelligent Maintenance Platform

Tran and Yang (2012) proposed architecture of Intelligent CBM (I-CBM) platform as shown in Fig. 10.1.

This platform contains modules with the aim of converting the rotating machinery signals into useful information for the maintainers to take remedial actions, inspect the conditions, and conduct a repair on the defect before the catastrophic failure occurs. In each module, many applicable algorithms could be appropriately selected to obtain the best result. As depicted in Fig. 10.1, data can be obtained from the sensors installed on the machinery for condition monitoring or manually input working conditions. Subsequently, these data are transformed into features by selecting appropriate algorithms for signal processing, feature extraction, and feature selection. In the feature space, a proper algorithm is employed for each task of classifying the type of faults, performing the degradation, and forecasting the remaining lifetime of machinery.

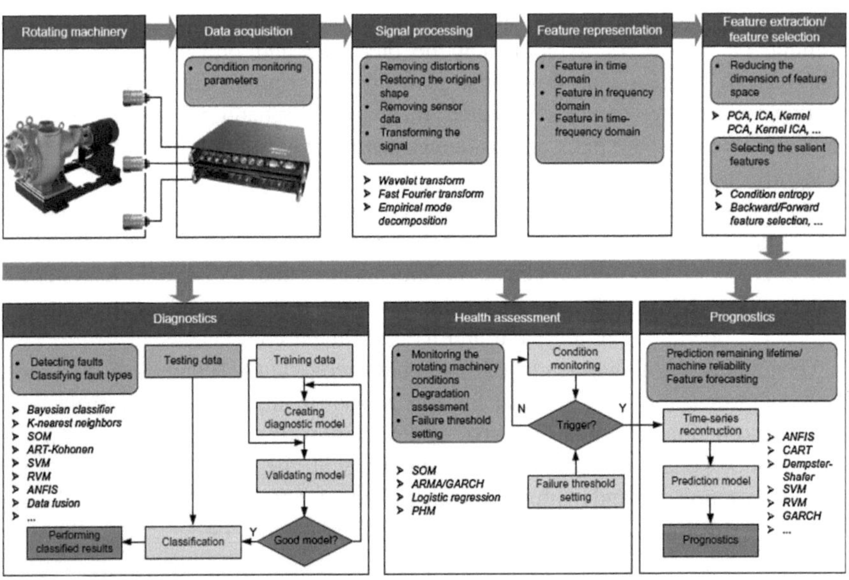

Fig. 10.1 The architecture of I-CBM platform (Tran and Yang 2012)

10.2.1 Data Acquisition

A major challenge confronted with CBM is that how the functional symptoms are monitored in terms of measurable machinery states. Data are the requirement for this challenge. Data acquisition is a process of collecting and storing useful data from target system to monitor the condition, diagnoses the faults, and prognosticate the future states and remaining lifetime. According to Jardine et al. (2006), data used for CBM could be categorized into two main types: event data and condition monitoring data. The former includes the information on what has happened (e.g., installation, breakdown, and overhaul) and what has been done (e.g., minor repair, preventive maintenance, and oil change) to the machinery. The latter is the measurements related to the health condition/states of the machinery. Condition monitoring data are very versatile which could be vibration, acoustic, oil analysis, temperature, pressure, moisture, humidity, weather, or environment data, etc. In this platform, vibration and current data are commonly used due to the easy-to-measure signals and analysis.

10.2.2 Signal Processing

Signal processing is a process of removing distortions and restoring the original shape of signals, removing sensor data which is not relevant, and transforming the signal to make relevant features more explicit. Many methods could be applied for this process in I-CBM platform, for example, wavelet transform and fast Fourier transforms (FFT).

10.2.3 Feature Representation

Data obtained from signal processing process are rarely usable in its raw form due to the huge dimensionality. The huge dimensionality causes not only difficulties of data storage but also data transfer. Therefore, representing data as features is the demand for reduce the huge dimensionality. Feature representation or feature calculation module is a submodule of feature-based techniques and plays a crucial role in attaining the performance of I-CBM platform.

Here, the represented features include time domain features (e.g., root mean square, variance, shape factor, skewness, kurtosis, and crest factor) and frequency domain features (e.g., content at the feature frequency and the amplitude of FFT spectrum). Figure 10.2a presents an example of feature representation module for vibration signals.

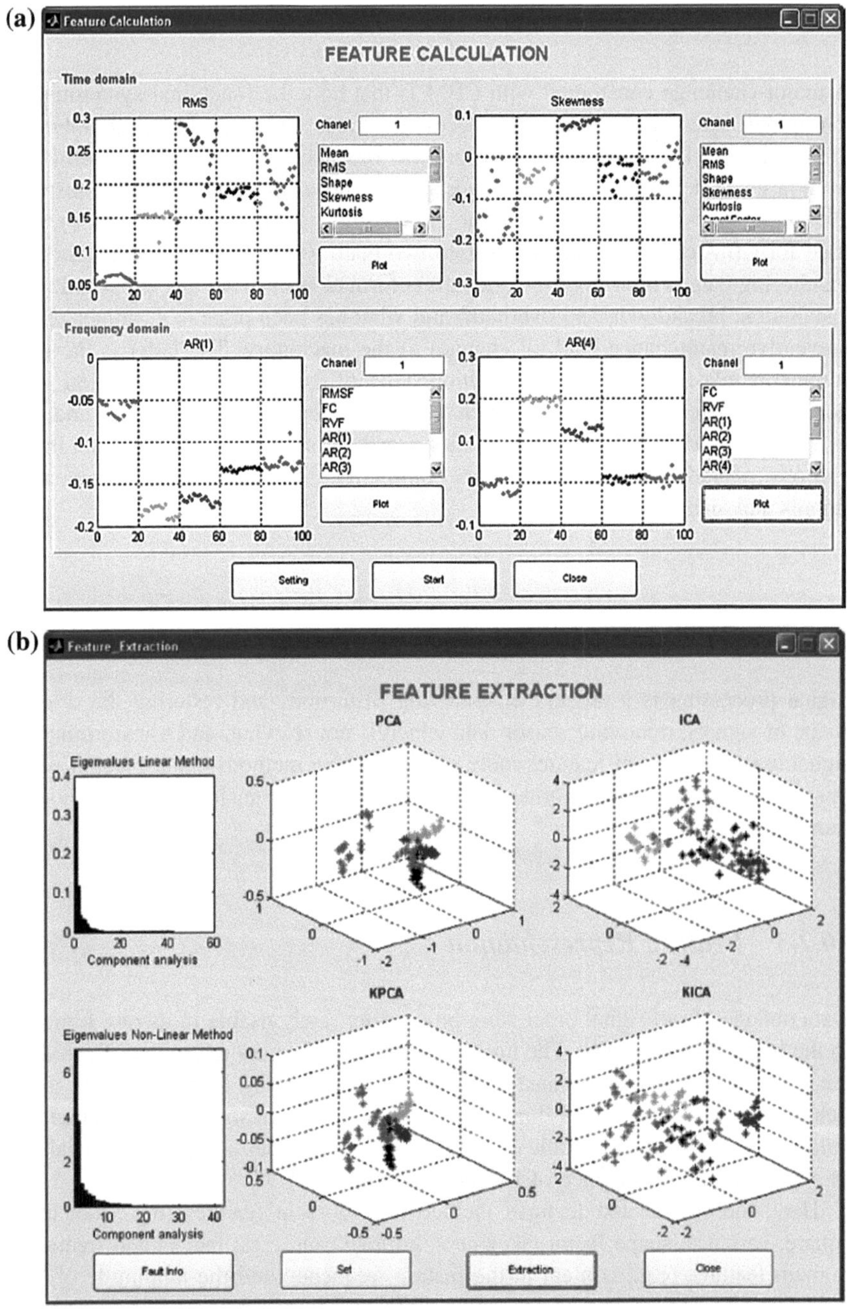

Fig. 10.2 Feature representation extraction and selection in I-CBM platform (Tran and Yang 2012). **a** Feature representation, **b** feature extraction, and **c** feature selection

(c)

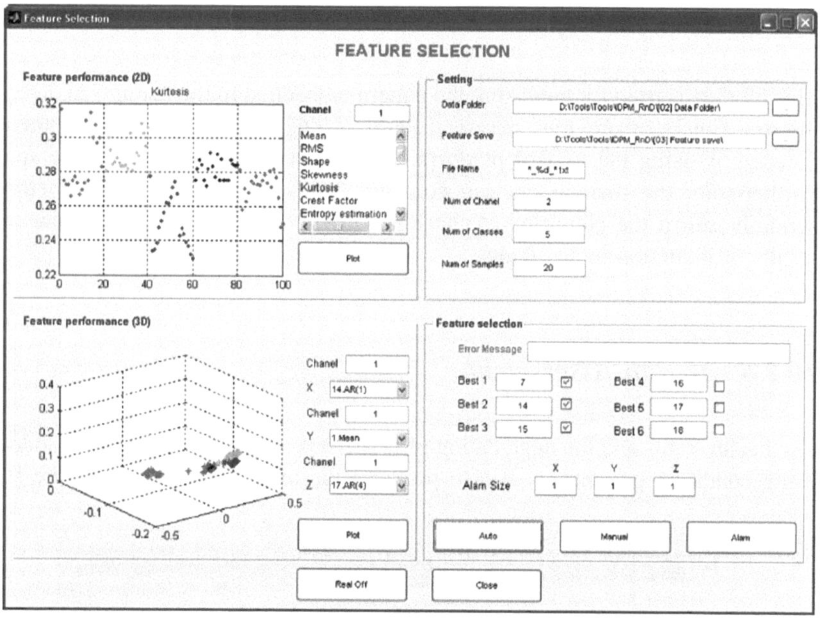

Fig. 10.2 (continued)

10.2.4 *Feature Extraction and/or Feature Selection*

Total features obtained in the previous process can cause a curse of dimensionality and peaking phenomenon that greatly degrade the classification accuracy. Feature extraction can be viewed as a prepruning process to choose a small subset of total features that is necessary and sufficient to describe the overall operations of machine systems. The importance of feature extraction is not only to reduce the search space but also to speed up the process of classification and also to improve the quality of classification. The extracted feature vectors will serve as one of the essential inputs to fault diagnosis and prognosis algorithms. In this platform, common algorithms used for feature extraction are principal component analysis (PCA), independent component analysis (ICA), kernel PCA, kernel ICA, linear discriminant analysis, etc. as shown in Fig. 10.2b.

Even though dimensionality is reduced by the feature extraction process, each feature set contains many redundant or irrelevant features as well as salient features in feature space. Consequently, feature selection process is of necessity to find an optimal subset of features that maximizes information content or predictive accuracy. Figure 10.2c gives the result obtained from feature selection process.

10.2.5 Diagnostics

This module is used for analyzing the pattern embedded in the features to determine the root causes of previous observed faults or degradation. In order to attain this purpose by using the I-CBM platform, several classification methods are applied. Furthermore, the maintainers can compare the performance of each method to certainly affirm the condition of machine. Figure 10.3a describes the diagnosing results of induction motor faults.

10.2.6 Health Assessment

The health status and the degradation of machinery are performed in this module by using condition parameters. It also provides the unacceptable level or the failure

(a) (b)

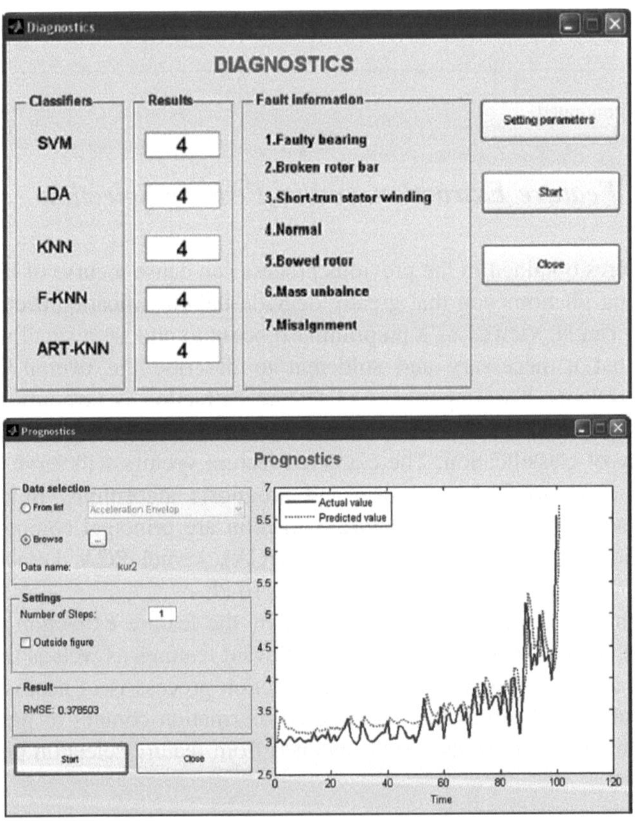

Fig. 10.3 An example of fault diagnostics and prognostics in I-CBM platform (Tran and Yang 2012). **a** Fault diagnostics, **b** failure prognostics

threshold for the operations, so that the appropriate actions will be taken to avoid the consequences of failure before the failure occurs.

10.2.7 Prognostics

Prognostics are the ability to predict the remaining lifetime, future health states, or reliability of machinery based on current health assessment and historical trends. Thus, there are two main functions of prognostics: failure prediction and remaining lifetime estimation. Failure prediction allows the pending failures to be identified before they come to a serious situation. Remaining lifetime is the time left before a particular fault will occur or any part needs to be replaced. The techniques related to prognostics can be classified as experience-based, model-based, and data-driven-based. The prognostics module of I-CBM platform addresses both these functions in which most of methods belong to data driven-based techniques. Figure 10.3b shows a forecasted result of kurtosis feature of bearing by using ARMA/GARCH model (Pham and Yang2010).

10.3 Autonomous Control for Safety Operation

A substantial benefit may be realized from the application of PHM technologies to critical systems by taking advantage of the diagnostic and prognostic pronouncements and exploiting them in a feedback configuration to reconfigure or restructure system controls, replan missions, and sustain stable operating conditions in the presence of non-catastrophic failures. Failures and damage accommodation of critical system components (aircraft actuators, control surfaces, etc.) has been the subject of intense research activities in recent years. Automated contingency management (ACM) aims to optimize available assets and control authority so that the system (unmanned aerial vehicles, engines, etc.) continues to operate within an acceptable and stable regime (operation envelope, stable state, etc.).

Concepts of *fault-tolerant control* (FTC) and *fault accommodation* are promoted to assist not only the operator but also the system designer in optimizing system design, sensor requirements, component redundancy, active controls, etc. so that modern critical assets exhibit attributes of high confidence. Furthermore, notions of active/adaptive control are investigated in an attempt to design an added control authority (not a part of the conventional controller design) that may extend the flight or stability envelope in critical situations. Consider, for example, the case of an unmanned aerial vehicle (UAV) that experiences a collective actuator failure in flight. Such an event invariably will drive the vehicle to an unstable state resulting in loss of the aircraft. If a controller is designed and implemented to regulate the RPMs of the helicopter's main rotor and is activated immediately after the failure is detected and isolated, the UAV will remain stable and land safely during the emergency. The system designer stands to benefit from PHM by gaining a better understanding of the physics of failure mechanisms, the need for monitoring and sensing, issues of hardware and analytic redundancy, and means to design and build high confidence systems.

Key ACM/FTC assumptions include:

- The failures or system parameter changes are unanticipated,
- After the appearance of a failure, the system operates in an emergency mode with another criterion until the failure is recovered,
- Nonstandard techniques, effectors, and configurations may be required,
- The available reaction time is small,
- The handling qualities of the restructured/reconfigured system may be degraded, and
- The failure, in general, may influence the system's behavior and stability.

Sensor, component, and operational failures are considered. The last category refers to such failure modes for a typical rotorcraft as rotor stall, engine stall and surge, rapid battery discharge, loss of data link, etc. Figure 10.4 shows an engine perspective that illustrates the application of PHM at various levels of the control hierarchy. At the higher echelons, mission replanning and midlevel "intelligent" controls are exercised to safeguard the engine's operational integrity, whereas gain scheduling is practiced at the lowest level to ascertain system stability.

Next, consider the case of a vertical takeoff and landing (VTOL) UAV suffering a fault condition in flight. The vehicle is equipped with appropriate PHM sensors and active controllers designed to accommodate certain fault modes. A fault detection and identification (FDI) routine receives periodic sensor information,

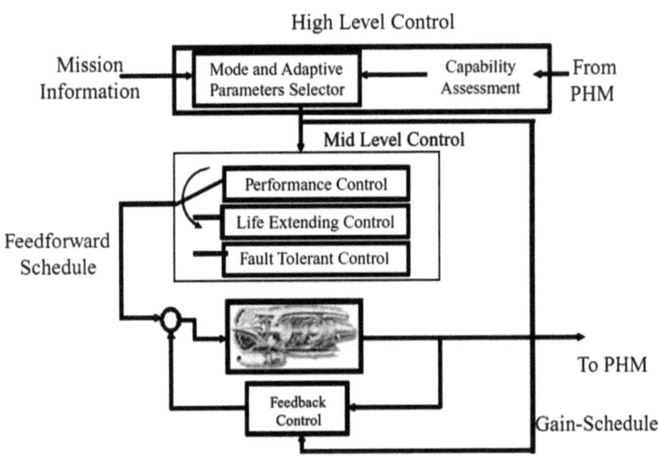

Fig. 10.4 ACM, an engine perspective (Vachtsevanos et al. 2006)

extracts "features" from the signals, and compares these features with known fault signatures. When a failure is identified, a fault-mode signal is generated that contains information about the nature of the fault. An FTC routine determines the reconfiguration actions necessary to maintain system stability and performance. These actions may include:

- Starting new algorithms (controllers),
- Changing the source of certain signals (i.e., sensor signals and actuator commands),
- Changing set points, and
- Modifying local controller parameters.

The FTC routine also sets new priority and quality-of-service (QoS) requirements for signals during the fault recovery process. Figure 10.5 is a schematic representation of the integration of PHM and control systems.

A three-level control reconfiguration hierarchy is suggested:

- Redistribution of control authority between subsystems,
- Updating of local controller set points (based on reconfigured overall system),
- Reconfiguration of local controllers.

A typical FDI/FTC example situation involves a stuck collective actuator while the helicopter is descending from 300 to 100 ft. Without reconfiguration, the vehicle would be unable to slow its descent, as shown in Fig. 10.6 and validated via computer simulations.

With FDI/FTC, the FDI routine detects the stuck collective actuator and outputs a fault-mode signal. The FTC routine receives the fault mode signal and proceeds to reconfigure the system; that is, it starts the RPM controller routine, stops the

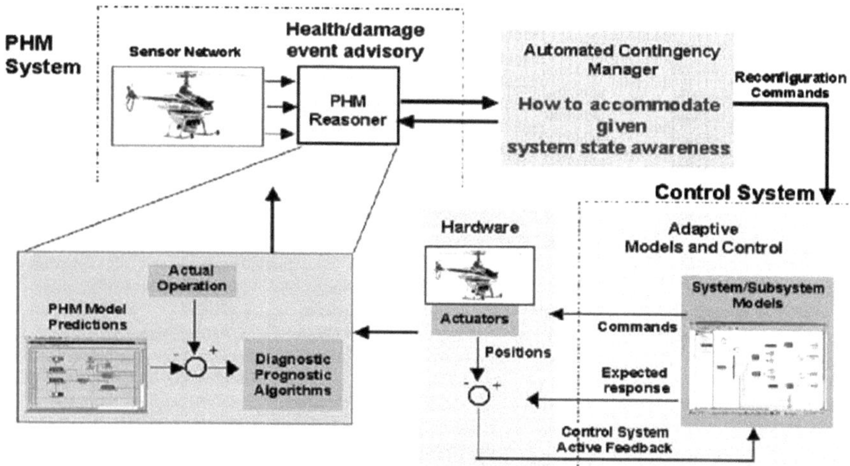

Fig. 10.5 Integration of PHM and control systems (Vachtsevanos et al. 2006)

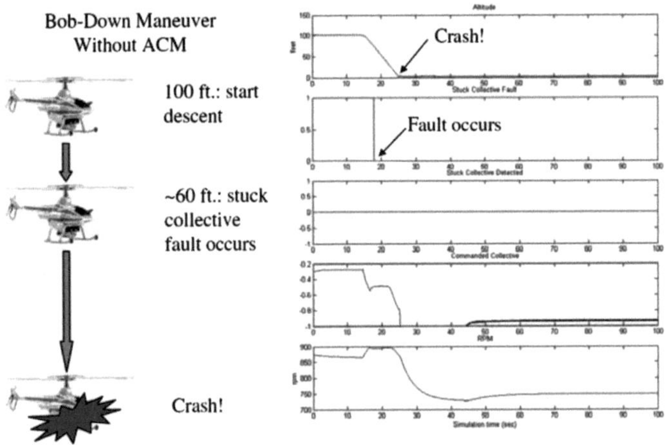

Fig. 10.6 UAV bob-down maneuver without FDI/FTC (Vachtsevanos et al. 2006)

nominal (collective) controller, and the RPM controller subscribes the sensor information and outputs low-level actuator commands. The vehicle hovers at 30 ft, now attaining a stable state, as shown in Fig. 10.7.

Researchers at the Georgia Institute of Technology have developed and validated/verified via simulation and flight testing this and other control reconfiguration algorithms under the Software Enabled Control (SEC) Program. They used for this purpose the GTMax, an experimental VTOL UAV instrumented and owned by Georgia Tech.

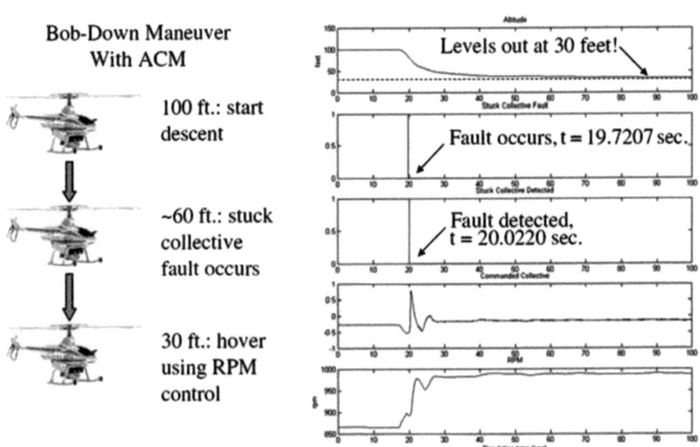

Fig. 10.7 UAV bob-down maneuver with FDI/FTC (Vachtsevanos et al. 2006)

10.4 Future PHM

PHM is a generic way of dealing with a certain degree of system uncertainty and complexity. In situations, in which system uncertainty and complexity further increase in future, *self-maintenance* abilities, *resilient systems* and *engineering immune systems* are necessary and should be developed (Lee et al. 2013), as shown in Fig. 10.8. Systems with high uncertainty and are applicable to nonintrusive approaches only require solutions that are more advanced than preventive maintenance in order to avoid untimely maintenance and nonoptimal cost. Self-maintenance could be a suitable approach to this problem. Self-maintenance refers to the ability to carry out regular quality and safety checking by machine itself, to detect anomaly, and to make immediate repairs when needed by using stocked spare parts to avoid potential catastrophic loss. Based on the PHM functions, such as current health assessment and RUL prediction, self-maintenance will be achievable. High complexity and dynamic are noticeable challenges for current PHM techniques. Resilient system could be the solution to this situation. Resilient systems can manage functions across multiple possible states, resist different types of disorder, and gradually return to the equilibrium state. Compared with the concept of robustness of a system which mainly refers to a static behavior of the system, resilience means the system is capable to dynamically survive different unforeseen impacts and adaptive to disturbance to reach a new stable status at a steady rate. Resilience is one of many qualities to be integrated into future system

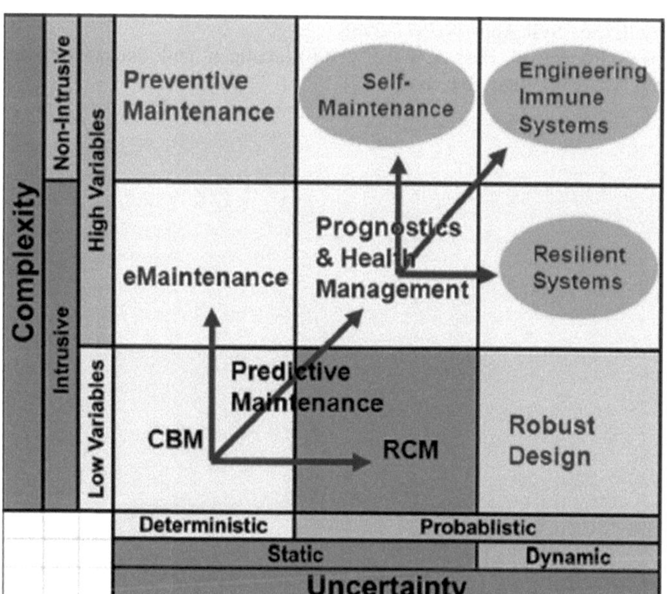

Fig. 10.8 PHM transformation and future trends (Lee et al. 2013)

with the support of advanced PHM techniques. Beyond self-maintenance and resilient system, *engineering immune system (EIS) will be the next-generation PHM* (Lee et al. 2013). EIS is an analogy of the biological immune system which protects against invasion and infection by identifying and killing the pathogens. It can address the machine maintenance issues in highly complex and uncertain environment. The goal of having an EIS is to achieve efficient near-zero breakdown performance with minimal human intervention. EIS should be *robust* in diverse and dynamic environment, *adaptive* to learn and respond to new infections, *adaptive* to retain memory to facilitate future responses, and *autonomous* for self-controlled ability with no requirement of external control. Currently, artificial immune system has already been developed to enable computing system to manage itself and adjust to accommodate varying circumstances with minimized interference from human operator. To improve the prognostics performance of the engineering system, further development of EIS is essential.

References

Jardine AKS, Lin D, Banjevic D (2006) A review on machinery diagnostics and prognostics implementing condition-based maintenance. Mech Syst Signal Process 20(7):1483–1510

Lee J et al (2013) Prognostics and health management design for rotary machinery systems-reviews, methodology and applications. Mech Syst Signal Process

Pham HT, Yang BS (2010) Estimation and forecasting of machine health condition using ARMA/GARCH model. Mech Syst Signal Process 24(2):546–558

Tran TV, Yang BS (2012) An intelligent condition-based maintenance platform for rotating machinery. Expert Syst Appl 39:2977–2988

Vachtsevanos G, Roemer M, Hess A, Wu B (2006) Intelligent fault diagnosis and prognosis for engineering systems. Wiley, Inc. ISBN: 978-0-471-72999-0

Index

G. Niu, *Data-Driven Technology for Engineering Systems Health Management*,
DOI 10.1007/978-981-10-2032-2

MIX
Papier aus verantwortungsvollen Quellen
Paper from responsible sources
FSC® C105338
FSC
www.fsc.org

If you have any concerns about our products,
you can contact us on
ProductSafety@springernature.com

In case Publisher is established outside the EU,
the EU authorized representative is:
Springer Nature Customer Service Center GmbH
Europaplatz 3, 69115 Heidelberg, Germany

Printed by Libri Plureos GmbH
in Hamburg, Germany